ONCOGENES

The Jones and Bartlett Series in Biology

AIDS Smartbook, Kopec/Wood/ Bennett

Anatomy and Physiology: An Easy Learner, Sloane

Aquatic Entomolgy, McCafferty/ Provonsha

Biochemistry, Abeles/Frey/Jencks

Biology, Ethics, and the Origins of Life, Rolston

Biology: Investigating Life on Earth, Second Edition, Avila

The Biology of AIDS, Third Edition, Fan/Conner/Villarreal

Biotechnology, Theory and Techniques, Volume I, Chirikjian

Biotechnology, Theory and Techniques, Volume II, Chirikjian

The Cancer Book, Cooper

Cell Biology: Organelle Structure and Function, Sadava

Cells: Principles of Molecular Structure and Function, Prescott

Creative Evolution?!, Campbell/Schopf

Early Life, Margulis

Electron Microscopy, Bozzola/Russell

Elements of Human Cancer, Cooper

Essentials of Molecular Biology, Second Edition, Freifelder/Malacinski

Essentials of Neurochemistry, Wild/Benzel

Evolution, Strickberger

Experimental Research Notebook, Jones and Bartlett Publishers

Experimental Techniques in Bacterial Genetics, Maloy

Genetics, Third Edition, Hartl

Genetics of Populations, Hedrick

The Global Environment, ReVelle/ ReVelle

Grant Application Writer's Handbook, Reif-Lehrer

Handbook of Protoctista, Margulis/ Corliss/Melkonian/Chapman

Human Anatomy and Physiology Coloring Workbook and Study Guide, Anderson

Human Biology, Farish

Human Genetics: The Molecular Revolution, McConkey

The Illustrated Glossary of Protoctista, Margulis/McKhann/Olendzenski

Laboratory Research Notebooks, Jones and Bartlett Publishers

Major Events in the History of Life, Schopf

Medical Biochemistry, Bhagavan

Methods for Cloning and Analysis of Eukaryotic Genes, Bothwell/ Yancopoulos/Alt

Microbial Genetics, Second Edition, Maloy/Cronan/Freifelder

Molecular Biology, Second Edition, Freifelder

Oncogenes, Second Edition, Cooper

100 Years Exploring Life, 1888-1988, The Marine Biological Laboratory at Woods Hole, Maienschein

The Origin and Evolution of Humans and Humanness, Rasmussen

Origins of Life: The Central Concepts, Deamer/Fleischaker

Plants, Genes, and Agriculture, Chrispeels/Sadava

Population Biology, Hedrick

Statistics: An Interactive Text for the Health and Life Sciences, Krishnamurty/ Kasovia-Schmitt/Ostroff

Vertebrates: A Laboratory Text, Wessels

❖ ONCOGENES

Second Edition

GEOFFREY M. COOPER
Dana-Farber Cancer Institute
Harvard Medical School
Boston, Massachusetts

Jones and Bartlett Publishers
Sudbury, Massachusetts
Boston London Singapore

Editorial, Sales, and Customer Service Offices

Jones and Bartlett Publishers
40 Tall Pine Drive
Sudbury, MA 01776
info@jbpub.com
http://www.jbpub.com

Jones and Bartlett Publishers International
Barb House, Barb Mews
London W6 7PA
UK

Library of Congress Cataloging-in-Publication Data

Cooper, Geoffrey M.
Oncogenes / Geoffrey M. Cooper. — 2nd ed.
p. cm. —(The Jones and Bartlett series in biology)
Includes bibliographical references and index.
ISBN 0-86720-937-2
1. Oncogenes. I. Title. II. Series.
[DNLM: 1. Oncogenes. 2. Genes, Suppressor, Tumor. QZ 202 C7762o 1995]
RC268.42.C66 1995
616.99'4071—dc20
DNLM/DLC
for Library of Congress 94-42582
CIP

Acquisitions Editor: Joseph E. Burns
Assistant Production Manager/Coordinator: Judy Songdahl
Manufacturing Buyer: Dana L. Cerrito
Production and Design: Colophon
Typesetting: LeGwin Associates
Cover Design: Marshall Henrichs
Cover Printing: New England Book Components
Printing and Binding: Braun-Brumfield

Cover Illustration: Molecular surface of the high affinity complex between the Src SH2 domain and a phosphotyrosine peptide. The solvent-accessible surface of the SH2 domain is indicated by green dots, and the three basic residues that interact with the phosphotyrosine are shown in blue. The fold of the peptide chain of the SH2 domain is shown as a white ribbon. The atoms of the peptide are shown as solid spheres, colored white, blue, red, and yellow for carbon, nitrogen, oxygen, and phosphorus, respectively. The figure was prepared by Chi-Hon Lee and John Kuriyan based on the work reported in Waksman, G., Shoelson, S.E., Pant, N., Cowburn, D., and Kuriyan, J. 1993. *Cell* 72:779–790.

Generously supplied by John Kuriyan, The Rockefeller University.

Printed in the United States of America
99 98 97 10 9 8 7 6 5 4

To Ann

❖ Contents

❖ Preface

Work in a large number of laboratories has led to the realization that the development of cancer is a consequence of abnormal expression or function of specific cellular genes—oncogenes and tumor suppressor genes. Oncogenes are mutated forms of normal cell genes (called proto-oncogenes) that stimulate cell proliferation. Tumor suppressor genes, on the other hand, normally inhibit cell proliferation, and it is their loss or inactivation in tumors that contributes to the unregulated growth of cancer cells. Proto-oncogenes and tumor suppressor genes are not only key to the development of cancer, but they also play central roles in regulatory systems governing the proliferation and differentiation of normal cells. Consequently, understanding oncogenes and tumor suppressor genes not only is important from the standpoint of cancer, but has become a key area for students and scientists interested in diverse aspects of biology and medicine.

This book is intended to serve the needs of advanced undergraduates, medical students, graduate students, physicians, and scientists by providing an introduction and broad overview of the oncogene/tumor suppressor gene field. As in the first edition, the approach I have taken throughout the book is to describe the key experiments that led to our current thinking, and then to summarize the results of subsequent studies within this framework. In so doing, I have frequently indicated the dates of key discoveries and the names of the scientists. My intent in doing this is to impart to the reader a flavor of the way in which work in this field has progressed, and of the large number of researchers who have made important contributions.

Since the first edition of *Oncogenes* appeared in 1990, major advances have been made in several areas. Many new tumor suppressor genes have been identified, and we have gained a clearer understanding of the role of tumor suppressor genes and oncogenes in the multistep development of human cancers. Striking progress has been made in understanding signal transduction pathways leading to cell proliferation as well as the mechanisms that control the progression of cells through the replicative cycle. Important advances have also been made in elucidating the roles of proto-oncogenes in controlling cell growth and differentiation during development of multicellular organisms.

Nonetheless, the study of oncogenes and tumor suppressor genes continues to be a fast-moving area of science, our understanding of which is far from complete. New results continue to emerge at a rapid pace, and there remain many unanswered questions. Perhaps foremost among them, posing a critical challenge for future research, is the problem of translating our progress in understanding the molecular biology of cancer into practical advances in cancer prevention and treatment. It is therefore hoped that the following overview will provide a conceptual framework to help the reader understand not only where we are now, but also where future research may lead.

❖ Acknowledgments

A number of people have made critical contributions to both the first and the current editions of this project, and I am pleased to acknowledge their help.

My wife, Ann Kiessling (Harvard Medical School), read and critiqued the current manuscript in its entirety, as she had the first edition. Her comments and advice were once again of great value.

The first edition benefited from the input of David Prescott (University of Colorado) as scientific editor and from the comments, criticisms, and suggestions of several reviewers: Karen Beemon (Johns Hopkins University), Walter Carney (Oncogene Sciences), Errol Friedberg (University of Texas), Michael Greenberg (Harvard Medical School), Tony Hunter (The Salk Institute), Andrew Laudano (University of New Hampshire), Paul Neiman (Fred Hutchinson Cancer Research Center), Eric Stanbridge (University of California), Howard Temin (University of Wisconsin), Peter Vogt (Scripps Research Institute), and Debra Wolgemuth (Columbia University). Their helpful input carried over to the current edition, and I remain grateful for their time and advice.

The current edition has likewise benefited from the additional comments and criticisms of Stuart Aaronson (Mount Sinai Medical Center), Doug Lowy (National Institutes of Health), Paul Neiman (Fred Hutchinson Cancer Research Center), Charles Sherr (St. Jude Children's Research Hospital), Bruce Spiegelman (Dana-Farber Cancer Institute), Tom Sturgill (University of Virginia), and Bert Vogelstein (Johns Hopkins Oncology Center). Their reviews of chapters covering recent progress were especially appreciated.

It is also once again a pleasure to thank the staff of Jones and Bartlett Publishers for their support throughout this undertaking. Joe Burns and Judy Songdahl merit special thanks for their interest, support, and help with *Oncogenes* in both its first and its second editions. The editorial assistance of Patty Zimmerman is also gratefully acknowledged.

Finally, it is with great personal sorrow that I must note the recent death of my mentor and friend, Howard Temin. His pioneering discovery of the replication of retroviruses through a DNA provirus changed the way that we think about the flow of genetic information and provided the foundation upon which the discovery of oncogenes rests.

❖ Part I

Introduction

❖ Chapter 1

The Cancer Cell

Cancer is a disease entity in which the fundamental rules of cell behavior break down. Whereas the growth of normal cells is carefully regulated to meet the needs of the whole organism, cancer cells replicate autonomously and continuously, ultimately invading and interfering with the function of normal tissues. The transformation of a normal cell into a cancer cell thus appears to be due to aberrations in regulatory systems that are central to normal cell physiology. Consequently, understanding the abnormalities that result in cancerous growth may also illuminate the mechanisms that control normal cell growth and development, much as the use of mutants in bacterial genetics has led to an understanding of normal gene function and regulation.

A fundamental feature of cancer cells is that they divide to form more cancer cells, virtually never giving rise to normal progeny. Thus, the abnormalities that result in cancerous growth are stably inherited at the cellular level. This suggests the hypothesis that genetic changes are responsible for neoplastic transformation and raises the prospect of identifying critical genes that play a causal role in tumor development. The application of molecular biology to the study of cancer has centered on finding such genes, now called *oncogenes* and *tumor suppressor genes,* and understanding their mechanisms of action in regulating cell proliferation.

Biology of Human Cancers

Neoplasms, literally "new growths," can develop from normal cells of any tissue. There are consequently a wide variety of tumors that differ with respect to their origin, growth, and prognosis for the patient. The most important biological distinction between tumors is that between benign and malignant neoplasms, of which only the latter are properly referred to as cancers. Both arise as a consequence of abnormal cell proliferation, but they differ in their ability to invade surrounding tissues. A *benign tumor*—for example, a wart or papilloma—is a growth that remains confined to its normal location and neither invades surrounding tissue nor spreads to other organs. In contrast, a *malignant tumor*, or cancer, is both invasive and metastatic, capable of invading and destroying adjacent normal tissue as well as spreading to distant organ sites through the circulatory system. Whereas benign tumors can usually be removed surgically, the ability of malignant tumors to metastasize frequently makes them refractory to such localized treatment. The distinction between benign and malignant tumors (cancers) is thus critical for therapy and prognosis.

More than one hundred different kinds of human cancer are recognized and historically classified according to their origin (Table 1.1 is a sample listing). Most of these neoplasms are divided into three major groups: carcinomas, sarcomas, and leukemias/lymphomas. *Carcinomas*, which comprise approximately 90% of human cancers, arise from epithelial cells of either endodermal or ectodermal origin. *Sarcomas* and the *leukemias* and *lymphomas* develop from cells of mesodermal origin, including muscle, bone, blood vessels, fibroblasts, and circulating cells of the blood and lymph systems. Tumors are further classified according to tissue of origin—for example, lung, breast, or colon carcinoma—and by histologic type. For example, lung carcinomas can be divided into several histologic types—squamous cell carcinomas, small or large cell undifferentiated carcinomas, and adenocarcinomas—that differ in their patterns of growth and metastasis and, therefore, in the clinical course of the disease.

Although cancer can be considered a diverse collection of different diseases, there are common features in the development of different tumors that form the basis for our current understanding of carcinogenesis. First, most tumors—whether benign or malignant—have been found to arise from a single cell of origin. Thus, neoplasms are usually clonal growths in which the tumor cells represent the descendants of a single progenitor cell that underwent an initial transforming event and began to proliferate abnormally. As noted earlier, the neoplastic phenotype is stably inherited at cell division; so continued proliferation of an initially transformed cell will give rise to a clonally derived cell population.

The clonal origin of neoplasms does not, however, imply that a malignant tumor originates from only a single transforming event. On the contrary, the development of malignant neoplasms is a complex multistep process in which cells acquire the fully neoplastic phenotype through a progressive series of alterations. One indication of the multistep nature of neoplastic disease is the fact that most cancers develop late in life with the incidence of disease increasing

TABLE 1.1
Examples of Human Tumors

Origin	Representative Tumors
Ectoderm Derivatives	
skin	squamous and basal cell carcinomas
mammary gland	breast carcinoma
neuronal cells	neuroblastoma
glial cells	glioblastoma
retinal cells	retinoblastoma
melanocytes	melanoma
adrenal medulla	pheochromocytoma
Mesoderm Derivatives	
bone	osteosarcoma
fibrous tissue	fibrosarcoma
cartilage	chondrosarcoma
muscle	rhabdomyosarcoma
blood vessel	hemangiosarcoma
fat cells	liposarcoma
erythrocytes	erythrocytic leukemia
granulocytes	myelocytic leukemia
lymphocytes	lymphocytic leukemias and lymphomas
plasma cells	myeloma
Endoderm Derivatives	
urinary bladder	bladder carcinoma
pancreas	pancreatic carcinoma
colon	colon carcinoma
lung	lung carcinoma
thyroid	thyroid carcinoma
liver	hepatic carcinoma

rapidly with age. For example, the death rate from colon carcinoma increases more than 1000-fold between the ages of thirty and eighty (Fig. 1.1). Such a dramatic nonlinear increase of cancer incidence with age is not consistent with the hypothesis that a neoplasm arises as the result of a single transforming

FIGURE 1.1
Annual rate of death from colon cancer in the United States. (From J. Cairns, *Cancer: Science and Society*, 1978.)

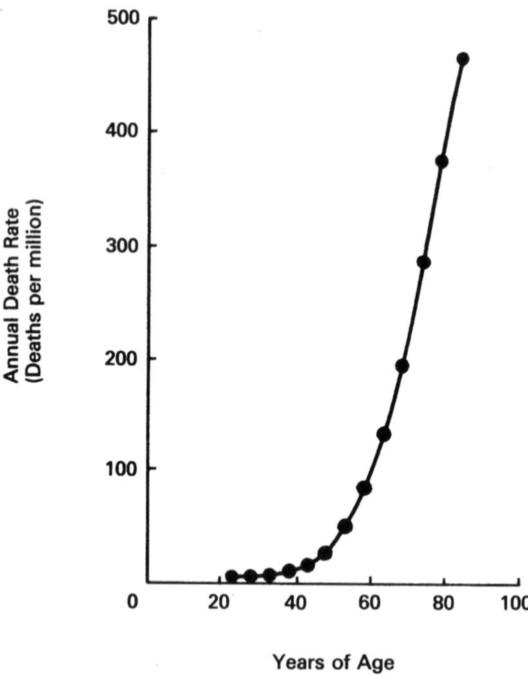

event. If this were the case, tumor incidence would be a linear function of age. Instead, the increasing incidence of most cancers with age suggests that neoplasms result from the accumulation of multiple abnormalities (for example, mutations in several different genes) during the life of an individual.

The multistep nature of neoplasm development is also illustrated by the identification of preneoplastic or premalignant lesions that can subsequently progress to fully malignant neoplasms. Such lesions appear to represent cell populations with partial loss of normal growth control and increasing malignant potential that correspond to progressive steps in neoplasm development. For example, an early preneoplastic stage in the development of carcinomas (Fig. 1.2) is recognized as *epithelial dysplasia.* These lesions may correspond to the initiation of neoplasm development, representing the result of an alteration in a single cell that leads to abnormal proliferation and the genesis of a preneoplastic cell population. A later stage in neoplasm progression is *carcinoma in situ,* corresponding to small localized tumors that have not invaded the surrounding normal tissue. These tumors likely result from further mutations within a preneoplastic cell, leading to increased proliferative potential and the more aggressive growth of neoplastic cells. Still further mutations are likely involved in the progression of carcinoma *in situ* to fully *malignant carcinomas,* which invade

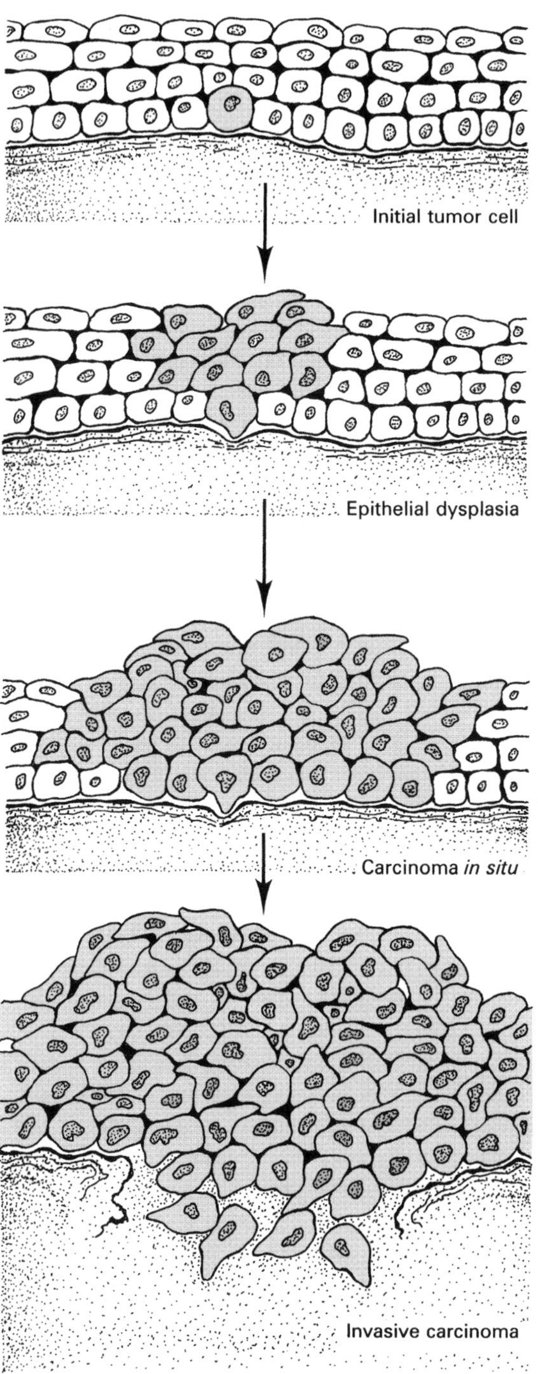

FIGURE 1.2
Stages of carcinoma development. A single initially transformed cell begins to proliferate, giving rise to a preneoplastic cell population (dysplasia) that progresses first to carcinoma *in situ* and then to invasive carcinoma.

adjacent tissues and eventually metastasize. The pathogenesis of neoplasia is thus commonly viewed as the evolution of increasingly malignant cell populations that arise through a series of mutations in the descendants of a single initially transformed cell. Understanding the development of cancer at the molecular level therefore requires identifying the progressive alterations that culminate in malignancy.

Experimental Induction of Cancer in Animals

Studies of the development of neoplastic disease require the use of experimental systems in which cancers can be reproducibly induced and systematically analyzed. This obviously implies the need for animal models in which neoplasms can be induced by controlled experimental manipulations. Such model systems have shown that cancers can be induced by a variety of carcinogenic chemicals, by radiation, and by some viruses. Chemical and viral carcinogens have been particularly useful as experimental tumor-inducing agents and have served to focus research efforts on identification of the genes critical to the development of neoplasia.

A wide variety of chemicals induce cancer in animals and some of these chemicals are undoubtedly also important as human carcinogens. The active forms of most carcinogenic chemicals are potent mutagens, consistent with the notion that genetic alterations are central to cell transformation. Most chemical carcinogens induce tumors in rodents only after long latent periods, typically many months after their administration. As discussed earlier, such long latent periods suggest that multiple changes must occur before a normal cell is converted into the fully malignant phenotype, and the reproducible induction of some tumors by chemical carcinogens has helped to define discrete preneoplastic and neoplastic stages of tumor development. However, the long time required for tumor development after chemical carcinogen application is also a complicating factor in designing experiments to identify the primary targets for carcinogen action. Moreover, carcinogens induce mutations at multiple random sites in cellular DNA; so the problem of identifying the biologically critical target genes for carcinogen-induced mutations is formidable. Only in recent years has the identification of cellular oncogenes allowed definition of some biologically significant targets for chemical carcinogen action (see chapter 11).

Studies of the viruses that cause cancer in experimental animals (tumor viruses) offer two major advantages over studies of chemical carcinogens. First, some tumor viruses are extremely potent inducers of cancer so that susceptible animals develop tumors within days or weeks after infection. Second, and most important, the small size of the genome of a virus (ca. 10 kb) compared with that of a typical animal cell (ca. 10^6 kb) has allowed molecular studies of tumor viruses to be readily undertaken. Such molecular characterizations of tumor viruses have led to the direct identification of the viral genes responsible for neoplastic transformation (viral oncogenes) and have provided the basis for our understanding of the cellular genes (cellular oncogenes) that are critical to the development of nonvirus-induced tumors, including human neoplasms.

Growth of Cells in Culture

Although induction of cancer can be studied in whole animals, such experiments are difficult to quantitate and suffer from a number of potentially uncontrolled variables in the context of the intact organism. Consequently, the ability to propagate animal cells in culture has been central to the experimental development of cancer research. *In vitro* cell culture systems have provided researchers with the ability to analyze the growth requirements and behavior of normal cells, to compare the properties of normal cells and cancer cells under controlled conditions, and to develop quantitative and reproducible assays for cell transformation. Although cell culture is indispensable as an experimental system, it must be remembered that it provides only an *in vitro* approximation to the environment of a cell in the intact organism. It is consequently important to be wary of potential artifacts of cell culture and to critically compare results of *in vitro* experiments with the "reality" of whole-animal studies.

Most cells are cultured in media consisting of physiologic salt solutions, amino acids, vitamins, glucose, and serum, which provides growth factors required for cell proliferation. A number of different polypeptides that support proliferation of different cell types have been identified. For example, one of the serum growth factors utilized by fibroblasts in culture is platelet-derived growth factor (PDGF), which is a dimeric polypeptide of ~30 kd. It is stored in blood platelets and released during the clotting reaction, suggesting that it normally plays a role in wound healing. Proliferation of fibroblasts requires other factors in addition to PDGF, including epidermal growth factor (EGF) and insulin or insulin-like growth factors (IGFs). These and other extracellular polypeptides are critical regulators of normal cell growth and differentiation, as will be discussed in detail in subsequent chapters.

In order to use cells in culture as a model to study neoplasia, we have to ask how the behavior of cells *in vitro* relates to their behavior *in vivo*. In particular, are there differences between normal and neoplastic cells in culture that can be correlated with the fundamental difference between these cells in an animal: the ability to form tumors? The basic distinction *in vivo* is that cancer cells proliferate under conditions in which normal cells do not. This lack of growth control that is characteristic of tumor cells in an animal is mimicked by a number of differences in the growth of normal cells and cancer cells in culture (Table 1.2). Although different kinds of tumor cells display considerable variability in their *in vitro* growth properties, they clearly differ from their normal counterparts in at least some of these general characteristics.

TABLE 1.2
Growth of Normal and Neoplastic Fibroblasts in Culture

Growth Characteristic	Normal Cells	Tumor Cells
density-dependent inhibition of growth	present	absent
growth factor requirements	high	low
anchorage dependence	present	absent
proliferative life span	finite	indefinite
contact inhibition of movement	present	absent
adhesiveness	high	low
morphology	flat	rounded

A primary distinction between normal and neoplastic cells in culture is that normal cells display density-dependent control of cell growth. That is, cultures of normal cells maintained in a constant level of serum growth factors proliferate to a finite cell density and then cease growth. Cell proliferation is resumed upon subculture, when the cells are replated at a lower density. In contrast, tumor cells continue growing to much higher cell densities.

A related difference between normal and neoplastic cells in culture is their requirement for serum growth factors. Neoplastic cells frequently display a reduced dependence upon extracellular growth factors, and some tumor cells proliferate in the absence of exogenous factors required for normal cell growth. In some cases, such tumor cells produce growth factors that drive their own proliferation. Because the availability of serum growth factors appears to determine, at least in part, the cell density at which proliferation of normal fibroblasts becomes arrested, the reduced growth factor requirement of neoplastic cells is closely correlated with their ability to proliferate to higher densities than normal cells.

A third difference between normal and neoplastic cells in culture is that normal fibroblasts and epithelial cells must attach to a solid matrix, such as the surface of a culture dish, in order to grow. In contrast, many transformed fibroblast and epithelial cells can grow without attachment to a surface and can therefore proliferate in suspension, a property referred to as *anchorage independence.* This distinction between normal and transformed cells does not apply to hematopoietic cells because these cells normally proliferate without attachment to a substrate.

Normal and neoplastic cells also frequently differ in their reproductive life span in culture. Most normal cells have limited proliferative capacity *in vitro,* which varies for different cell types. For example, normal human fibroblasts can usually be cultured for fifty to one hundred cell divisions, after which they stop growing and die (a phenomenon called *senescence*). In contrast, cells derived from many rodent and human tumors will continue to grow indefinitely in culture. Such permanent cultures are called *established, continuous,* or *immortal cell lines.* The capacity for continuous proliferation, "immortality," is thus another property that frequently distinguishes neoplastic from normal cells. It should be noted, however, that many tumor cells (for example, cells cultured from chicken sarcomas) are not immortal. Conversely, not all immortalized cell lines are neoplastic. For example, it is possible to isolate cell lines from normal rodent fibroblasts that proliferate indefinitely in culture but otherwise display the *in vitro* growth characteristics of normal cells and are not tumorigenic when inoculated into susceptible animals. Cell lines of this type are therefore immortalized but not neoplastic, indicating that immortality can be separated from tumorigenicity. Such cells have been particularly useful in many studies because they afford a uniform and continuously available source of cells that display many aspects of normal growth control, although they have clearly undergone at least one change characteristic of tumor cells.

Another general characteristic of tumor cells is abnormal differentiation. Such defective differentiation is closely related to their increased proliferation, because most fully differentiated normal cells either cease division or divide only slowly. Tumor cells, however, are generally blocked at an early stage of differentiation consistent with their active proliferation. In addition, tumor cells frequently fail to undergo programmed cell death (*apoptosis*), which is a normal part of the differentiation program of many cell types with limited life spans *in vivo.*

In addition to these changes in proliferative capacity, several other aspects of cell behavior distinguish neoplastic cells from their normal counterparts. Normal fibroblasts move across the surface of a culture dish but cease such movement upon contact with a neighboring cell. Such "contact inhibition of movement" is not displayed by tumor cells, which continue moving after cell contact. Probably related to this is the fact that neoplastic cells are less adhesive than normal cells and are thus less firmly attached to either neighboring cells or the surface of culture dishes. In addition, many tumor cells differ morphologically from normal cells, generally displaying a rounder appearance. These differences in cell interactions and morphology are correlated with several characteristics common to neoplastic cells, including increased secretion of proteases, decreased levels of surface proteins taking part in cell adhesion, and disorganization of the cytoskeleton.

The behavior of neoplastic cells in culture thus mimics several features of their behavior in the intact animal. In contrast to normal cells in culture, tumor cells generally proliferate indefinitely, are not stringently regulated by cell density or growth factor availability, are able to grow without attachment to a surface, and are less responsive to cell contact. It is important to note, however, that the foregoing characteristics constitute only a general description of the neoplastic phenotype. The behavior of different kinds of cells in culture is complex, and no single feature serves to distinguish all neoplastic cells from all normal cells. Nevertheless, these general characteristics displayed by many tumor cells *in vitro* appear to be directly related to the continuous unregulated growth and invasiveness characteristic of cancer cells *in vivo*.

Growth Control and the Cell Cycle

In contrast to bacteria, which synthesize DNA continuously throughout their division cycle, animal cells undergo a more complex cell cycle, which has classically been divided into four stages based on the timing of DNA synthesis (Fig. 1.3). After mitosis (M), the cell enters the G_1 (gap 1) stage, during which the cell is metabolically highly active but not engaged in DNA synthesis. At this time, the cell is diploid (2N) in DNA content. The cell progresses from G_1 to the S (synthesis) phase, in which DNA replication occurs and DNA content increases from 2N to 4N. Following S, the cell, now tetraploid (4N) in DNA content, enters the G_2 (gap 2) stage prior to mitosis, which results in the formation of two diploid daughter cells. The total cell cycle time varies with different cell types, with most of the variability occurring in G_1. As an example, a typical proliferating cell in culture might display a total cycle time of 20 hours, with 1 hour for M, 8 hours for G_1, 8 hours for S, and 3 hours for G_2.

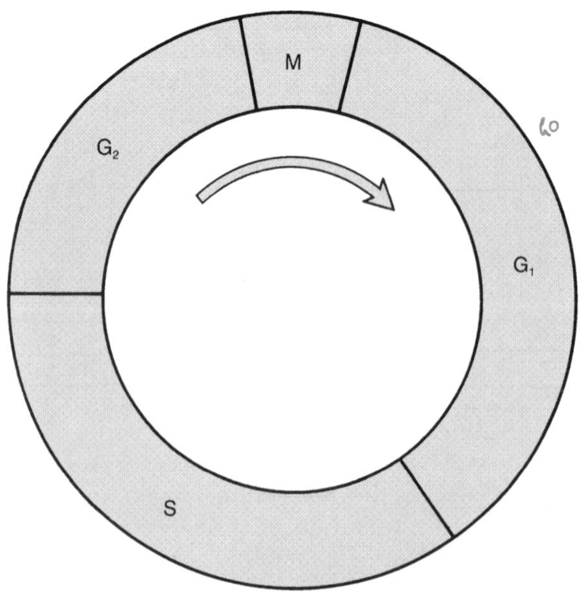

FIGURE 1.3
The cell cycle. M, mitosis; G_1, gap 1; S, DNA synthesis; G_2, gap 2.

Normal fibroblasts that cease proliferation as a consequence of growth factor deprivation or density-dependent arrest display a 2N DNA content, indicating that proliferation is arrested in G_1 or at the G_1/S interphase. This is consistent with growth control *in vitro* occurring at a physiologically normal state, because most nondividing cells *in vivo* also have a 2N DNA content. If such growth-arrested fibroblasts are stimulated by growth factor addition, there is a significant lag (8–12 hours) before DNA synthesis begins. The arrested cells thus appear to have entered a quiescent phase of the cell cycle, which is generally distinguished from the G_1 period of actively cycling cells and is referred to as G_0. One model to account for growth control kinetically is that cycling cells reach a decision point in G_1. At this point, normal cells either proceed through G_1 and the rest of the cell cycle or enter G_0, depending on the presence or absence of the appropriate growth factors. Transformed cells, on the other hand, proceed through G_1 rather than entering G_0, even in the absence of the growth factors that would be required by their normal counterparts.

Cell Transformation *in Vitro*

In addition to facilitating studies of normal and neoplastic cell growth, the ability to propagate animal cells in culture has allowed the development of *in vitro* assays for the carcinogenic activity of chemicals and viruses. Such assays of cell transformation in culture detect the conversion of a cell with normal growth properties into one with *in vitro* growth characteristics of a neoplastic cell. Insofar as the *in vitro* growth characteristics of normal and neoplastic cells are only approximations to the phenotype of cancer cells *in vivo,* the transformation of cells in culture should be considered only an approximation to the formation of cancer cells in the intact organism. Nevertheless, transformation of cultured cells clearly provides a simpler and more quantitative assay for the biological activity of chemical and viral carcinogens than induction of tumors in experimental animals. In fact, the availability of sensitive and reproducible assays for cell transformation *in vitro* has been critical to developing an experimental understanding of the molecular basis of neoplasia. Several such assays are now commonly used, based on the characteristic loss of growth control and morphological alterations displayed by tumor cells *in vitro.*

The first and most widely used assay of transformation *in vitro* depends on the ability to recognize a group of transformed cells as a morphologically distinct "focus" against a background of normal cells on the surface of a culture dish (Fig. 1.4). The focus assay takes advantage of three properties of transformed cells: altered morphology, loss of contact inhibition of movement, and loss of density-dependent inhibition of growth. The result is an overgrowth of morphologically altered cells that tend to "pile up" over a normal cell monolayer. The disadvan-

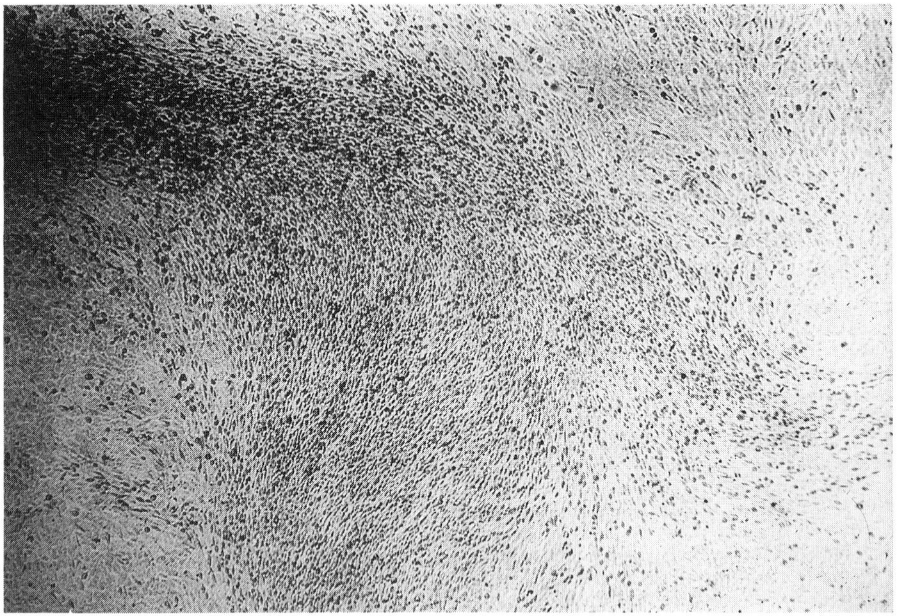

FIGURE 1.4
A focus of NIH 3T3 mouse cells transformed by a *ras*H oncogene from human tumor DNA. (From T.G. Krontiris and G.M. Cooper, *Proc. Natl. Acad. Sci. USA* 78:1181–1184, 1981.)

tage of the focus assay is that it is relatively nonselective, because both normal and transformed cells grow under the culture conditions used.

Two commonly used selective assays of transformation in culture take advantage of the reduced growth factor requirements and anchorage independence of transformed cells. In the first of these assays, cells are plated in growth factor deficient medium: under these conditions transformed cells will proliferate, whereas normal cells will not. In the second assay, cells are plated in suspension in a semisolid medium to prevent attachment to a surface: only anchorage-independent transformed cells grow to form colonies (Fig. 1.5).

A variety of *in vitro* assays can thus be employed to assess several distinct parameters of neoplastic transformation. Importantly, the ability of cells transformed *in vitro* to form tumors *in vivo* can also be determined by animal inoculation; so alterations in particular parameters of *in vitro* cell growth can be correlated with tumorigenicity. An important test system for such *in vivo* assays is the nude mouse, which carries a mutation resulting in thymic deficiency. As a consequence, thymus-derived (T) lymphocytes do not develop, and the mouse is immunologically incompetent. Inoculation of these mice

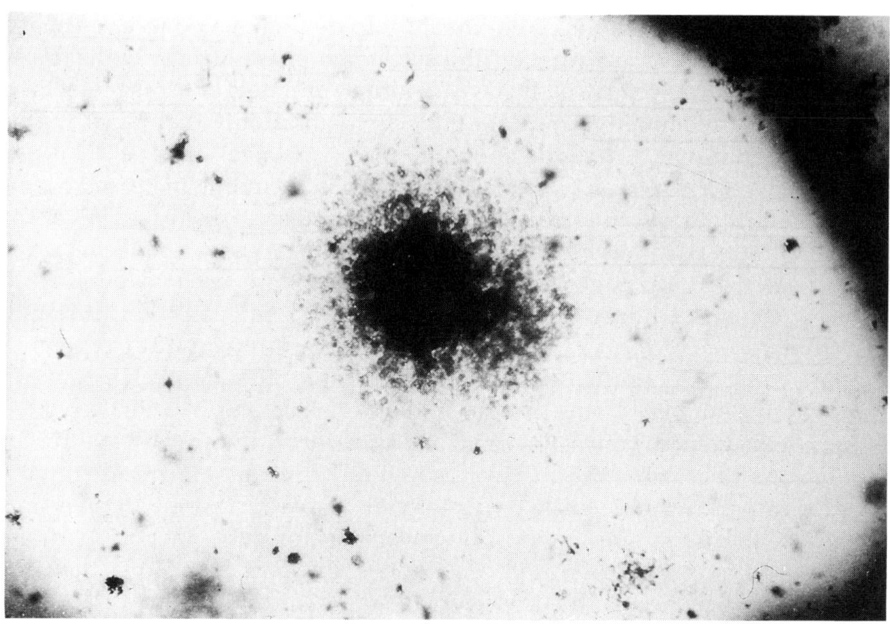

FIGURE 1.5
A colony of chicken embryo fibroblasts growing in suspension in agar medium following transformation by Rous sarcoma virus.

therefore provides an assay for tumorigenicity that is not complicated by the possibility of immune rejection. This is particularly useful because it allows testing tumorigenicity of a cell line in a nonsyngeneic host that would otherwise reject the "foreign" tumor graft. Such studies of tumorigenicity indicate a good overall correlation between *in vitro* transformation and the ability to form tumors *in vivo*, supporting the validity of *in vitro* transformation assays. The correlation between transformation *in vitro* and tumorigenicity *in vivo* is not, however, absolute. Many cells that are transformed by one or more criteria *in vitro* still fail to form tumors in animals. Such nontumorigenic transformed cells have apparently acquired only some of the alterations needed to convert a normal cell into the fully neoplastic phenotype.

Biochemical Alterations in Transformed Cells

A great deal of work has been devoted to comparative analysis of the biochemistry of transformed and normal cells with the hope of identifying differences that are central to the transformed phenotype. This approach, however, has suffered from an inherent inability to distinguish between primary causes of neoplastic transformation and secondary consequences of the transformed state. Cancer cells usually have abnormal numbers or arrangements of chromosomes (aneuploidy), indicating a genetic instability that plays an important role in tumor development. In addition, cancer cells frequently differ from normal cells in several biochemical properties, including increased glycolysis, secretion of growth factors, secretion of proteases, decreased levels of surface proteins required for cell adhesion, and disorganization of the cytoskeleton. As discussed previously, some of these differences may be directly related to the altered growth properties of transformed cells. For example, growth factor secretion can lead to reduced requirements for exogenous growth factors, reduced levels of cell adhesion molecules are likely related to loss of contact inhibition, and changes in the cytoskeleton correlate with morphological differences between normal and transformed fibroblasts. Secretion of proteases and growth factors as well as reduced cell adhesion are also important determinants of the ability of cancer cells to invade normal tissues and metastasize to distant sites *in vivo*. However, none of these phenotypic alterations are unique to transformed cells; nor is it apparent whether they are causes or consequences of the transformed state. For this reason, productive attempts to understand the molecular basis of neoplastic disease have focused on specific genes that can directly induce transformation rather than on comparative phenotypic analyses of neoplastic and normal cells.

Summary

Malignant neoplasms develop through a series of stepwise progressive changes that culminate in the loss of growth control characteristic of cancer cells: continuous unregulated proliferation, the ability to invade surrounding normal tissues, and the ability to metastasize to distant organ sites. The abnormal proliferation characteristic of cancer cells *in vivo* is paralleled by their loss of growth control in culture, and studies of cells *in vitro* have allowed more carefully controlled experimental analyses than is possible in whole animals. Such *in vitro* studies have defined the parameters that characterize the growth of normal and neoplastic cells and have led to the identification and isolation of specific polypeptide factors that control cell growth and differentiation. In addition, the characteristic differences between normal and neoplastic cells in culture have allowed the development of quantitative *in vitro* assays for neoplastic transformation. The ability to study cell transformation in controlled and quantitative experimental systems has in turn allowed detailed analysis of the transformation process, leading to the identification of specific genes capable of inducing the transformed cell phenotype. As we will see in subsequent chapters, such genes were first identified in tumor viruses and, more recently, in a wide variety of neoplasms, including human malignancies.

References

General References

Brugge, J., Curran, T., Harlow, E., and McCormick, F., eds. 1991. *Origins of human cancer: a comprehensive review*. Cold Spring Harbor Laboratory Press, Plainview, NY.

Cairns, J. 1978. *Cancer: science and society*. W.H. Freeman, San Francisco.

Cooper, G.M. 1992. *Elements of human cancer*. Jones and Bartlett Publishers, Boston.

Ruddon, R.W. 1987. *Cancer biology*, 2d ed. Oxford Univ. Press, New York.

Biology of Human Cancers

Fialkow, P.J. 1979. Clonal origin of human tumors. *Ann. Rev. Med.* 30:135–143.

Foulds, L. 1969. *Neoplastic development*, Vol. 1. Academic Press, London.

Liotta, L.A., and Stetler-Stevenson, W.G. 1991. Cancer metastasis and angiogenesis: an imbalance of positive and negative regulation. *Cell* 64:327–336.

Nowell, P.C. 1989. Chromosomal and molecular clues to tumor progression. *Sem. Oncol.* 16:116–127.

Peto, R. 1977. Epidemiology, multistage models and short-term mutagenicity tests. *In Origins of human cancer*, ed. Hiatt, H.H., Watson, J.D., and Winsten, J.A. Cold Spring Harbor Laboratory, New York. pp. 1403-1428.

Wainscott, J.S., and Fey, M.F. 1990. Assessment of clonality in human tumors: a review. *Cancer Res.* 50:1355–1360.

Experimental Induction of Cancer in Animals

Ames, B.N. 1979. Identifying environmental chemicals causing mutations and cancer. *Science* 204:587–593.

Ames, B.N., Durston, W.E., Yamasaki, E., and Lee, F.D. 1973. Carcinogens are mutagens: a simple test system combining liver homogenates for activation and bacteria for detection. *Proc. Natl. Acad. Sci. USA* 70:2281–2285.

Boutwell, R.K. 1974. The function and mechanism of promoters of carcinogenesis. *CRC Crit. Rev. Toxicol.* 2:419–431.

Farber, E., and Cameron, C. 1980. The sequential analysis of cancer development. *Adv. Cancer Res.* 31:125–226.

Gross, L. 1983. *Oncogenic viruses*, 3d ed. Pergammon Press, New York.

Miller, J.A. 1970. Carcinogenesis by chemicals: an overview—G.H.A. Clowes memorial lecture. *Cancer Res.* 30:559–576.

Growth of Cells in Culture

Abercrombie, M., and Heaysman, J.E.M. 1954. Observations on the social behaviour of cells in tissue culture II. "Monolayering" of fibroblasts. *Exptl. Cell Res.* 6: 293–306.

Carpenter, G., and Cohen, S. 1979. Epidermal growth factor. *Ann. Rev. Biochem.* 48:193–216.

Cross, M., and Dexter, T.M. 1991. Growth factors in development, transformation, and tumorigenesis. *Cell* 64:271–280.

Dulbecco, R. 1970. Topoinhibition and serum requirement of transformed and untransformed cells. *Nature* 227:802–806.
Eagle, H. 1955. Nutrition needs of mammalian cells in tissue culture. *Science* 122:501–504.
Folkman, J., and Klagsbrun, M. 1987. Angiogenic factors. *Science* 235:442–447.
Goldstein, S. 1990. Replicative senescence: the human fibroblast comes of age. *Science* 249:1129–1133.
Hayflick, L., and Moorhead, P.S. 1961. The serial cultivation of human diploid cell strains. *Exptl. Cell Research* 25:585–621.
Holley, R.W., and Kiernan, J.A. 1968. "Contact inhibition" of cell division in 3T3 cells. *Proc. Natl. Acad. Sci. USA* 60:300–304.
Metcalf, D. 1991. Control of granulocytes and macrophages: molecular, cellular, and clinical aspects. *Science* 254:529–533.
Raff, M.C. 1992. Social controls on cell survival and cell death. *Nature* 356:397–400.
Ross, R., Raines, E.W., and Bowen-Pope, D.F. 1986. The biology of platelet-derived growth factor. *Cell* 46:155–169.
Sporn, M.B., and Roberts, A.B. 1988. Peptide growth factors are multifunctional. *Nature* 332:217–219.
Todaro, G.J., and Green, H. 1963. Quantitative studies of the growth of mouse embryo cells in culture and their development into established lines. *J. Cell Biol.* 17:299–313.

Growth Control and the Cell Cycle

Baserga, R. 1985. *The biology of cell reproduction.* Harvard University Press, Cambridge.
Murray, A., and Hunt, T. 1993. *The cell cycle: an introduction.* W.H. Freeman, New York.
Pardee, A.B. 1989. G_1 events and regulation of cell proliferation. *Science* 246:603–608.

Cell Transformation *in Vitro*

Freedman, V.H., and Shin, S.-I. 1974. Cellular tumorigenicity in nude mice: correlation with cell growth in semi-solid medium. *Cell* 3:355–359.
Macpherson, I., and Montagnier, L. 1964. Agar suspension culture for the selective assay of cells transformed by polyoma virus. *Virology* 23:291–294.
Temin, H.M., and Rubin, H. 1958. Characteristics of an assay for Rous sarcoma virus and Rous sarcoma cells in tissue culture. *Virology* 6:669–688.

Biochemical Alterations in Transformed Cells

Gottesman, M. 1990. The role of proteases in cancer. *Sem. Cancer Biol.* 1:97–160.
Plantfaber, L.C., and Hynes, R.O. 1989. Changes in integrin receptors on oncogenically transformed cells. *Cell* 56:281–290.
Pollack, R., Osborn, M., and Weber, K. 1975. Patterns of organization of actin and myosin in normal and transformed cultured cells. *Proc. Natl. Acad. Sci. USA* 72:994–998.
Ruoslahti, E. 1988. Fibronectin and its receptors. *Ann. Rev. Biochem.* 57:375–413.
Sporn, M.B., and Roberts, A.B. 1985. Autocrine growth factors and cancer. *Nature* 313:745–747.

❖ PART II

Viral Oncogenes

❖ Chapter 2

Tumor Viruses

Tumor viruses have been particularly important models for studying neoplastic transformation because, as discussed in chapter 1, they efficiently and reproducibly induce tumors in experimental animals and transform cells in culture. Moreover, the relatively small size of viral compared with cellular genomes has stimulated molecular biological studies to identify specific viral genes responsible for induction of the transformed state. Consequently, genes responsible for the induction of neoplastic transformation were first identified in tumor viruses, ultimately leading to the identification of cellular oncogenes.

Animal viruses can be divided broadly into two groups: those with DNA genomes and those with RNA genomes. The DNA viruses with oncogenic potential constitute members of six distinct virus families: hepatitis B viruses, simian virus 40 (SV40) and polyomavirus, papillomaviruses, adenoviruses, herpesviruses, and poxviruses (Table 2.1). In contrast, members of only one family of RNA viruses, the retroviruses, are capable of inducing oncogenic transformation. However, it is studies of this family of viruses that have been most revealing in terms of our understanding the genes that are pivotal to the transformation process.

Hepatitis B Viruses

The hepatitis B viruses (hepadnaviruses) are the least complex genetically of the animal DNA viruses, with genomes of only ~3 kb. Members of this virus family have been isolated from several species, including ducks, squir-

TABLE 2.1
Oncogenic Viruses

Virus Family	Approximate Genome Size (kb)
DNA Viruses	
hepatitis B viruses	3
SV40 and polyomavirus	5
papillomaviruses	8
adenoviruses	35
herpesviruses	100–200
poxviruses	200
RNA Viruses	
retroviruses	9

rels, woodchucks, and human beings. They all share a common specificity for infection of liver cells, and they are of considerable interest because of their ability to induce malignant neoplasms in their natural hosts. For example, infection of woodchucks with woodchuck hepatitis virus has been clearly demonstrated to result in the development of hepatocellular carcinoma with frequencies of nearly 100% after latent periods of two to four years. Human hepatitis B virus infection is also strongly associated with the development of human hepatocellular carcinoma, such that the frequency of this disease is increased more than 100-fold in chronically infected individuals.

In spite of the fact that hepatitis B virus (HBV) is the causative agent of a common human cancer, the molecular basis for its oncogenic activity is incompletely understood. One HBV gene, called the *X* gene, has been implicated in the transformation process because it is consistently expressed in human hepatocellular carcinomas and is able to induce liver cancer in transgenic mice. The HBV X protein appears to act by stimulating the transcription of host cell genes that drive cell proliferation. In addition, the integration of HBV DNA into the genome of infected cells may induce mutations or chromosomal abnormalities that affect critical host cell genes. Such mutations could result in either the activation of cellular oncogenes or the inactivation of tumor suppressor genes by mechanisms similar to those described in chapters 6 through 10. Moreover, HBV infection results in chronic tissue damage leading to continual cell proliferation, which may in itself contribute to tumor development.

Simian Virus 40 And Polyomavirus

Simian virus 40 (SV40) and polyomavirus are particularly important from the standpoint of experimental cancer research in that they are probably the best understood DNA tumor viruses and serve as models for DNA tumor virus-induced cell transformation. The small size of their genomes (~5 kb) has facilitated molecular analysis, and the availability of good *in vitro* culture systems to study both virus replication and transformation has allowed extensive biological experimentation.

Importantly, neither polyomavirus nor SV40 induces tumors in its natural host species (mice and monkeys, respectively). Rather, their oncogenicity is purely a laboratory phenomenon that is revealed by infection of heterologous species in which the viruses cannot replicate. This situation is mimicked in tissue culture (Fig. 2.1). Infection of permissive cell cultures (mouse cells for polyomavirus and monkey cells for SV40) results in a lytic infection in which the viruses replicate, resulting in cell death and the release of progeny virus particles. The consequence of infection of cells of the natural host species is virus replication and cell death, not transformation. The transforming potential of these viruses is revealed by infection of a nonpermissive species in which virus replication is blocked (for example, hamster cells for polyomavirus and mouse cells for SV40). However, the efficiency of transformation following infection of nonpermissive cells is much lower (at least 100-fold) than the efficiency of lytic infection of permissive cells. Transformation by these viruses is therefore a rare event that occurs when their normal life cycle is aborted.

In spite of the fact that polyomavirus and SV40 are not natural transforming agents, they have proved to be excellent models and a great deal has been learned about the molecular basis for their transforming activities. Transformation of cells by these viruses is a direct result of stable integration of viral DNA followed by expression of the same viral genes that function at early stages of virus replication in permissive cells. The genomes of polyomavirus and SV40 are divided into two regions: the early region and the late region (Fig. 2.2). The early region encodes proteins expressed immediately after viral infection and required for viral DNA synthesis. The late region is not expressed until after synthesis of viral DNA has started and includes genes encoding the viral structural proteins. The early region of SV40 encodes two proteins, called small t and large T (for tumor) antigens, of ~17 kd and 94 kd, respectively. The mRNAs for these proteins are generated by alternative splicing of a single early region primary transcript, such that two different donor splice sites are joined to a common acceptor site to yield the large T and small t mRNAs. The early region of polyomavirus likewise encodes small and large T antigens similar to those of SV40, as well as a third protein designated middle T (~55 kd). In SV40, large T

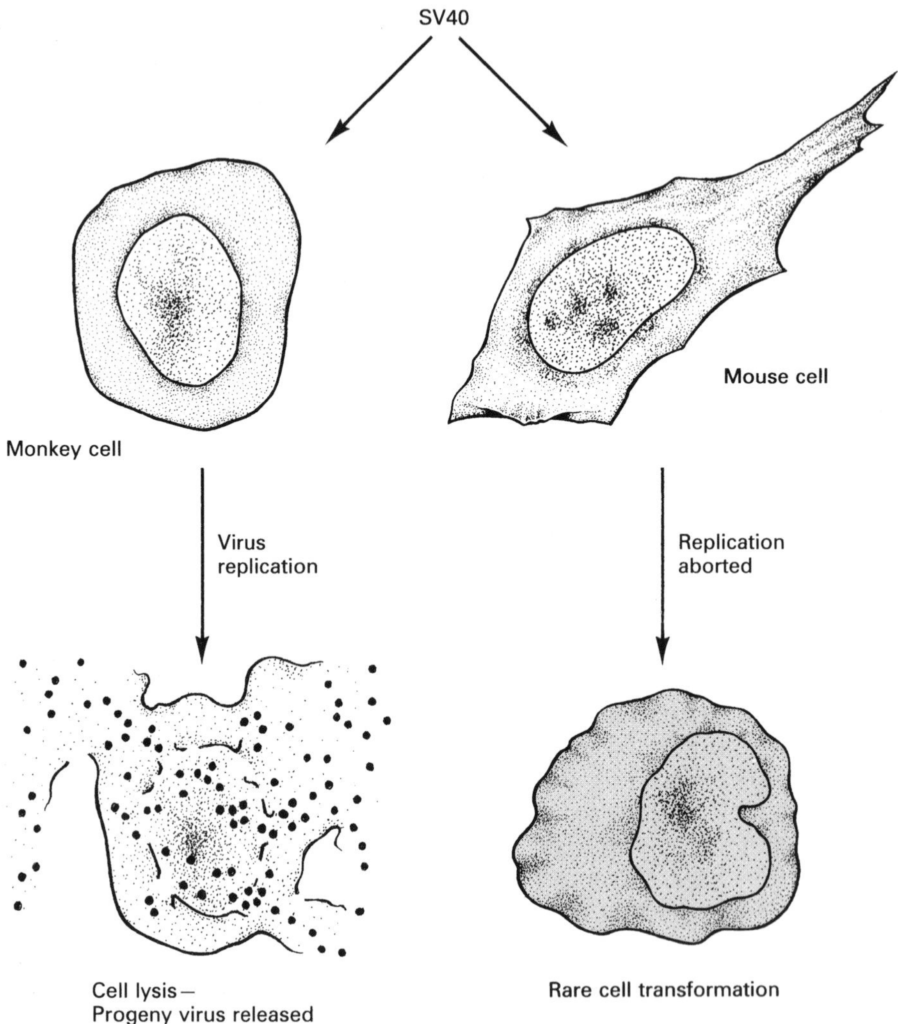

Figure 2.1
SV40 replication and transformation. Infection of a permissive cell (e.g., monkey) results in virus replication, cell lysis, and release of progeny virus particles. In a nonpermissive cell (for example, mouse), virus replication is blocked and a small fraction of the infected cells stably integrate SV40 DNA and become permanently transformed.

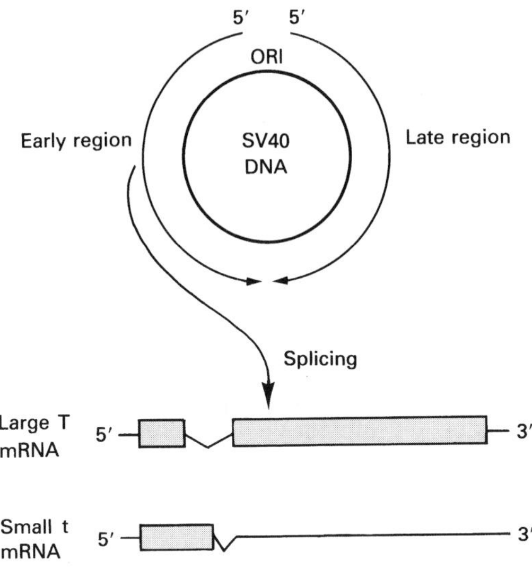

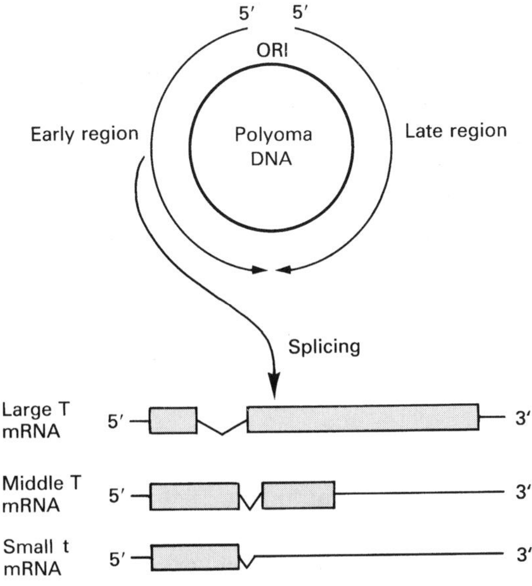

FIGURE 2.2
The SV40 and polyomavirus genomes and early region transcripts. Multiple early region mRNAs are produced by alternative splicing. Coding sequences are indicated as filled boxes.

alone is sufficient to induce transformation. Transformation by polyomavirus primarily results from the action of middle T, although the large T antigen cooperates with middle T in the transformation of primary cells. The small t antigens are not required for transformation of cultured cells by either SV40 or polyomavirus, although they appear to provide a helper function that contributes to the full expression of the neoplastic phenotype.

During lytic infection, these early region proteins fulfill multiple functions required for virus replication, including stimulation of host cell DNA synthesis, initiation of viral DNA synthesis, activation of transcription of viral late genes, and activation of transcription of some genes of the host cell. Stimulation of host cells is a critical event early in the virus life cycle, because viral DNA synthesis is dependent on the activity of host cell enzymes (for example, DNA polymerases). Most normal target cells for viral infection *in vivo* are nonproliferating cells (arrested in G_0) that do not express DNA synthetic enzymes. Therefore, the SV40 and polyomavirus early genes must stimulate host cell proliferation in order to induce the host cell machinery needed for viral DNA replication in the normal lytic cycle. Cell transformation apparently results from stable integration of the early region of viral DNA and continual expression of the early gene products in a cell that is nonpermissive for virus replication.

Both the polyomavirus and the SV40 T antigens induce cell transformation, at least in part, by interacting with cellular proteins encoded by proto-oncogenes or tumor suppressor genes. Both SV40 and polyomavirus large T antigens bind and inactivate the protein encoded by the *Rb* tumor suppressor gene, which acts as a negative regulator of normal cell proliferation (see chapter 9). The SV40 large T antigen, which alone is sufficient to transform cells, also binds and interferes with the action of a second tumor suppressor gene product, p53 (chapter 10). In contrast, polyomavirus middle T stimulates cell growth by activating the proteins encoded by *src* and other proto-oncogenes that participate in stimulating cell proliferation in response to growth factors (see chapters 13 and 17).

Papillomaviruses

The papillomaviruses, although comparable in genetic complexity (genomes of ~8 kb), have been far more difficult to study than polyomavirus and SV40, largely owing to the lack of suitable cell culture systems for papillomavirus propagation *in vitro*. However, the papillomaviruses are important etiologic agents of neoplastic disease in a variety of species, including human beings. In fact, the first DNA tumor virus isolated was the Shope papillomavirus. This virus induces benign papillomas, which occasionally progress to malignant carcinomas, in cottontail rabbits. Bovine papillomavirus

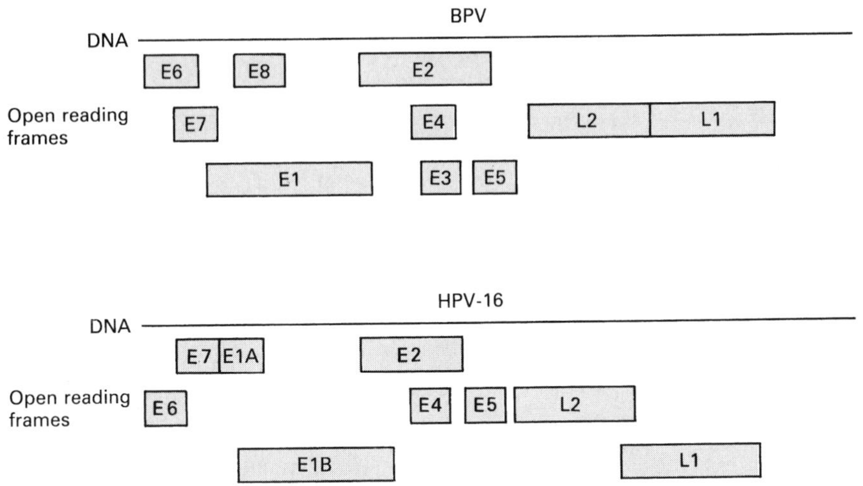

FIGURE 2.3
The genomes of bovine papillomavirus (BPV) and human papillomavirus type 16 (HPV-16). Protein coding regions (open reading frames) are designated E (for early) or L (for late). The transforming genes are *E5/E6* for BPV and *E6/E7* for HPV-16.

likewise induces skin papillomas in its natural host and has been well studied because of its ability to induce transformation of cultured cells. In humans, more than sixty different types of papillomaviruses have been isolated. Some of these viruses induce only benign neoplasms—for example, warts. Other human papillomaviruses, however, are causative agents of common malignancies, particularly cervical carcinoma and other anogenital cancers. More than 90% of anogenital carcinomas have been found to contain integrated human papillomavirus DNA sequences, primarily of human papillomavirus types 16 and 18, indicating that these viruses play a causal role in tumor development.

Molecular analysis has identified two genes of bovine papillomavirus, *E5* and *E6*, that induce transformation of cultured cells (Fig. 2.3). The major transforming gene, *E5*, encodes a small polypeptide of only forty-four amino acids that stimulates the activity of growth factor receptors on the cell surface. Transformation by the oncogenic human papillomaviruses also is mediated by two viral genes, *E6* and *E7*, which are regularly expressed in cervical carcinomas. The E6 and E7 proteins of these viruses act similarly to SV40 large T antigen by interfering with the activity of cellular proteins encoded by the *Rb* and *p53* tumor suppressor genes. In particular, E7 binds to Rb and E6 stimulates degradation of p53.

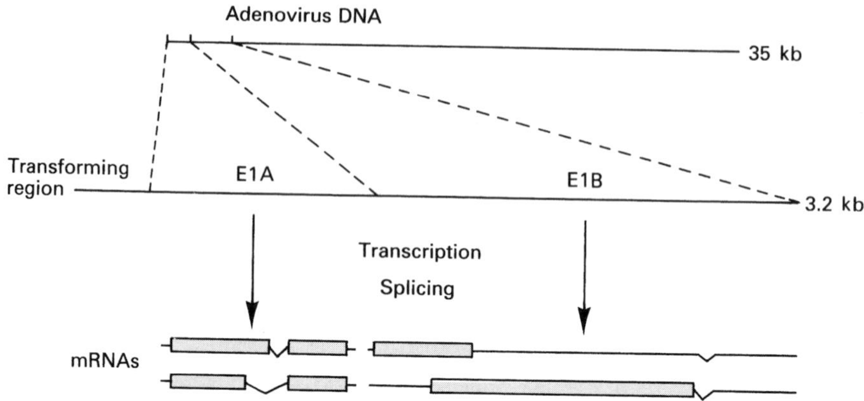

FIGURE 2.4
The adenovirus genome and transforming region. Two adenovirus genes (*E1A* and *E1B*) function to induce transformation. Both *E1A* and *E1B* encode multiple mRNAs and proteins as a consequence of alternative splicing. Only those mRNAs found in transformed cells are shown, with coding regions depicted as filled boxes.

Adenoviruses

The adenoviruses are a large and diverse virus family with genomes of ~35 kb. Like SV40 and polyomavirus, the adenoviruses are lytic in cells of their natural host but induce transformation of heterologous cells that are nonpermissive for virus replication. Thus, human adenoviruses replicate productively in human cells and have no known association with human neoplastic disease. However, the human adenoviruses will transform rat or hamster cells, nonpermissive hosts that do not support virus replication. As with SV40 and polyomavirus, transformation by adenoviruses is a rare event with an efficiency much lower than the efficiency of productive infection of permissive cells.

The molecular biology of adenovirus replication and transformation, like that of polyomavirus and SV40, has been intensively studied. Transformation is mediated by the combined action of two early region genes, *E1A* and *E1B*, both of which encode multiple protein products as a result of alternative splicing (Fig. 2.4). Two *E1A* mRNAs are expressed in transformed cells and translated to yield overlapping proteins of ~26 kd and 32 kd that, like SV40 large T, function to activate transcription of both viral and cellular genes. The *E1B* region likewise encodes two major proteins in transformed cells, of ~19 kd and 55 kd.

The adenovirus E1A and E1B proteins, like SV40 large T antigen and the human papillomavirus transforming proteins, specifically interact with cellular proteins encoded by the *Rb* and *p53* tumor suppressor genes. In particular, the adenovirus E1A proteins interact with Rb and the E1B 55 kd protein interacts with p53. It is noteworthy that the adenovirus E1A and E1B proteins—which, like the human papillomavirus E6 and E7 proteins, are needed in combination to induce transformation—interact separately with Rb and p53. This contrasts with SV40 large T, which can alone induce cell transformation and interacts with both Rb and p53. The interactions of the transforming proteins of SV40, human papillomaviruses, and adenoviruses with the same tumor suppressor gene products clearly indicate that these small DNA tumor viruses induce cell transformation by a common pathway in which inactivating the proteins encoded by the cellular *Rb* and *p53* tumor suppressor genes plays a central role.

Herpesviruses

The herpesviruses are among the most complex animal viruses, with genomes ranging from 100 to 200 kb. Several members of the herpesvirus family are of particular interest because they induce tumors in their natural hosts. Such naturally occurring herpesvirus-induced neoplasms in animals include Lucké's carcinoma in frogs and Marek's disease (a T-cell lymphoma) in chickens. Another highly oncogenic member of this family is herpesvirus saimiri, which is nonpathogenic in its natural host (squirrel monkeys) but induces virulent T-cell lymphomas in other monkey species.

A human herpesvirus, Epstein-Barr virus (EBV), has been identified as a causative agent of African Burkitt's lymphoma, B- cell lymphomas in immunosuppressed individuals (for example, AIDS patients), and nasopharyngeal carcinoma. The epidemiological association of EBV with these lymphomas is further supported by the ability of EBV to transform human lymphocytes in culture. In spite of this important biology, molecular studies have been hindered both by the large size of the EBV genome (~180 kb) and by the lack of cell culture systems for virus propagation. Nonetheless, several EBV genes have been found to be required for transformation of primary human lymphocytes. Moreover, one of these genes (*LMP*1) is itself capable of inducing transformation of fibroblasts and altering the growth properties of lymphoid cell lines. This gene encodes a plasma membrane protein of currently unknown function.

Poxviruses

The last family of DNA viruses with transforming activity comprises the poxviruses. Like herpesviruses, the poxviruses are extremely complex animal viruses with genomes of ~200 kb. Unlike all of the other DNA viruses, which replicate in the cell nucleus, the poxviruses replicate in the cytoplasm of infected cells. Poxviruses do not induce malignant tumors, but they do induce benign neoplasms. Two examples are the Shope fibroma virus, which causes benign fibromas in rabbits, and the Yaba monkey virus, which induces transient superficial tumors on the limbs of rhesus monkeys. The molecular basis of poxvirus transformation is obscure, because of both the large genome size and the lack of *in vitro* transformation systems.

Retroviruses

In contrast with the diversity of DNA viruses with oncogenic potential, only one family of RNA viruses has the capacity to induce neoplastic transformation. The unique feature of this family, the retroviruses, is that they replicate in infected cells through a DNA intermediate, called a *provirus,* which is integrated into chromosomal DNA of the host (Fig. 2.5). The virion RNA consists of two identical subunits of ~9 kb. In infected cells, this RNA is copied into DNA by the viral reverse transcriptase, which is carried in virus particles. The viral DNA is integrated into host DNA to form the DNA provirus. Proviral DNA is replicated along with cellular DNA in dividing cells. In addition, the proviral DNA is transcribed by host cell RNA polymerase to yield progeny viral RNA genomes as well as mRNAs, which are translated into viral proteins (see chapter 3).

In most cases, productive infection with retroviruses is not cytocidal. Progeny virus particles are released by budding from the plasma membrane without cell lysis. In contrast to most of the DNA tumor viruses, oncogenic retroviruses are consequently able to transform the same cells in which replication proceeds efficiently. Perhaps as a corollary, many retroviruses induce cancer in nature in their normal host species. They are a large virus family found in a wide variety of animals, including reptiles, chickens, mice, cats, monkeys, and human beings. One human retrovirus, human T-lymphotropic virus type I (HTLV I), is the causative agent of adult T cell leukemia and a second, HTLV II, may cause hairy cell leukemia. Although not directly oncogenic, a third human retrovirus, human immunodeficiency virus (HIV), is of major interest as the causative agent of AIDS.

The efficiency with which retroviruses induce transformation varies with different viruses. Some viruses of this family are the most potent known carcinogens, both in intact animals and in cell culture. A good example is provided by Rous sarcoma virus, which was the first clearly accepted tumor virus, isolated from a chicken tumor by Peyton Rous in 1911. Inoculation of chickens with this virus results in the formation of large sarcomas within one to two weeks. Likewise, Rous sarcoma virus transforms chicken embryo fibroblasts in culture with high efficiency, such that nearly every infected cell is converted into the transformed phenotype.

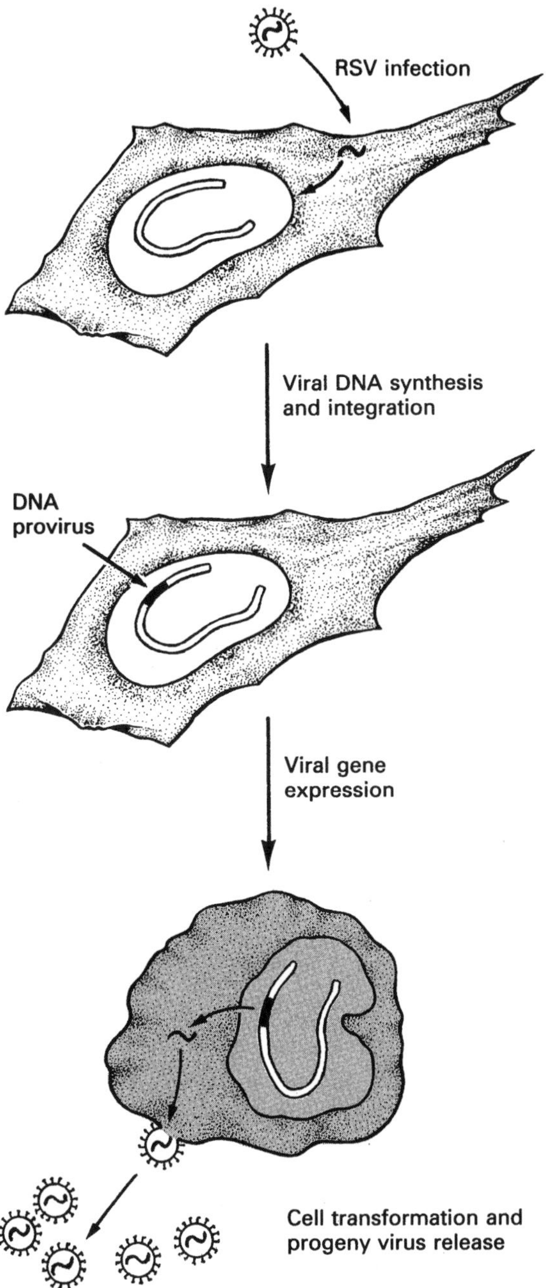

FIGURE 2.5
Cycle of replication and cell transformation by Rous sarcoma virus (RSV). Viral DNA is synthesized and integrated into chromosomal DNA of the host to form the DNA provirus. Viral gene expression results in both cell transformation and production of progeny virus particles, which are released by budding from the plasma membrane without cell lysis.

The oncogenic retroviruses have proved to be extraordinarily informative objects of study for the molecular biologist interested in cancer. A few such retroviruses, including the HTLVs, induce neoplasia as a result of the activity of viral gene products that also function directly in virus replication. These gene products (for example, the HTLV I Tax protein) apparently serve as transcriptional activators of both the viral genome and some genes of the host cell. The induction of neoplastic transformation by most retroviruses, however, is not mediated by genes that also function in virus replication. As discussed in the next chapter, the retroviral oncogenes are instead a unique set of genes that function specifically to induce transformation without playing a direct role in virus replication. Most importantly, as will be appreciated throughout this text, the identification of retroviral oncogenes has served to elucidate multiple different aspects of carcinogenesis and cellular oncogene activation.

Summary

Members of seven distinct families of animal viruses have the ability to induce tumors either in their natural hosts or in heterologous species. Studies of these tumor viruses have made major contributions, both directly and indirectly, to our understanding of human neoplastic disease. Some tumor viruses, such as hepatitis B virus, human papillomaviruses, Epstein-Barr virus, and HTLV, are causative agents of human neoplasms. Other tumor viruses, including SV40, polyomavirus, and adenovirus, are not natural agents of neoplastic disease but have provided simple model systems for molecular studies of cell transformation. The retroviruses in particular have provided important models for both molecular analysis and for biological studies of the role of tumor viruses in naturally occurring neoplasms. Most importantly, understanding the molecular biology of retroviruses has led not only to the identification of a diverse set of viral oncogenes, but also to the discovery of cellular oncogenes and their roles in both virus- and nonvirus-induced cancers.

References

General References

Botchan, M., Grodzicker, T. and Sharp, P.A., eds. 1986. *Cancer cells,* Vol. 4: *DNA tumor viruses.* Cold Spring Harbor, New York.

Fields, B.D., Knipe, D.M., Chanock, R.M., Hirsch, M.S., Melnick, J.L., Monath, T.P., and Roizman, B., eds. 1990. *Virology,* 2d ed. Raven Press, New York.

Gross, L. 1983. *Oncogenic viruses,* 3d ed. Pergamon Press, New York.

Tooze, J., ed. 1981. *Molecular biology of tumor viruses: DNA tumor viruses,* 2d ed. Cold Spring Harbor, New York.

Weiss, R., Teich, N., Varmus, H., and Coffin, J., eds. 1985. *Molecular biology of tumor viruses: RNA tumor viruses,* 2d ed. Cold Spring Harbor, New York.
Zur Hausen, H. 1991. Viruses in human cancer. *Science* 254:1167–1173.

Hepatitis B Viruses

Chisari, F.V., Klopchin, K., Moriyama, T., Pasquinelli, C., Dunsford, H.A., Sell, S., Pinkert, C.A., Brinster, R.L., and Palmiter, R.D. 1989. Molecular pathogenesis of hepatocellular carcinoma in hepatitis B virus transgenic mice. *Cell* 59:1145–1156.
Hansen, L.J., Tennant, B.C., Seeger, C., and Ganem, D. 1993. Differential activation of *myc* gene family members in hepatic carcinogenesis by closely related hepatitis B viruses. *Mol. Cell. Biol.* 13:659–667.
Kekule, A.S., Lauer, U., Weiss, L., Luber, B., and Hofschneider, P.H. 1993. Hepatitis B virus transactivator HBx uses a tumour promoter signalling pathway. *Nature* 361:742–745.
Kim, C.-M., Koike, K., Saito, I., Miyamura, T., and Jay, G. 1991. *HBx* gene of hepatitis B virus induces liver cancer in transgenic mice. *Nature* 351:317–320.
Summers, J., Smolec, J.M., and Snyder, R.A. 1978. A virus similar to human hepatitis B virus associated with hepatitis and hepatoma in woodchucks. *Proc. Natl. Acad. Sci. USA* 75:4533–4537.

Simian Virus 40 and Polyomavirus

Courtneidge, S.A., and Smith, A.E. 1983. Polyoma virus transforming protein associates with the product of the c-*src* cellular gene. *Nature* 303:435–439.
DeCaprio, J.A., Ludlow, J.W., Figge, J., Shew, J.-Y., Huang, C.- M., Lee, W.-H., Marsilio, E., Paucha, E., and Livingston, D.M. 1988. SV40 large tumor antigen forms a specific complex with the product of the retinoblastoma susceptibility gene. *Cell* 54:275–283.
Dilworth, S.M., Brewster, C.E.P., Jones, M.D., Lanfrancone, L., Pelicci, G., and Pelicci, P.G. 1994. Transformation by polyoma virus middle T-antigen involves the binding and tyrosine phosphorylation of Shc. *Nature* 367:87–90.
Dyson, N., Bernards, R., Friend, S.H., Gooding, L.R., Hassell, J.A., Major, E.O., Pipas, J.M., Vandyke, T., and Harlow, E. 1990. Large T antigens of many polyomaviruses are able to form complexes with the retinoblastoma protein. *J. Virol.* 64:1353–1356.
Gross, L. 1953. A filterable agent, recovered from Ak leukemic extracts, causing salivary gland carcinomas in C3H mice. *Proc. Soc. Exp. Biol. Med.* 83:414–421.
Kriegler, M., Perez, C.F., Hardy, C., and Botchan, M. 1984. Transformation mediated by the SV40 T antigens: separation of the overlapping SV40 early genes with a retroviral vector. *Cell* 38:483–491.
Lane, D.P., and Crawford, L.V. 1979. T antigen is bound to a host protein in SV40-transformed cells. *Nature* 278:261–263.
Linzer, D.I.H., and Levine, A.J. 1979. Characterization of a 54K dalton cellular SV40 tumor antigen present in SV40-transformed cells and uninfected embryonal carcinoma cells. *Cell* 17:43–52.

Rassoulzadegan, M., Cowie, A., Carr, A., Glaichenhaus, N., Kamen, R., and Cuzin, F. 1982. The role of individual polyoma virus early proteins in oncogenic transformation. *Nature* 300: 713–718.

Sweet, B.H., and Hilleman, M.R. 1960. The vacuolating virus, SV 40. *Proc. Soc. Exp. Biol. Med.* 105:420–427.

Vogt, M., and Dulbecco, R. 1960. Virus-cell interaction with a tumor-producing virus. *Proc. Natl. Acad. Sci. USA* 46:365–370.

Papillomaviruses

Durst, M., Gissman, L., Ikenberg, H., and zur Hausen, H. 1983. A papillomavirus DNA from a cervical carcinoma and its prevalence in cancer biopsy samples from different geographic regions. *Proc. Natl. Acad. Sci. USA* 80:3812-3816.

Dyson, N., Howley, P.M., Munger, K., and Harlow, E. 1989. The human papilloma virus-16 E7 oncoprotein is able to bind to the retinoblastoma gene product. *Science* 243:934–937.

Martin, P., Vass, W.C., Schiller, J.T., Lowy, D.R., and Velu, T.J. 1989. The bovine papillomavirus E5 transforming protein can stimulate the transforming activity of EGF and CSF-1 receptors. *Cell* 59:21–32.

Munger, K., Phelps, W.C., Bubb, V., Howley, P.M., and Schlegel, R. 1989. The *E6* and *E7* genes of the human papillomavirus type 16 together are necessary and sufficient for transformation of primary human keratinocytes. *J. Virol.* 63:4417–4421.

Nakabayashi, Y., Chattopadhyay, S.K., and Lowy, D.R. 1983. The transformation function of bovine papilloma virus DNA. *Proc. Natl. Acad. Sci. USA* 80:5832–5836.

Nilson, L.A., and DiMaio, D. 1993. Platelet-derived growth factor receptor can mediate tumorigenic transformation by the bovine papillomavirus E5 protein. *Mol. Cell. Biol.* 13:4137–4145.

Scheffner, M., Werness, B.A., Huibregtse, J.M., Levine, A.J., and Howley, P.M. 1990. The E6 oncoprotein encoded by human papillomavirus types 16 and 18 promotes the degradation of p53. *Cell* 63:1129–1136.

Schiller, J.T., Vass, W.C., and Lowy, D.R. 1984. Identification of a second transforming region in bovine papillomavirus DNA. *Proc. Natl. Acad. Sci. USA* 81:7880–7884.

Shope, R.E. 1933. Infectious papillomatosis of rabbits. *J. Exp. Med.* 58:607–624.

Adenoviruses

Berk, A.J. 1986. Adenovirus promoters and E1A transactivation. *Ann. Rev. Genet.* 20:45–79.

Graham, F.L., van der Eb, A.J., and Heijneker, H.L. 1974. Size and location of the transforming region in human adenovirus type 5 DNA. *Nature* 251:687–691.

Hilleman, M.R., and Werner, J.R. 1954. Recovery of new agent from patients with acute respiratory illness. *Proc. Soc. Exp. Biol. Med.* 85:183–188.

Rowe, W.P., Huebner, R.J., Gillmore, L.K., Parrott, R.H., and Ward, T.G. 1953. Isolation of a cytopathogenic agent from human adenoids undergoing spontaneous degeneration in tissue culture. *Proc. Soc. Exp. Biol. Med.* 84:570–573.

Sarnow, P., Ho, Y.S., Williams, J., and Levine, A.J. 1982. Adenovirus E1B-58kd tumor antigen and SV40 large tumor antigen are physically associated with the same 54 kd cellular protein in transformed cells. *Cell* 28:387–394.

Trentin, J.J., Yabe, Y., and Taylor, G. 1962. The quest for human cancer viruses. *Science* 137:835–841.

Van den Elsen, P., Houweling, A., and van der Eb, A.J. 1983. Expression of region E1B of human adenovirus in the absence of region E1A is not sufficient for complete transformation. *Virology* 128:377–390.

Whyte, P., Buchkovich, K.J., Horowitz, J.M., Friend, S.H., Raybuck, M., Weinberg, R.A., and Harlow, E. 1988. Association between an oncogene and an anti-oncogene: the adenovirus E1A proteins bind to the retinoblastoma gene product. *Nature* 334:124–129.

Herpesviruses

Biggs, P.M., Churchill, A.E., Rootes, D.G., and Chubb, R.C. 1968. The etiology of Marek's disease—an oncogenic herpes-type virus. *Perspect. Virol.* 6:211–234.

De Thé G., Geser, A., Day, N.E., Tukei, P.M., Williams, E.H., Beri, D.P., Smith, P.G., Dean, A.G., Bornkamm, G.W., Feorinio, P., and Henle, W. 1978. Epidemiological evidence for causal relationship between Epstein-Barr virus and Burkitt's lymphoma from Ugandan prospective study. *Nature* 274:756–761.

Epstein, M.A., Achong, B.G., and Barr, Y.M. 1964. Virus particles in cultured lymphoblasts from Burkitt's lymphoma. *Lancet* 1:702–703.

Fahreus, R., Rymo, L., Rhim, J.S., and Klein, G. 1990. Morphological transformation of human keratinocytes expressing the *LMP* gene of Epstein-Barr virus. *Nature* 345:447–449.

Hammerschmidt, W., and Sugden, B. 1989. Genetic analysis of immortalizing functions of Epstein-Barr virus in human B lymphocytes. *Nature* 340:393–397.

Kaye, K.M., Izumi, K.M., and Kieff, E. 1993. Epstein-Barr virus latent membrane protein 1 is essential for B-lymphocyte growth transformation. *Proc. Natl. Acad. Sci. USA* 90:9150–9154.

Lucké, B. 1938. Carcinoma in the leopard frog: its probable causation by a virus. *J. Exp. Med.* 68:457–468.

Melendez, L.V., Hunt, R.D., Daniel, M.D., Blake, B.J., and Garcia, F.G. 1971. Acute lymphocytic leukemia in owl monkeys inoculated with Herpesvirus saimiri. *Science* 171:1161–1163.

Wang, D., Liebowitz, D., and Kieff, E. 1985. An EBV membrane protein expressed in immortalized lymphocytes transforms established rodent cells. *Cell* 43:831–840.

Poxviruses

Niven, J.S.F., Armstrong, J.A., Andrews, C.H., Pereira, H.G., and Valentine, R.C. 1961. Subcutaneous "growths" in monkeys produced by a poxvirus. *J. Pathol. Bacteriol.* 81:1–14.

Shope, R.E. 1932. A filtrable virus causing a tumor-like condition in rabbits and its relationship to virus myxomatosum. *J. Exp. Med.* 56:803–822.

Retroviruses

Baltimore, D. 1970. RNA-dependent DNA polymerase in virions of RNA tumour viruses. *Nature* 226:1209–1211.

Nerenberg, M., Hinrichs, S.H., Reynolds, R.K., Khoury, G., and Jay, G. 1987. The *tat* gene of human T-lymphotropic virus type 1 induces mesenchymal tumors in transgenic mice. *Science* 237:1324–1329.

Poiesz, B.J., Ruscetti, F.W., Gazdar, A.F., Bunn, P.A., Minna, J.D., and Gallo, R.C. 1980. Detection and isolation of type C retrovirus particles from fresh and cultured lymphocytes of a patient with cutaneous T-cell lymphoma. *Proc. Natl. Acad. Sci. USA* 77: 7415–7419.

Rous, P. 1911. A sarcoma of the fowl transmissible by an agent separable from the tumor cells. *J. Exp. Med.* 13:397–411.

Tanaka, A., Takahashi, C., Yamaoka, S., Nosaka, T., Maki, M., and Hatanaka, M. 1990. Oncogenic transformation by the *tax* gene of human T-cell leukemia virus type I *in vitro. Proc. Natl. Acad. Sci. USA* 87:1071-1075.

Temin, H.M. 1964. Nature of the provirus of Rous sarcoma. *Natl. Cancer Inst. Monogr.* 17:557–570.

Temin, H.M., and Mizutani, S. 1970. RNA-dependent DNA polymerase in virions of Rous sarcoma virus. *Nature* 226:1211–1213.

Varmus, H. 1988. Retroviruses. *Science* 240:1427–1435.

❖ Chapter 3

Retroviral Oncogenes

The discovery of retroviral oncogenes was the first step toward understanding the function of genes specifically responsible for induction of abnormal cell proliferation. As noted in the preceding chapter, different retroviruses vary substantially in their effects on infected cells, particularly the efficiency with which they induce oncogenic transformation. Some retroviruses, the prototype of which is Rous sarcoma virus (RSV), very rapidly induce tumors in infected animals and efficiently transform cells in culture (Fig. 3.1). Viruses of this type are called *acutely transforming viruses.* About forty such viruses have now been isolated from chickens, turkeys, mice, rats, cats, and monkeys.

Other oncogenic retroviruses, however, are much less efficient inducers of oncogenic transformation than the acutely transforming viruses (Fig. 3.1). These *weakly oncogenic viruses* induce tumors only after long latent periods (several months) in infected animals and do not transform cells in culture. The prototype example of this class of viruses is the avian leukosis virus (ALV), which induces B-cell lymphomas in chickens. It will replicate in cultures of chicken embryo fibroblasts but, in contrast with its close relative RSV, does not induce cell transformation. It is weakly oncogenic viruses of this type that are commonly found in nature.

The differences in transforming potential between the acutely transforming viruses and their much less oncogenic relatives led a number of investigators to focus their attention on attempting to elucidate the molecular basis for this difference in virus pathogenicity. These studies led to the identification of retroviral oncogenes.

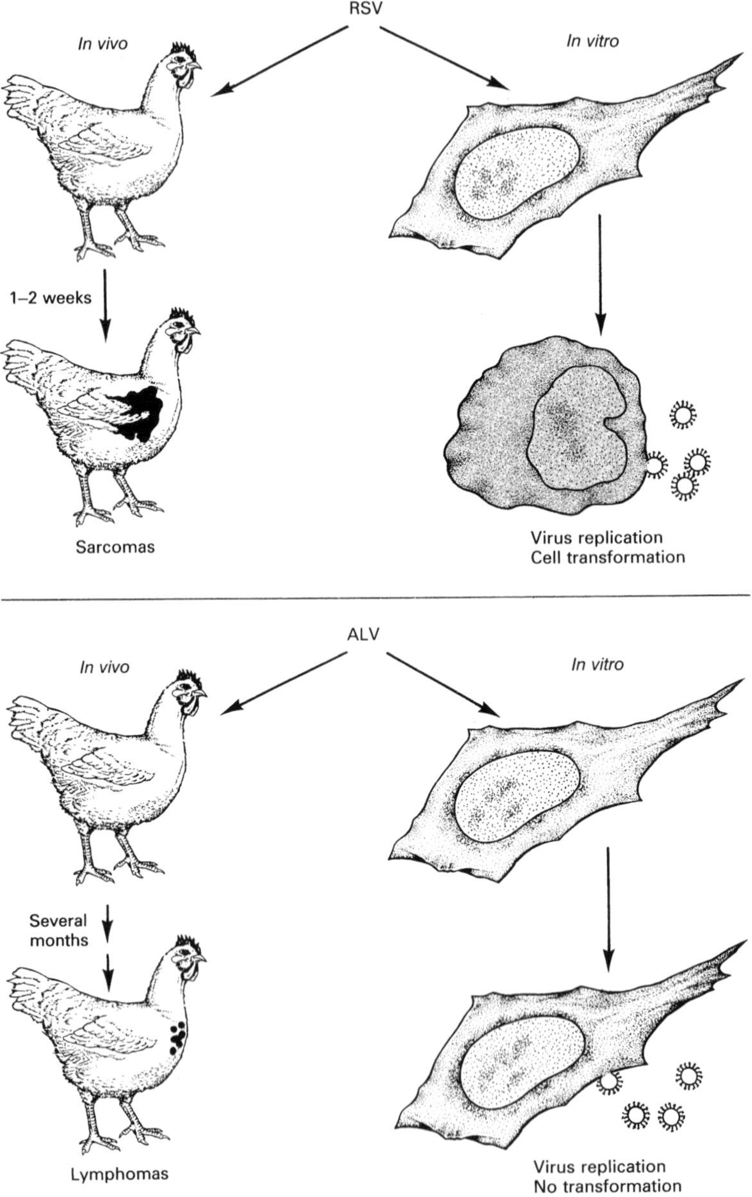

FIGURE 3.1
Neoplasm induction and cell transformation by Rous sarcoma virus (RSV) and avian leukosis virus (ALV). RSV induces sarcomas rapidly in infected chickens and efficiently transforms fibroblasts in culture. In contrast, ALV induces lymphomas only after long latent periods in infected birds and does not transform cells in culture.

The *src* Gene of RSV

The difference in transforming activity of RSV and ALV carried a critical implication. These two viruses were closely related and shared a common mode of replication in the same host cells, chicken embryo fibroblasts. Yet, only infection with RSV resulted in cell transformation. It thus appeared possible that RSV contained genetic information that was responsible for transformation and that was lacking in ALV. Identification of this information in the genome of RSV was the first step in the study of retroviral oncogenes.

Analysis of the sizes of genomic RNAs of RSV and ALV strains revealed an important difference. In 1970, Peter Duesberg and Peter Vogt found that the genomic RNA of RSV was ~10 kb, whereas that of ALV was smaller, ~8.5 kb. The suggestion was that the extra 1.5 kb of RSV RNA contained the information responsible for its unique biological activity: induction of transformation. However, substantiating this hypothesis required the isolation of RSV mutants that were defective in transforming ability. Two types of RSV mutants have contributed to our understanding of transformation: (1) deletion mutants that are totally defective in transforming activity and (2) point mutants that are temperature sensitive for transformation.

Transformation-defective deletion mutants were first isolated by Peter Vogt in 1971 (Fig. 3.2). Vogt infected microwell cultures of chick embryo fibroblasts with a limiting dilution of a clonal RSV stock such that each microwell culture received on average only a single transforming virus particle. As expected from the statistics of random sampling distributions, approximately two-thirds of the microwell cultures developed foci of transformed cells. The remaining cultures could have received no virus and been uninfected. Alternatively, some of these nontransformed cultures could have been infected with a nontransforming mutant virus of the type that Vogt hoped to isolate. To distinguish these possibilities, Vogt tested all of the nontransformed cultures for the presence of replicating nontransforming virus. He was able to isolate such transformation-defective viruses at high frequency, thus demonstrating that RSV could lose its transforming ability by mutation. These transformation-defective RSV mutants had also lost the ability to induce sarcomas in infected birds; so the effect of the mutation was not limited to tissue culture. Because the transformation-defective mutants were capable of normal replication, an important conclusion of these experiments was that wild-type RSV contained genetic information that was specifically required for transformation but not for virus replication.

Comparative analysis of the genomic RNAs of the transformation-defective mutants indicated that their genomes were ~8.5 kb. Thus, these mutants had arisen by deletion of about 15% of the wild-type RSV genome and were now similar in size to ALV. RSV therefore apparently contained about fifteen hundred nucleotides that were not required for virus replication but which

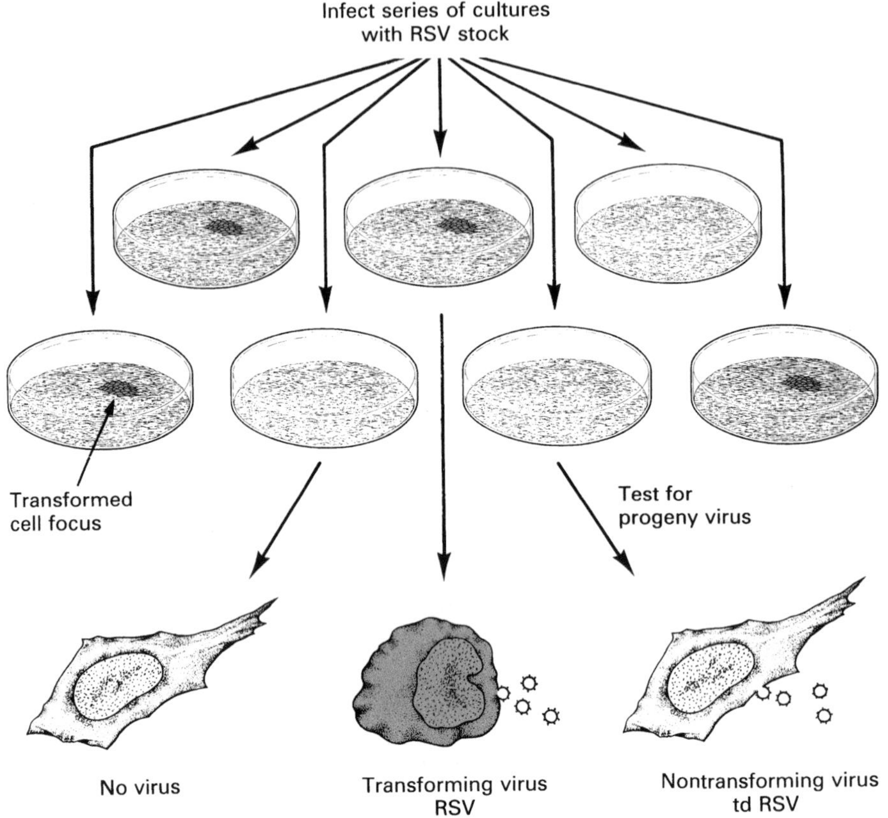

FIGURE 3.2
Isolation of transformation-defective (td) mutants of RSV. Cell cultures were infected with a limiting dilution of RSV such that each culture received on average a single transforming particle. Approximately two-thirds of the cultures developed foci of transformed cells and produced transforming progeny virus. Of the remaining cultures, some produced no progeny virus, indicating that they had not been infected. Others, however, produced nontransforming progeny (transformation-defective mutants) that were capable of replication but had lost the ability to induce cell transformation.

were necessary, at least in part, to induce both transformation of fibroblasts in culture and formation of sarcomas in chickens.

Although the isolation of transformation-defective deletion mutants defined transformation-specific genetic information, it did not establish whether this information was required continuously to maintain the transformed phenotype or only early in infection to initiate transformation. Resolution of this question required analysis of mutants that allowed control of gene function by

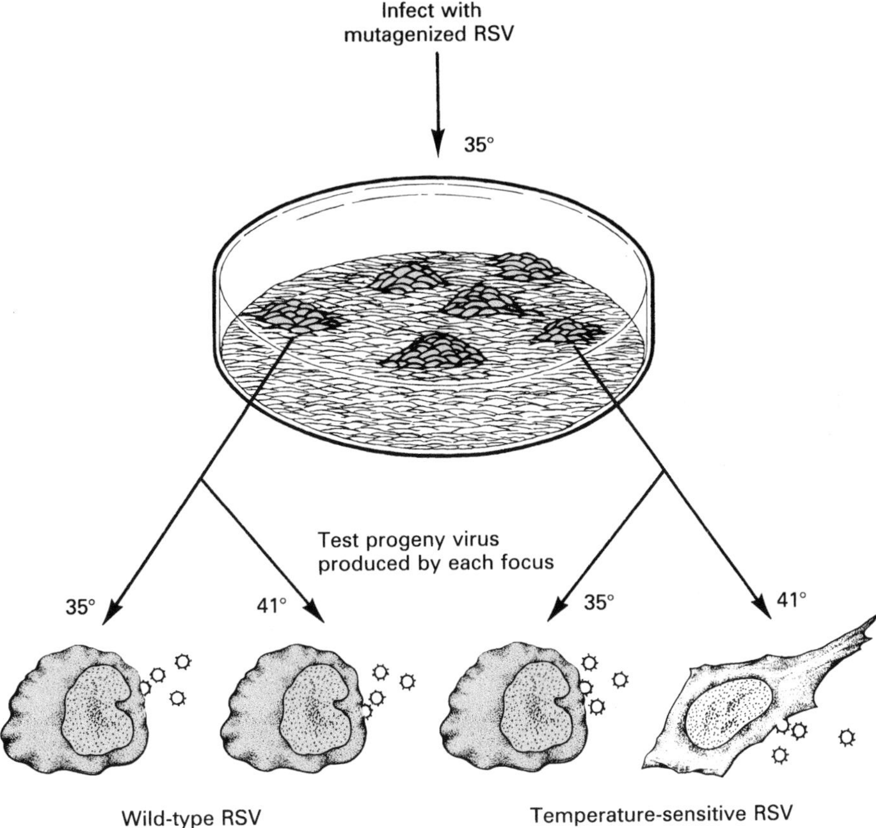

FIGURE 3.3
Isolation of temperature-sensitive RSV mutants. Cell cultures were infected with a mutagenized RSV stock at 35°C, the permissive temperature. Individual clones of progeny virus were isolated and tested for transformation at both 35° and 41°C, the nonpermissive temperature. Wild-type RSV induces transformation at both temperatures. In contrast, a temperature-sensitive mutant was isolated that induced transformation at 35° but not at 41°C. This RSV mutant replicated at both temperatures, indicating that the mutation affected only transformation.

experimental manipulation. This was achieved by the isolation of RSV mutants that were temperature sensitive for transformation (Fig. 3.3).

The first RSV mutant that was temperature sensitive specifically for induction of transformation was isolated by Steven Martin, following treatment of RSV with the potent chemical mutagen nitrosoguanidine. The mutagenized virus stock was plated on chick embryo fibroblasts at 35°C, the chosen permissive temperature for transformation, and virus clones produced by individual foci of RSV-transformed cells were isolated. Each of these individual virus clones were then tested for the ability to induce transformation at both 35° and 41°C, the chosen nonpermissive temperature. This protocol resulted in isolation of a mutant that induced focus formation at 35° but not at 41° and was thus temperature sensitive for transformation. This mutant virus was able to replicate equally well at both 35° and 41°; so the mutation affected only transformation, not virus replication. Thus, like the transformation-defective deletion mutants, Martin's temperature-sensitive mutant indicated that RSV contained genetic information specifically related to its transforming activity.

The question of whether this information was required for maintenance or initiation of transformation could then be resolved by temperature-shift experiments with infected cells. If a nontransformed culture at 41° was shifted to 35°, the cells rapidly became morphologically transformed. Conversely, if a transformed culture at 35° was shifted to 41°, the cells reverted to normal morphology. The temperature-sensitive mutation therefore defined a gene that was required for continuous maintenance of the transformed phenotype in infected cells.

A large number of RSV mutants that are temperature sensitive for maintenance of transformation have since been isolated. The standard genetic approaches of complementation and recombination were initially employed to ask whether RSV contained one or more than one transforming gene. Coinfection of cells with different combinations of temperature-sensitive transformation mutants did not result in cell transformation. The mutants therefore did not complement each other, indicating that they all affected a single gene. Moreover, coinfection with a temperature-sensitive mutant and a transformation-defective deletion mutant resulted in neither complementation nor recombination. Both types of mutants therefore defined a single gene of RSV that was required for transformation of fibroblasts in culture and for induction of sarcomas *in vivo*. Because RSV induces sarcomas, its transforming gene, which was the first oncogene to be identified, was called *src*.

The RSV And ALV Genomes

The genomic organization of retroviruses has been analyzed by a variety of molecular biological approaches, including cloning and sequencing. The nontransforming and weakly transforming retroviruses share a common genome structure, of which ALV is a representative example (Fig. 3.4).

Typical simple retroviral genomes contain three coding segments, each of which gives rise to multiple proteins as a consequence of posttranslational cleavage. The *gag* gene encodes core proteins, which are the major structural proteins of the virus particles. The *pol* gene encodes virion reverse transcriptase and integrase, enzymes that function in synthesis and integration of proviral DNA. The *env* gene encodes envelope glycoproteins, which interact with receptors on the surface of susceptible host cells. The viral genome also contains a

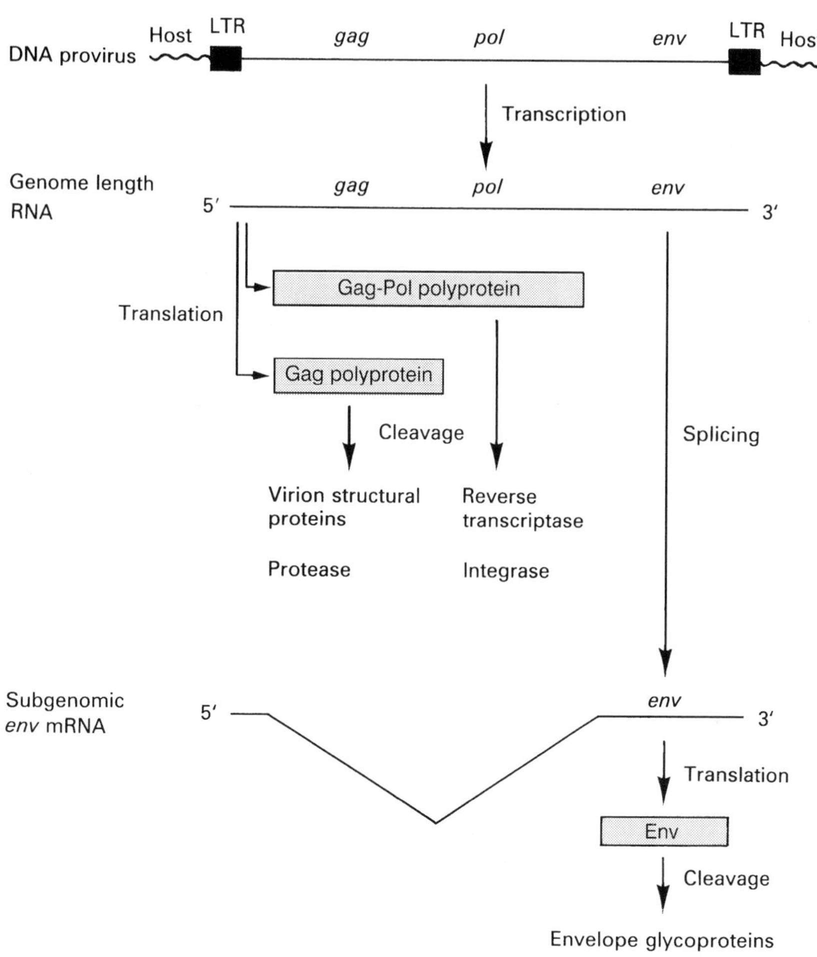

FIGURE 3.4
Retrovirus genome organization and expression. ALV is shown as an example of a typical nontransforming retrovirus. Proteins are depicted as filled boxes. The DNA provirus, integrated into the host chromosome, is transcribed to yield full length viral RNA. This primary transcript is the genomic RNA for progeny virus particles, as well as serving as the mRNA for the *gag* and *pol* genes. In addition, the full length primary transcript is spliced to yield mRNA for *env*.

number of *cis*-acting sequences that play important roles at several stages of the virus life cycle. These sequences include signals for packaging RNA into virus particles, sites for initiation of DNA synthesis, specific sequences that recombine with host DNA during viral DNA integration, and signals for transcription and processing of viral RNA.

The viral transcriptional regulatory sequences, which lead to efficient transcription of proviral DNA in infected cells, are contained within the long terminal repeats (LTRs). The LTRs are generated in the course of reverse transcription by duplication of several hundred nucleotides present at the termini of viral RNA. Sequences at the ends of the LTRs also direct the efficient integration of viral DNA into the host chromosome, which occurs specifically at the termini of the LTRs. Integrated proviral DNA is therefore flanked by these direct repeat sequences, which are referred to as the 5' and 3' LTRs. The LTRs contain three types of transcriptional regulatory sequences: (1) the promoter, at which transcription initiates in the 5' LTR; (2) enhancers, which result in high level transcription of the viral genome; and (3) sequences in the 3' LTR that signal polyadenylation of the viral RNA. Transcription is initiated in the 5' LTR and terminated in the 3' LTR to yield a primary transcript that is a complete copy of the viral RNA genome. Little transcription is normally initiated in the 3' LTR, even though it is identical in sequence with the 5' LTR. The full length primary transcript is the mRNA for *gag* and *pol* as well as the genomic RNA of progeny virus particles. In addition, subgenomic mRNAs (for example, for the *env* gene) are derived from this primary transcript by splicing.

Comparison of the RSV genome with those of ALV and transformation-defective RSV deletion mutants indicates that the RSV *src* gene is an additional gene located near the 3' end of RSV RNA, internal to the 3' LTR (Fig. 3.5). It is expressed in infected cells as a spliced mRNA whose transcription is initiated, like that of the viral replicative genes, in the 5' LTR. Gene transfer experiments with subgenomic fragments of RSV DNA indicate that *src* is not only necessary but also sufficient to induce cell transformation. Nucleotide sequence analysis indicates that the *src* gene contains a single open reading frame that encodes a protein of ~60 kd. The *src* gene product has been identified and extensively studied: it will be discussed in detail in chapter 13.

Other Retroviral Oncogenes

All of the other acutely transforming viruses, like RSV, contain at least one viral oncogene (and, in some cases, two) that does not have a role in virus replication but is responsible for the viruses' tumorigenicity *in vivo* and transforming activity *in vitro*. In several cases, the same oncogene is found in different viruses, sometimes including viruses isolated from different species. The current list (Table 3.1) comprises a total of more than two dozen distinct retroviral oncogenes, all of which are potent inducers of neoplastic transformation. These oncogenes have been designated by three-letter names that in most cases refer to the strain, species of isolation, or type of neoplasm induced by the acutely transforming virus in which the oncogene was first identified.

The genomic organization of the other acutely transforming viruses differs

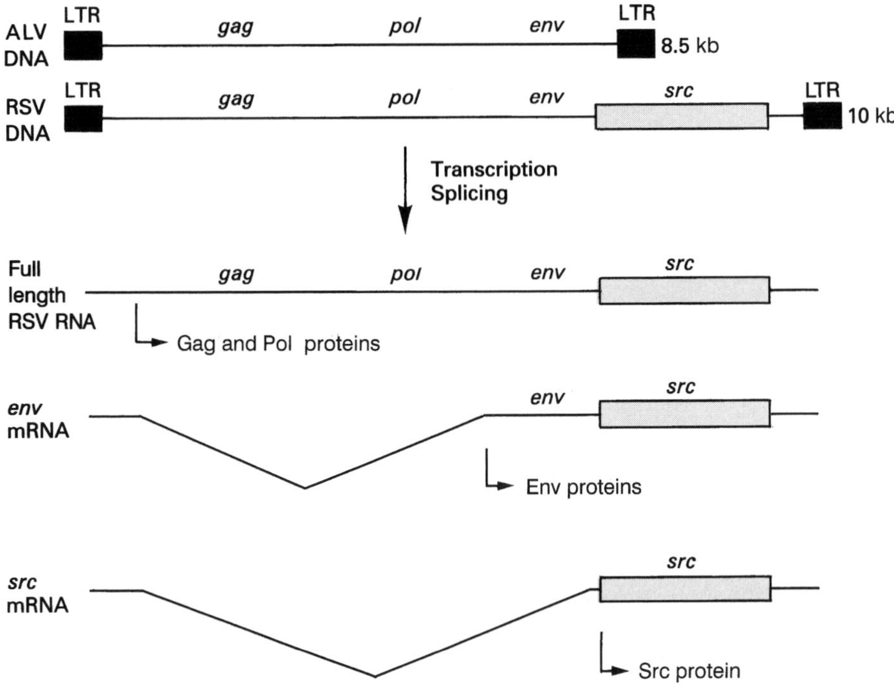

FIGURE 3.5
Organization and expression of the RSV genome. RSV contains an additional gene, *src,* that is not present in ALV. The RSV provirus is transcribed to yield a full length RNA, which serves as progeny genomic RNA and mRNA for *gag* and *pol.* The RSV primary transcript is spliced to yield two subgenomic mRNAs, one for *env* and the other for *src.*

from that of RSV. Most currently studied strains of RSV contain all of the viral replicative genes, in addition to the *src* oncogene, and are therefore replication competent as well as acutely transforming. All other acutely transforming viruses identified so far, however, are defective for replication and must be propagated by coinfection of the host cell with a helper virus that contains functional *gag, pol,* and *env* replicative genes. These genes of the helper virus complement the replication defects of the acutely transforming virus, thereby allowing its propagation.

The replication defectiveness of most acutely transforming viruses is a consequence of deletion of one or more replication genes, which have been replaced by the viral oncogenes. Consequently, the genomes of most acutely transforming viruses are smaller than their nontransforming counterparts. RSV is the exception, containing a larger genome than ALV, because it has not sustained deletions of replicative genes.

TABLE 3.1
Retroviral Oncogenes

Oncogene	Virus	Species	Oncogene Protein
abl	Abelson leukemia virus	mouse	$p120^{gag\text{-}abl}$
akt	AKT8 virus	mouse	$p105^{gag\text{-}akt}$
cbl	Cas NS-1 virus	mouse	$p100^{gag\text{-}cbl}$
crk	avian sarcoma virus CT10	chicken	$p47^{gag\text{-}crk}$
*erb*A	avian erythroblastosis virus-ES4	chicken	$p75^{gag\text{-}erbA}$
*erb*B	avian erythroblastosis virus-ES4	chicken	$p72^{erbB}$
ets	avian erythroblastosis virus-E26	chicken	$p135^{gag\text{-}myb\text{-}ets}$
fes	Gardner-Arnstein feline sarcoma virus	cat	$p110^{gag\text{-}fes}$
fgr	Gardner-Rasheed feline sarcoma virus	cat	$p70^{gag\text{-}fgr}$
fms	McDonough feline sarcoma virus	cat	$p180^{gag\text{-}fms}$
fos	FBJ murine osteogenic sarcoma virus	mouse	$p55^{fos}$
fps[1]	Fujinami sarcoma virus	chicken	$p140^{gag\text{-}fps}$
jun	avian sarcoma virus 17	chicken	$p55^{gag\text{-}jun}$
kit	Hardy-Zuckerman-4 feline sarcoma virus	cat	$p80^{gag\text{-}kit}$
maf	AS42 sarcoma virus	chicken	$p100^{gag\text{-}maf}$
mos	Moloney sarcoma virus	mouse	$p37^{env\text{-}mos}$
mpl	myeloproliferative leukemia virus	mouse	$p31^{env\text{-}mpl}$
myb	avian myeloblastosis virus	chicken	$p45^{gag\text{-}myb\text{-}env}$
myc	avian myelocytomatosis virus-29	chicken	$p110^{gag\text{-}myc}$
qin	avian sarcoma virus 31	chicken	$p85^{gag\text{-}qin}$
raf	3611 murine sarcoma virus	mouse	$p75^{gag\text{-}raf}$
	avian carcinoma virus MH2	chicken	$p100^{gag\text{-}raf}$
*ras*H	Harvey sarcoma virus	rat	$p21^{rasH}$
	Rasheed sarcoma virus	rat	$p29^{gag\text{-}rasH}$
*ras*K	Kirsten sarcoma virus	rat	$p21^{rasK}$
rel	reticuloendotheliosis virus	turkey	$p56^{env\text{-}rel}$
ros	UR2 sarcoma virus	chicken	$p68^{gag\text{-}ros}$
sea	avian erythroblastosis virus S13	chicken	$p137^{env\text{-}sea}$
sis	simian sarcoma virus	monkey	$p28^{env\text{-}sis}$

TABLE 3.1 (CON'T)
Retroviral Oncogenes

Oncogene	Virus	Species	Oncogene Protein
ski	avian SK virus	chicken	$p110^{gag\text{-}ski\text{-}gag}$
src	Rous sarcoma virus	chicken	$p60^{src}$
yes	Y73 sarcoma virus	chicken	$p90^{gag\text{-}yes}$

Note: Representative acutely transforming viruses bearing each of the indicated oncogenes are listed. The oncogene proteins are designated by approximate size in kilodaltons.

[1]Oncogenes *fes* and *fps* are homologous genes of cat and chicken origin.

All of these viruses share a similar basic strategy of expression of their oncogenes: they are transcribed from the proviral 5' LTR as part of the retroviral genome. However, the viruses of this type vary as to the details by which this is accomplished. Some oncogenes (including *ras, fos,* and *erb*B) are, like *src,* expressed as independent translation products that do not contain sequences encoded by viral replicative genes (Fig. 3.6A). In contrast, many of the other oncogenes are expressed as fusion proteins that contain amino acid sequences encoded by viral replicative genes in addition to a transformation-specific sequence. The most common genome organization results in expression of the viral oncogene as a fusion with viral *gag* sequences (Fig. 3.6B). In these viruses, most of *gag* is deleted and replaced with oncogene sequences (such as *abl, myc, erb*A, *jun, raf,* and several others). The oncogene coding sequences are inframe with the remaining 5' *gag* sequences; so the gene is transcribed and translated to yield an oncogene protein with amino acid sequences encoded by *gag* at its amino terminus. Variants of this basic fusion protein theme include the synthesis of recombinant proteins with a short stretch of amino acid sequences encoded by *env* instead of *gag* at the amino terminus of the oncogene product. Finally, some acutely transforming retroviruses contain two distinct oncogenes (such as *erb*A + *erb*B, *myb* + *ets,* and *myc* + *raf*), both of which contribute to the induction of neoplasms (Fig. 3.6C).

Summary

The acutely transforming retroviruses rapidly induce tumors in animals and efficiently transform cells in culture. The transforming potential of these viruses is a direct result of the activity of specific viral genes, the retroviral oncogenes. At present, more than two dozen different oncogenes have been identified in the genomes of more than forty acutely transforming

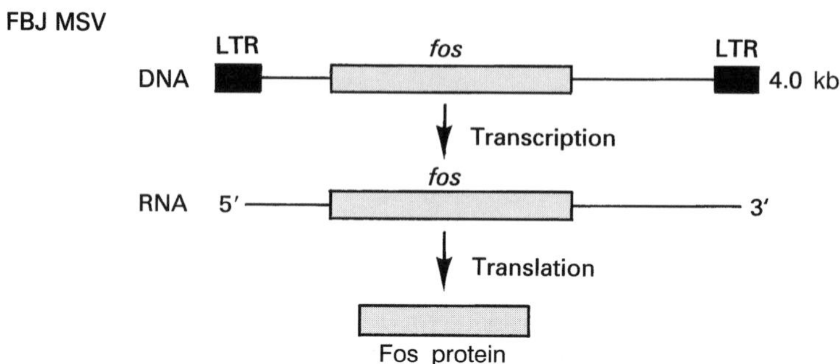

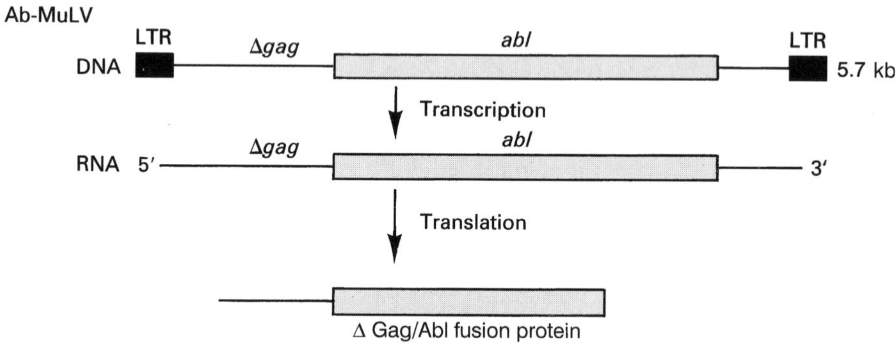

FIGURE 3.6
Organization and expression of defective acutely transforming virus genomes. Three representative strategies of viral oncogene expression are illustrated. A: The *fos* oncogene of FBJ MSV is expressed as an independent translation product of full length viral RNA. B: The *abl* oncogene of Ab-MuLV is expressed as a fusion protein formed by recombination with viral *gag* sequences (Δ*gag*). C: AEV ES4 has two oncogenes, *erb*A and *erb*B. The *erb*A oncogene is expressed as a *gag* fusion protein. *erb*B is translated independently from a spliced mRNA.

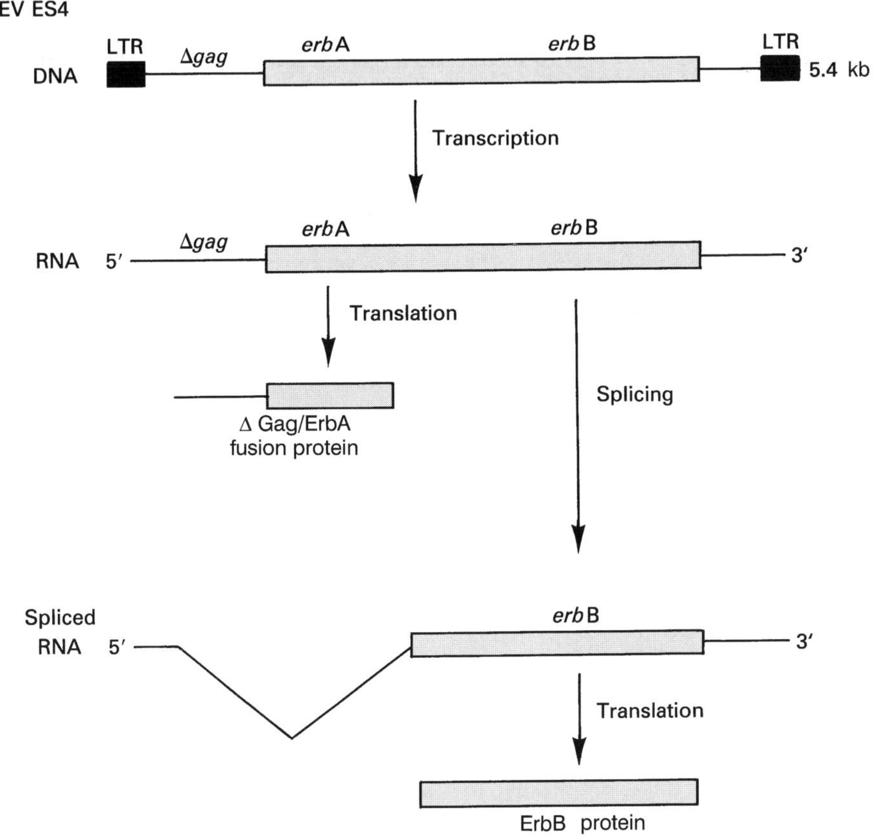

viruses isolated from chickens, turkeys, mice, rats, cats, and monkeys. Despite differences in genome organization and strategies of gene expression, the common theme underlying the biological properties of these potent transforming viruses is the efficient expression of an oncogene as an integrated part of the viral genome. In all cases, the oncogenes are specific genes responsible for cell transformation without taking part in retrovirus replication. The retroviral oncogenes have thus provided clear evidence for the role of specific genes in cell transformation and neoplasia.

References

General Reference

Weiss, R., Teich, N., Varmus, H., and Coffin, J., eds. 1985. *Molecular biology of tumor viruses: RNA tumor viruses*, 2d ed. Cold Spring Harbor, New York.

The *src* Gene of RSV

Bernstein, A., MacCormick, R., and Martin, G.S. 1976. Transformation-defective mutants of avian sarcoma viruses: the genetic relationship between conditional and nonconditional mutants. *Virology* 70:206–209.

Duesberg, P.H., and Vogt, P.K. 1970. Differences between the ribonucleic acids of transforming and nontransforming avian tumor viruses. *Proc. Natl. Acad. Sci. USA* 67:1673–1680.

Duesberg, P.H. and Vogt, P.K. 1973. RNA species obtained from clonal lines of avian sarcoma and from avian leukosis virus. *Virology* 54:207–219.

Kawai, S., and Hanafusa, H. 1971. The effects of reciprocal changes in temperature on the transformed state of cells infected with a Rous sarcoma virus mutant. *Virology* 46:470–479.

Lai, M.M.C., Duesberg, P.H., Horst, J., and Vogt, P.K. 1973. Avian tumor virus RNA: a comparison of three sarcoma viruses and their transformation-defective derivatives by oligonucleotide fingerprinting and DNA-RNA hybridization. *Proc. Natl. Acad. Sci. USA* 70:2266–2270.

Martin, G.S. 1970. Rous sarcoma virus: a function required for the maintenance of the transformed state. *Nature* 227:1021–1023.

Martin, G.S., and Duesberg, P.H. 1972. The α subunit in the RNA of transforming avian tumor viruses: I. Occurrence in different virus strains II. Spontaneous loss resulting in nontransforming variants. *Virology* 47:494–497.

Toyoshima, K., Friis, R.R., and Vogt, P.K. 1970. The reproductive and cell-transforming capacities of avian sarcoma virus B77: inactivation with UV light. *Virology* 42:163–170.

Vogt, P.K. 1971. Spontaneous segregation of nontransforming viruses from cloned sarcoma viruses. *Virology* 46:939–946.

Wyke, J.A., Bell, J.G., and Beamand, J.A. 1975. Genetic recombination among temperature-sensitive mutants of Rous sarcoma virus. *Cold Spring Harbor Symp. Quan. Biol.* 39:897–905.

The RSV and ALV Genomes

Coffin, J.M., and Billeter, M.A. 1976. A physical map of the Rous sarcoma virus genome. *J. Mol. Biol.* 100:293–318.

Eisenman, R.N., and Vogt, V.M. 1978. The biosynthesis of oncovirus proteins. *Biochim. Biophys. Acta* 473:187–239.

Hayward, W.S. 1977. Size and genetic content of viral RNAs in avian oncovirus-infected cells. *J. Virol.* 24:47–63.

Hughes, S.H., Shank, P.R., Spector, D.H., Kung, H.-J., Bishop, J.M., Varmus, H.E., Vogt,

P.K., and Breitman, M.L. 1978. Proviruses of avian sarcoma virus are terminally redundant, coextensive with unintegrated linear DNA and integrated at many sites. *Cell* 15:1397–1410.

Schwartz, D.E., Tizard, R., and Gilbert, W. 1983. Nucleotide sequence of Rous sarcoma virus. *Cell* 32:853–869.

Shank, P.R., Hughes, S.H., Kung, H.-J., Majors, J.E., Quintrell, N., Guntaka, R.V., Bishop, J.M., and Varmus, H.E. 1978. Mapping unintegrated avian sarcoma virus DNA: termini of linear DNA bear 300 nucleotides present once or twice in two species of circular DNA. *Cell* 15:1383–1395.

Varmus, H.E. 1982. Form and function of retroviral proviruses. *Science* 216:812–820.

Weiss, S.R., Varmus, H.E., and Bishop, J.M. 1977. The size and genetic composition of virus-specific RNAs in the cytoplasm of cells producing avian sarcoma-leukosis viruses. *Cell* 12:983–992.

Other Retroviral Oncogenes

Aaronson, S.A., Jainchill, J.L., and Todaro, G.J. 1970. Murine sarcoma virus transformation of BALB/3T3 cells: lack of dependence on murine leukemia virus. *Proc. Natl. Acad. Sci. USA* 66:1236–1243.

Bellacosa, A., Testa, J.R., Staal, S.P., and Tsichlis, P.N. 1991. A retroviral oncogene, *akt*, encoding a serine-threonine kinase containing an SH2-like region. *Science* 254:274–277.

Hartley, J.W., and Rowe, W.P. 1966. Production of altered cell foci in tissue culture by defective Moloney sarcoma virus particles. *Proc. Natl. Acad. Sci. USA* 55:780–786.

Langdon, W.Y., Hartley, J.W., Klinken, S.P., Ruscetti, S.K., and Morse, H.C. III. 1989. v-*cbl*, an oncogene from a dual-recombinant murine retrovirus that induces early B-lineage lymphomas. *Proc. Natl. Acad. Sci. USA* 86:1168–1172.

Li, J., and Vogt, P.K. 1993. The retroviral oncogene *qin* belongs to the transcription factor family that includes the homeotic gene *fork head*. *Proc. Natl. Acad. Sci. USA* 90:4490–4494.

Maki, Y., Bos, T.J., Davis, C., Starbuck, M., and Vogt, P.K. 1987. Avian sarcoma virus 17 carries the *jun* oncogene. *Proc. Natl. Acad. Sci. USA* 84:2848–2852.

Mayer, B.J., Hamaguchi, M., and Hanafusa, H. 1988. A novel viral oncogene with structural similarity to phospholipase C. *Nature* 332:272–275.

Nishizawa, M., Kataoka, K., Goto, N., Fujiwara, K.T., and Kawai, S. 1989. v-*maf*, a viral oncogene that encodes a "leucine zipper" motif. *Proc. Natl. Acad. Sci. USA* 86:7711–7715.

Reddy, E.P., Skalka, A.M., and Curran, T., eds. 1988. *The oncogene handbook.* Elsevier, New York.

Smith, D.R., Vogt, P.K., and Hayman, M.J. 1989. The v-*sea* oncogene of avian erythroblastosis retrovirus S13: another member of the protein-tyrosine kinase gene family. *Proc. Natl. Acad. Sci. USA* 86:5291–5295.

Souyri, M., Vigon, I., Penciolilli, J.-F., Heard, J.-M., Tambourin, P., and Wendling, F. 1990. A putative truncated cytokine receptor gene transduced by the myeloproliferative leukemia virus immortalizes hematopoietic progenitors. *Cell* 63:1137–1147.

❖ Chapter 4
The Origin of Retroviral Oncogenes

The oncogenes of acutely transforming retroviruses are a distinct group of genes that are responsible for viral pathogenesis but not required for viral replication. The presence of such genes in a viral genome seems paradoxical because they do not function directly in virus growth. Where did the oncogenes originate and how did they become incorporated into retroviruses?

Isolation of Acutely Transforming Retroviruses

The first clue to the origin of oncogenes derives from the way in which the acutely transforming viruses were isolated. The early isolates of these viruses—for example, RSV—were obtained from tumors of animals that were chronically infected with nontransforming retroviruses (for example, ALV). Such animals occasionally developed tumors from which new acutely transforming viruses could be isolated and then maintained by either animal or tissue culture passage. More recent and informative examples are provided by the laboratory isolation of several murine acutely transforming viruses—for example, Harvey sarcoma virus and Abelson leukemia virus (Fig. 4.1). These viruses were isolated following inoculation of rats or mice with weakly transforming murine retroviruses that, like ALV, did not contain oncogenes. A small fraction of such animals developed tumors from which new, more highly oncogenic viruses could be isolated. These new acutely transforming viruses contained oncogenes that were responsible for their increased oncogenic potential: they now induced neoplasms rapidly in mice or rats and efficiently transformed cells in culture. The common element to the isolation of acutely

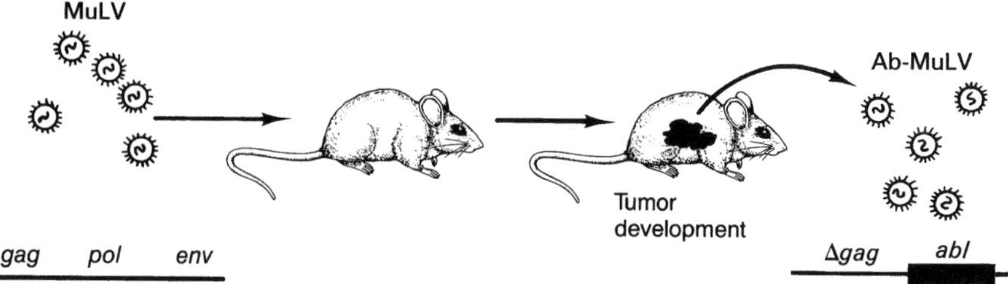

FIGURE 4.1
Isolation of Abelson leukemia virus. The acutely transforming virus Ab-MuLV was isolated from a rare tumor that developed in a mouse that had been inoculated with Moloney murine leukemia virus (MuLV). MuLV is a nonacutely transforming virus, which contains only the viral replicative genes *gag, pol,* and *env.* In contrast, Ab-MuLV has acquired a new viral oncogene, *abl,* which is responsible for its transforming activity.

transforming viruses, illustrated by these examples, is the rare isolation of a new virus from a tumor that developed in an animal infected by a nontransforming retrovirus. The acutely transforming viruses isolated from such tumors contained new oncogenes that were not present in the viruses that initially induced the neoplasms.

Normal Cell Homologs of Retroviral Oncogenes

The scenario of acutely transforming virus isolation suggested the hypothesis that the retroviral oncogenes were derived from host cells and incorporated into the genomes of nontransforming viruses to yield new recombinants with strikingly increased pathogenicity. The critical prediction of this hypothesis is that normal cells contain DNA sequences that are closely related to the oncogenes of acutely transforming retroviruses. In the early 1970s, the technology of nucleic acid hybridization permitted direct experimental evaluation of this hypothesis. Such experiments were initially complicated, however, by the fact that the genomes of many animal species were found to contain DNA sequences that were homologous to genomic RNAs of both transforming and nontransforming retroviruses. These sequences are endogenous retrovirus genomes, all of which have been found to derive from nontransforming retroviruses without oncogenes. They apparently resulted from infections of early embryos that led to incorporation of a provirus into the germ line, where it could then be chromosomally transmitted like any other cell gene.

The presence of such endogenous proviruses made it necessary to develop specific nucleic acid probes for the oncogenes of the acutely transforming viruses.

Otherwise, it was unclear whether hybridization between viral and cell DNA sequences was to be attributed to viral replication genes or to oncogenes. For example, hybridization of a radioactive probe representing the entire genome of either RSV or ALV to an excess of normal chicken DNA indicated that most of the sequences in both viral genomes were present in normal cell DNA. However, these experiments did not provide direct information about the possible presence of chicken DNA sequences that were specifically homologous to *src*.

The first direct experimental results to elucidate the origin of retroviral oncogenes were reported in 1976 from the laboratory of Harold Varmus and Michael Bishop, in collaboration with Peter Vogt. Their approach (Fig. 4.2) was to use RSV and its transformation-defective deletion mutants (discussed in chapter 3) to derive a nucleic acid hybridization probe that was specific for the RSV *src* gene. These workers initially synthesized a radioactive probe composed of DNA fragments that were representative of and complementary to the entire RSV genome. The probe was then hybridized with an excess of RNA of a transformation-defective deletion mutant so that only those DNA fragments

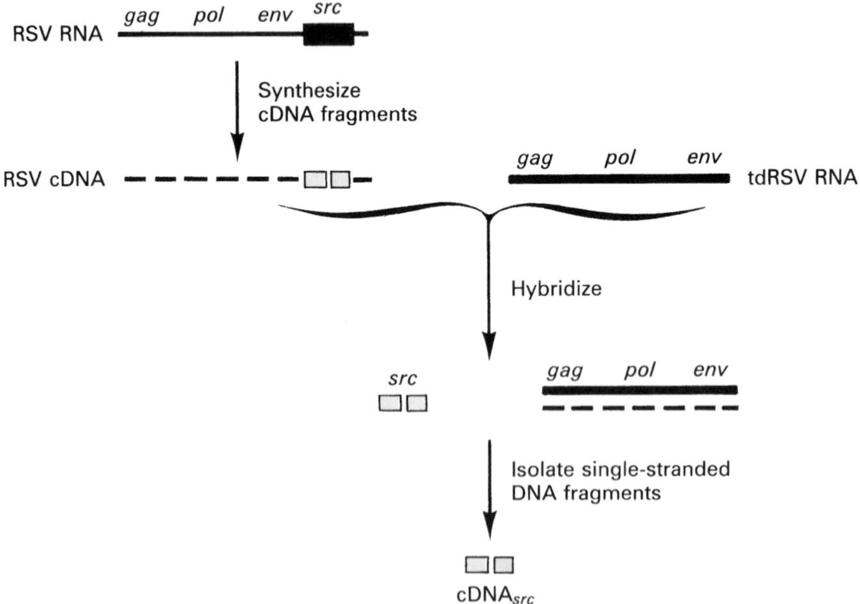

FIGURE 4.2
Preparation of cDNA$_{src}$. A cDNA copy of RSV RNA was hybridized to RNA of transformation-defective (td) RSV. Those cDNA fragments corresponding to the *src* gene failed to hybridize and were recovered as single-stranded DNA.

that were complementary to *src* failed to hybridize. The *src*-specific probe was then recovered as single-stranded unhybridized DNA fragments. Control experiments showed that the *src* probe hybridized to RSV RNA but not to RNAs of transformation-defective RSV mutants or to ALV strains. It therefore constituted a probe for only the *src* sequences, as defined genetically by the deletions of transformation-defective mutants.

With the availability of *src*-specific probe, it became possible to test directly for the presence of endogenous oncogene-related sequences in the genomes of normal animals. The results obtained by Varmus, Bishop, and their collaborators were striking. The *src*-specific probe hybridized extensively to DNAs of normal chickens, demonstrating that chickens regularly inherited DNA sequences that were closely related to the *src* oncogene. Not only were *src*-homologous sequences found to be part of the chicken genome, but related sequences were also detected in other vertebrate species, including human. These experiments thus established that normal cells contained genetic information closely related to a viral oncogene, providing the first definitive experimental support for the now well-substantiated hypothesis that the acutely transforming viruses originated by the incorporation of cellular genes into viral genomes. Moreover, these results provided the first clear indication that cells normally contained genes that might have the potential for inducing neoplastic transformation.

Analogous experiments with probes for the oncogenes of other acutely transforming viruses yielded results comparable to those obtained for *src*. In all cases, the retroviral oncogenes were homologous to DNA sequences that were normally found in the species from which the acutely transforming virus was isolated. Furthermore, as with *src*, the cellular homologs of retroviral oncogenes were conserved throughout vertebrates (and in some cases in lower organisms such as *Drosophila* and even yeast). It thus appeared that each of the retroviral oncogenes had originated from sequences that were part of the genetic complement of normal cells.

The normal cellular sequences from which the viral oncogenes were derived are now called *proto-oncogenes* in order to distinguish them from the oncogenes in the viruses. As we shall see throughout subsequent chapters, this distinction is critical to understanding the biology of both oncogene and proto-oncogene function. One of the central concepts that has developed is that the proto-oncogenes are normal cell genes that function in a number of aspects of cell growth and differentiation. The oncogenes are abnormally expressed or mutated forms of their normal cell progenitors. As a consequence of such alterations (including point mutations, deletions, recombination, or amplification), the oncogenes have acquired a new biological activity—the ability to induce cell transformation.

Structure and Expression of Proto-oncogenes

The discovery of proto-oncogenes as normal cell DNA sequences that were homologous to viral oncogenes immediately raised a number of questions concerning their structure and possible functions. Were the proto-oncogenes functional genetic elements that were expressed in normal cells? If so, what was their activity in normal cell physiology? Finally, did they contribute to the pathogenesis of cancer?

Understanding of the structure of proto-oncogenes in detail has come from their isolation and characterization as molecular clones. A typical example continues to be provided by the *src* gene of RSV. The *src* proto-oncogene was isolated from genomic libraries of chicken DNA by hybridization with viral *src* probe. The cellular sequences homologous to *src* constitute a single locus in the chicken genome that is not linked to sequences related to endogenous nontransforming retroviruses. In addition, the cellular sequences homologous to *src* are interrupted by several blocks of nonhomologous sequences (Fig. 4.3). These nonhomologous sequences are introns in the proto-oncogene (like those present in almost all cell genes) that have been removed from the viral oncogene as a result of RNA splicing. The structure of *src* and other proto-oncogenes, as well as their conservation in evolution, is typical of functional genes of normal cells.

Expression of proto-oncogenes has been investigated by analysis of both mRNA and protein in a variety of different types of cells. For example, such studies have identified cellular *src* mRNA in both normal cells and chemically

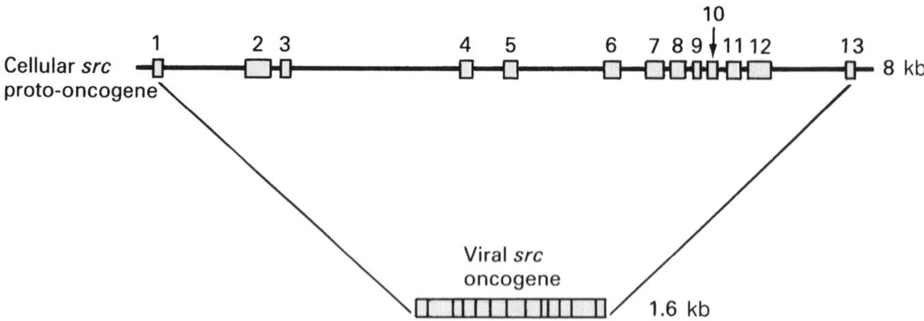

FIGURE 4.3
Comparison of the chicken *src* proto-oncogene and the RSV *src* oncogene. Exons of the cellular *src* proto-oncogene (shown as filled boxes) are separated by intervening sequences (introns), which were removed from the viral *src* oncogene by splicing. The sequence designated exon 13 is not a coding exon for cellular *src*, which terminates in exon 12: the exon 13 sequences were incorporated into viral *src* by a recombination event that led to deletion of the normal cellular *src* carboxy terminus.

induced tumors. A cellular Src protein, closely related in structure to the product of the viral *src* gene, also has been found in normal cells of both avian and mammalian species. Both the cellular and viral Src proteins are related in biochemical function: as discussed in detail in chapter 13, both the cellular *src* proto-oncogene (c-*src*) and the viral *src* oncogene (v-*src*) encode plasma membrane-associated proteins with protein-tyrosine kinase activity. Analogous investigations have found that the other proto-oncogenes are likewise expressed in various types of normal cells as well as in some nonvirus-induced tumors. The normal functions of proto-oncogenes in cell proliferation, differentiation, and development will be extensively discussed in later chapters. In any case, the proto-oncogenes appear to constitute a group of functional genes of normal cells that, when incorporated into the genome of retroviruses as oncogenes, have the capacity to induce neoplastic transformation.

Capture of Proto-oncogenes by Retroviruses

The life cycle of retroviruses, particularly their stable and noncytocidal association with their host cells, is a major factor in the ability of these viruses to acquire unique genes from the host through recombination with the viral genome. Recombination is known to occur with very high frequencies when two different retroviruses coinfect the same host cell. The first step in recombination between two viral genomes is the packaging of two different viral RNAs into the same virus particle (Fig. 4.4). This occurs frequently because the virus particles normally contain two copies of the genomic RNA molecules. Stable recombination between the two different viral RNA genomes then occurs when such a heterozygote virus infects a new cell, as a result of the polymerase switching templates in the process of proviral DNA synthesis.

The capture of cellular proto-oncogenes by retroviruses is a consequence of this high frequency of recombination between viral RNAs, as well as other features of the retroviral life cycle. The first step in the process is integration of a nontransforming provirus adjacent to a cellular proto-oncogene (Fig. 4.5). This is frequently followed by readthrough transcription, in which transcripts initiated in the 5' viral LTR extend into the adjacent proto-oncogene instead of terminating normally in the 3' LTR of the provirus. Cleavage and polyadenylation in the 3' LTR is an inefficient process, failing nearly 20% of the time; so the formation of such readthrough transcripts is a moderately frequent event in virus-infected cells. Such readthrough transcription may be followed by abnormal splicing between a donor site in the virus and an acceptor site in the proto-oncogene, resulting in a fusion of viral and cellular sequences at the RNA level. Alternatively, deletion of DNA sequences extending from within the provirus to the proto-oncogene also may result in formation of a new fused transcription unit, in which viral and adjacent cellular proto-oncogene sequences are tran-

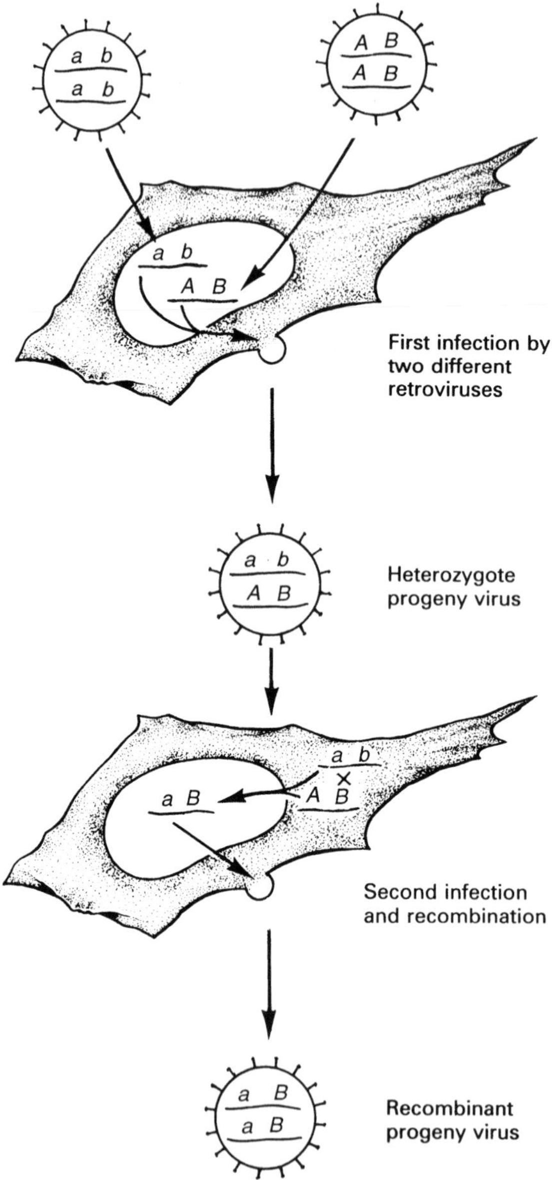

FIGURE 4.4
Recombination between retroviruses through formation of heterozygotes. The first infection results in a cell that expresses two different virus genomes, leading to formation of heterozygote progeny. Recombination then occurs in a second round of infection by such a heterozygote virus.

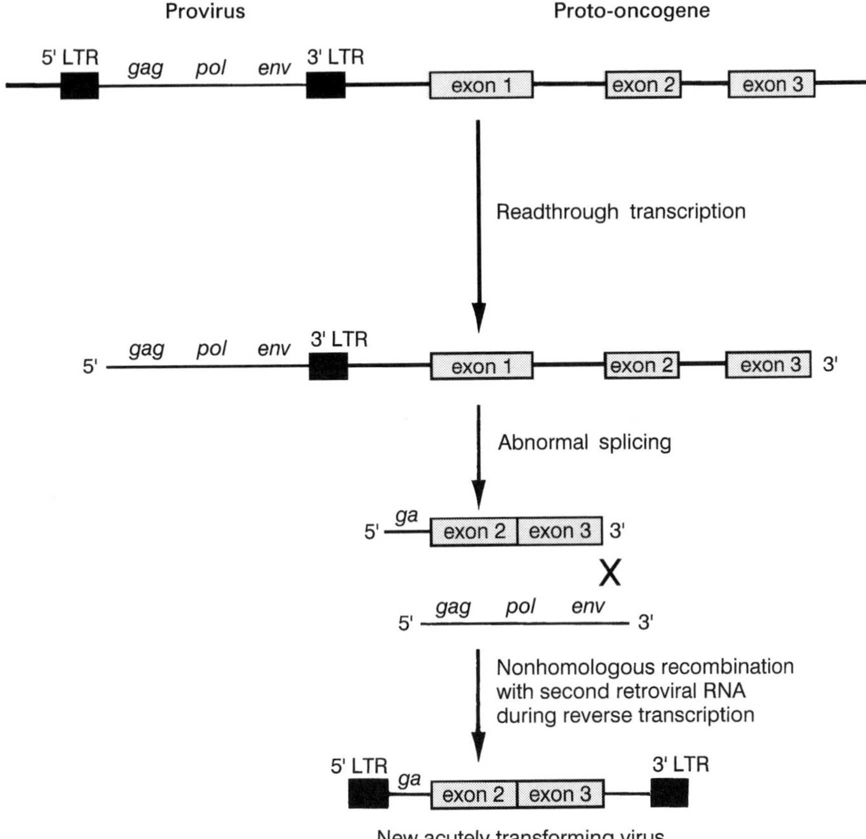

FIGURE 4.5
Model for generation of acutely transforming retroviruses. The provirus of a nontransforming retrovirus is initially integrated upstream of a cellular proto-oncogene. Readthrough transcription then generates a fusion transcript containing both viral and cellular sequences, followed by abnormal splicing. The fusion transcript is then packaged with another retrovirus genome to form a heterozygote and a new acutely transforming virus is generated by nonhomologous recombination during reverse transcription.

scribed from the retroviral LTR into a single RNA molecule. In either case, the new fused proto-oncogene transcript would contain a retroviral LTR as well as other viral sequences, including 5' RNA packaging signals. Consequently, it could be efficiently incorporated into progeny virus particles and used as a template for reverse transcription in the next round of infection. During this process, strand switching by reverse transcriptase has been shown to result in nonhomologous recombination events, leading to the incorporation of cellular proto-oncogene sequences into retroviral genomes.

The generation of acutely transforming viruses by such a mechanism would clearly be a rare event, requiring several low probability steps. This is consistent with the rarity of isolation of acutely transforming viruses. It is important to note that any cell gene could be recombined into a viral genome by these mechanisms. However, most such recombinants would lack distinctive new biological properties and consequently would not be detected. In contrast, incorporation of a proto-oncogene is easily detected when the progeny virus has the ability to induce rapid tumor formation and cell transformation.

Viral Oncogenes are Altered Versions of Proto-oncogenes

How do the proto-oncogenes of normal cells relate in structure and function to the oncogenes of retroviruses? Whereas the viral oncogenes are potent inducers of neoplastic transformation, their progenitors, the proto-oncogenes, are normal components of cellular genomes. Several differences between the viral oncogenes and their proto-oncogene homologs account for these differences in biological activity.

The first obvious consequence of incorporation of a proto-oncogene into a retroviral genome is that the resulting oncogene is now expressed as part of the virus. Consequently, the mechanisms that regulated normal expression of the proto-oncogene are abrogated. For example, the oncogene is transcribed under the control of a strong promoter and enhancer in the viral LTR, whereas transcription of the corresponding proto-oncogene is regulated by its normal cellular transcriptional control sequences. Because the viral LTRs are adapted for high-level transcription of the viral genome, most viral oncogenes are transcribed at much higher levels and can be expressed in different tissues compared with their normal cell counterparts.

The viral oncogenes lack introns, which are present in nearly all of the proto-oncogenes. Consequently, any transcriptional regulatory sequences present in introns of the proto-oncogene are lost in the viral oncogenes. In addition, most of the viral oncogenes are truncated versions of their proto-oncogene homologs, which lost either one or both of their normal ends when they were incorporated into viral genomes. Loss of the normal 5' and 3' untranslated regions of the cellular transcripts, and their replacement by viral sequences, can affect both stability and translational efficiency of the viral mRNAs compared with the cellular mRNAs.

In addition to these differences in regulation of expression, most of the proteins encoded by oncogenes differ significantly in structure from the corresponding proto-oncogene proteins. For example, as noted in chapter 3, a common genomic organization of acutely transforming viruses results in expression of the oncogene as a fusion protein with viral *gag* sequences at its amino terminus. In many of these instances, amino acids at the normal amino terminus of the proto-oncogene protein have been deleted. These truncated oncogene proteins are thus very different in structure, and potentially in function, from their proto-oncogene precursors. In many cases, sequences encoding amino acids at the carboxy terminus of the proto-oncogene have also been deleted from the oncogene.

Finally, most of the oncogenes have accumulated multiple point mutations relative to their proto-oncogene homologs. Some of these mutations may be neutral. However, because the acutely transforming viruses have been selected for transforming activity during multiple passages both in animals and cell culture, it is to be expected that some of these mutations result in amino acid substitutions that increase transforming potency of the oncogene products.

The specific roles of different changes in the altered biological and biochemical activities of oncogenes versus proto-oncogenes will be the subject of later chapters. However, comparison of the RSV *src* oncogene and proto-oncogene serves as an example to illustrate some of these alterations and their functional significance. Replacement of cellular regulatory sequences with the RSV LTR leads to levels of *src* mRNA and protein from tenfold to 100-fold higher in RSV-infected chicken embryo fibroblasts than in normal cells. For some proto-oncogenes—for example, *mos*—increased expression of the normal proto-oncogene protein is sufficient to induce cell transformation. However, although efficient expression of viral *src* is required for its potent transforming activity, elevated expression of the *src* proto-oncogene is not by itself sufficient to induce cell transformation. One structural alteration responsible for the transforming activity of viral *src* is deletion of the nineteen carboxy-terminal amino acids of the proto-oncogene protein. This region of the *src* proto-oncogene encodes a negative regulatory domain that modulates the protein's tyrosine kinase activity: its deletion from viral *src* thus releases the oncogene protein from normal controls and enhances its catalytic activity. In addition, RSV *src* differs from the *src* proto-oncogene by several point mutations resulting in amino acid substitutions, some of which increase both the transforming potency and the protein kinase activity of the oncogene protein.

Viral oncogenes thus differ significantly in several respects from their proto-oncogene homologs in normal cells. The viral oncogenes can therefore be viewed as mutant forms of normal cell genes that have acquired the ability to induce neoplastic transformation as a result of alterations in both gene expression and the structure and function of their protein products.

Summary

The retroviral oncogenes originated from normal cell genes, called proto-oncogenes, that were incorporated into retroviral genomes by virus-cell recombination events. The proto-oncogenes are widely conserved in evolution and, as discussed in later chapters, appear to function in a number of aspects of normal cell growth and differentiation. The oncogenes are altered versions of their normal cell progenitors, frequently differing from the corresponding proto-oncogenes in both regulatory and protein-coding sequences. These alterations are responsible for the ability of the oncogenes to efficiently induce neoplastic transformation. The proto-oncogenes thus constitute a group of normal cell genes with potential transforming activity. Their discovery paved the way to identifying the genetic abnormalities that contribute to the development of human cancers.

References

General Reference

Weiss, R., Teich, N., Varmus, H., and Coffin, J., eds. 1985. *Molecular biology of tumor viruses: RNA tumor viruses,* 2d ed. Cold Spring Harbor, New York.

Isolation of Acutely Transforming Retroviruses

Abelson, H.T., and Rabstein, L.S. 1970. Influence of prednisolone on Moloney leukemogenic virus in BALB/c mice. *Cancer Res.* 30:2208–2212.

Abelson, H.T., and Rabstein, L.S. 1970. Lymphosarcoma: virus induced thymic-independent disease in mice. *Cancer Res.* 30:2213–2222.

Harvey, J.J. 1970. An unidentified virus which causes the rapid production of tumours in mice. *Nature* 204:1104–1105.

Rous, P. 1911. A sarcoma of the fowl transmissible by an agent separable from the tumor cells. *J. Exp. Med.* 13:397–411.

Normal Cell Homologs of Retroviral Oncogenes

Ellis, R.W., DeFeo, D., Maryak, J.M., Young, H.A., Shih, T.Y., Chang, E.H., Lowy, D.R., and Scolnick, E.M. 1980. Dual evolutionary origin for the rat genetic sequences of Harvey murine sarcoma virus. *J. Virol.* 36:408–420.

Frankel, A.E., and Fischinger, P.J. 1976. Nucleotide sequences in mouse DNA and RNA specific for Moloney sarcoma virus. *Proc. Natl. Acad. Sci. USA* 73:3705–3709.

Scolnick, E.M., Rands, E., Williams, D., and Parks, W.P. 1973. Studies on the nucleic acid sequences of Kirsten sarcoma virus: a model for formation of a mammalian RNA-containing sarcoma virus. *J. Virol.* 12:458–463.

Shilo, B.-Z., and Weinberg, R.A. 1981. DNA sequences homologous to vertebrate oncogenes are conserved in *Drosophila melanogaster*. *Proc. Natl. Acad. Sci. USA* 78:6789–6792.

Spector, D.H., Varmus, H.E., and Bishop, J.M. 1978. Nucleotide sequences related to the transforming gene of avian sarcoma virus are present in the DNA of uninfected vertebrates. *Proc. Natl. Acad. Sci. USA* 75:4102–4106.

Stehelin, D., Guntaka, R.V., Varmus, H.E., and Bishop, J.M. 1976. Purification of DNA complementary to nucleotide sequences required for neoplastic transformation of fibroblasts by avian sarcoma viruses. *J. Mol. Biol.* 101:349–365.

Stehelin, D., Varmus, H.E., Bishop, J.M., and Vogt, P.K. 1976. DNA related to the transforming gene(s) of avian sarcoma viruses is present in normal avian DNA. *Nature* 260:170–173.

Structure and Expression of Proto-oncogenes

Collett, M.S., Brugge, J.S., and Erikson, R.L. 1978. Characterization of a normal avian cell protein related to the avian sarcoma virus transforming gene product. *Cell* 15:1363–1369.

Collett, M.S., Erikson, E., Purchio, A.F., Brugge, J.S., and Erikson, R.L. 1979. A normal cell protein similar in structure and function to the avian sarcoma virus transforming gene product. *Proc. Natl. Acad. Sci. USA* 76:3159–3163.

Golden, A., and Brugge, J.S. 1988. The *src* oncogene. In *The oncogene handbook*, ed. Reddy, E.P., Skalka, A.M., and Curran, T. Elsevier, Amsterdam. pp. 149–173.

Oppermann, H., Levinson, A.D., Varmus, H.E., Levintow, L., and Bishop, J.M. 1979. Uninfected vertebrate cells contain a protein that is closely related to the product of the avian sarcoma virus transforming gene (*src*). *Proc. Natl. Acad. Sci. USA* 76:1804–1808.

Parker, R.C., Varmus, H.E., and Bishop, J.M. 1981. Cellular homologue (c-*src*) of the transforming gene of Rous sarcoma virus: isolation, mapping and transcriptional analysis of c-*src* and flanking regions. *Proc. Natl. Acad. Sci. USA* 78:5842–5846.

Shalloway, D., Zelenetz, A.D., and Cooper, G.M. 1981. Molecular cloning and characterization of the chicken gene homologous to the transforming gene of Rous sarcoma virus. *Cell* 24:531–541.

Spector, D.H., Smith, K., Padgett, T., McCombe, P., Roulland-Dussoix, D., Moscovici, C., Varmus, H.E., and Bishop, J.M. 1978. Uninfected avian cells contain RNA related to the transforming gene of avian sarcoma viruses. *Cell* 13:371–379.

Takeya, T., and Hanafusa, H. 1981. Comparison between the viral transforming gene (*src*) of recovered avian sarcoma virus and its cellular homolog. *Mol. Cell. Biol.* 1:1024–1037.

Takeya, T., and Hanafusa, H. 1983. Structure and sequence of the cellular gene homologous to the RSV *src* gene and the mechanism for generating the transforming virus. *Cell* 32:881–890.

Wang, S.Y., Hayward, W.S., and Hanafusa, H. 1977. Genetic variation in the RNA transcripts of endogenous virus genes in uninfected chicken cells. *J. Virol.* 24:64–73.

Capture of Proto-oncogenes by Retroviruses

Hu, W.-S., and Temin, H.M. 1990. Genetic consequences of packaging two RNA genomes in one retroviral particle: pseudodiploidy and high rate of genetic recombination. *Proc. Natl. Acad. Sci. USA* 87:1556–1560.

Hu, W.-S., and Temin, H.M. 1990. Retroviral recombination and reverse transcription. *Science* 250:1227–1233.

Kawai, S., and Hanafusa, H. 1972. Genetic recombination with avian tumor virus. *Virology* 49:37–44.

Sugden, B. 1993. How some retroviruses got their oncogenes. *Trends Biochem. Sci.* 18:233–235.

Swain, A., and Coffin, J.M. 1992. Mechanism of transduction by retroviruses. *Science* 255:841–845.

Vogt, P.K. 1971. Genetically stable reassortment of markers during mixed infection with avian tumor viruses. *Virology* 46:947–952.

Weiss, R.A., Mason, W.S., and Vogt, P.K. 1973. Genetic recombinants and heterozygotes derived from endogenous and exogenous avian RNA tumor viruses. *Virology* 52:535–552.

Wyke, J.A., Bell, J.G., and Beamand, J.A. 1975. Genetic recombination among temperature-sensitive mutants of Rous sarcoma virus. *Cold Spring Harbor Symp. Quan. Biol.* 39:897–905.

Zhang, J., and Temin, H.M. 1993. Rate and mechanism of nonhomologous recombination during a single cycle of retroviral replication. *Science* 259:234–238.

Viral Oncogenes are Altered Versions of Proto-oncogenes

Blair, D.G., Oskarsson, M., Wood, T.G., McClements, W.L., Fischinger, P.J., and Vande Woude, G.F. 1981. Activation of the transforming potential of a normal cell sequence: a molecular model for oncogenesis. *Science* 212:941–943.

Bolen, J.B. 1993. Nonreceptor tyrosine protein kinases. *Oncogene* 8:2025–2031.

Cartwright, C.A., Eckhart, W., Simon, S., and Kaplan, P.L. 1987. Cell transformation by $pp60^{c\text{-}src}$ mutated in the carboxy-terminal regulatory domain. *Cell* 49:83–91.

Iba, H., Takeya, T., Cross, F.R., Hanafusa, T., and Hanafusa, H. 1984. Rous sarcoma virus variants that carry the cellular *src* gene instead of the viral *src* gene cannot transform chicken embryo fibroblasts. *Proc. Natl. Acad. Sci. USA* 81:4424–4429.

Kato, J., Takeya, T., Grandori, C., Iba, H., Levy, J., and Hanafusa, H. 1986. Amino acid substitutions sufficient to convert the nontransforming $p60^{c\text{-}src}$ protein to a transforming protein. *Mol. Cell. Biol.* 6:4155–4160.

Kmiecik, T.E., and Shalloway, D. 1987. Activation and suppression of $pp60^{c\text{-}src}$ transforming ability by mutation of its primary sites of tyrosine phosphorylation. *Cell* 49:65–73.

Levy, J.B., Iba, H., and Hanafusa, H. 1986. Activation of the transforming potential of $p60^{c\text{-}src}$ by a single amino acid change. *Proc. Natl. Acad. Sci. USA* 83:4228–4232.

Oskarsson, M., McClements, W.L., Blair, D.G., Maizel, J., and Vande Woude, G.F. 1980. Properties of a normal mouse cell DNA sequence (*sarc*) homologous to the *src* sequence of Moloney sarcoma virus. *Science* 207:1222–1224.

Parker, R.C., Varmus, H.E., and Bishop, J.M. 1984. Expression of v-*src* and chicken c-*src* in rat cells demonstrates qualitative differences between pp60$^{v\text{-}src}$ and pp60$^{c\text{-}src}$. *Cell* 37:131–139.

Piwnica-Worms, H., Saunders, K.B., Roberts, T.M., Smith, A.E., and Cheng, S.H. 1987. Tyrosine phosphorylation regulates the biochemical and biological properties of pp60$^{c\text{-}src}$. *Cell* 49:75– 82.

Reynolds, A.B., Vila, J., Lansing, T.J., Potts, W.M., Weber, M.J., and Parsons, J.T. 1987. Activation of the oncogenic potential of the avian cellular *src* protein by specific structural alteration of the carboxy terminus. *EMBO J.* 6:2359–2364.

Shalloway, D., Coussens, P.M., and Yaciuk, P. 1984. Overexpression of the c-*src* protein does not induce transformation of NIH 3T3 cells. *Proc. Natl. Acad. Sci. USA* 81:7071–7075.

❖ Part III

Cellular Oncogenes and Tumor Suppressor Genes

❖ Chapter 5

Identification of Cellular Oncogenes by Gene Transfer

The identification of retroviral oncogenes clearly established the fact that specific genes could induce cell transformation. Moreover, the cellular origin of retroviral oncogenes identified the proto-oncogenes as a group of normal cell genes with potential transforming activity. These considerations raised the question whether nonvirus-induced tumors, including human neoplasms, could be caused by alterations in normal cell genes. Such a hypothesis was also consistent with several aspects of the biology of neoplastic disease, discussed in chapter 1, particularly the stable inheritance of the transformed phenotype and the mutagenic activity of many carcinogens. Evaluation of this hypothesis depended on identifying cellular genes capable of inducing transformation and specifically altered in tumor cells.

Transformation by Cellular DNA

The transforming activity of retroviral oncogenes could be readily assessed because the viruses provided a vehicle for introduction of these genes into normal cells. Analogous direct study of the potential transforming activity of cellular genes is dependent on the ability to assay the biological activity of cellular DNA. Transfer of biologically active DNA in bacteria was first demonstrated in 1944 by Oswald Avery, Colin MacLeod, and Maclyn McCarty in classical experiments that established that DNA was the genetic material. Successful transfer of biologically active eucaryotic cell DNA was achieved by Miroslav Hill and Jana Hillova in 1971. These investigators demonstrated that the DNA of RSV-transformed cells could be transferred to recipient cultures of chicken embryo fibroblasts: the biological activity of the RSV provirus was indicated both by transformation of the recipient cells and by production of infectious RSV. These experiments appeared remarkable in that the activity of a single RSV genome could be detected in the presence of nearly a millionfold excess of unrelated cellular DNA. However, technical refinements in this gene transfer assay have allowed the detection and study of a number of single copy genes of either viral or cellular origin, as long as their biological activity is detectable by virtue of inducing an altered phenotype in appropriate recipient cells. For example, transfer of subgenomic fragments of RSV proviral DNA established that the RSV *src* gene was by itself sufficient to induce cell transformation. As long as *src* was associated with transcriptional regulatory sequences that allowed its efficient expression, none of the viral replicative genes was required.

The ability to detect viral oncogenes by gene transfer set the stage for identifying biologically active cellular oncogenes. Could cellular genes with the potential of inducing neoplastic transformation be detected in the same gene transfer assays as were used for viral oncogenes? The initial approaches to this question were reported in late 1979 and early 1980.

Robert Weinberg and colleagues assayed the transforming potential of DNAs isolated from fifteen lines of chemically transformed rodent cells (Fig. 5.1, left side). They found that DNAs from five of these donor cell lines induced transformation of recipient NIH 3T3 cells (an immortalized but nontransformed mouse cell line). The efficiency of transformation induced by the positive donor DNAs was similar to that induced by DNA of murine sarcoma virus-transformed cells. Thus, some chemically transformed cells contained active cellular oncogenes with biological transforming properties analogous to those of retroviruses.

Experiments in our laboratory indicated that normal cells also contained genes with a latent potential for inducing transformation (Fig. 5.1, right side). A low level of transformation was detected following transfer of normal chicken and mouse donor DNAs to recipient NIH 3T3 cell cultures. The efficiency of transformation induced by normal cell DNA was about 100-fold lower than that induced by an active viral oncogene. However, when colonies of cells that had been transformed by normal DNA were grown up and used as DNA donors in a second round of gene transfer, their DNAs were found to induce transformation with high efficiencies, comparable to the transforming efficiency of retroviral oncogenes. These results suggested that some normal cell genes could be activated to function as oncogenes as a consequence of DNA rearrangements that occurred during the process of gene transfer.

The implication of these combined results was threefold: (1) some chemically transformed rodent cell lines contained potent transforming oncogenes

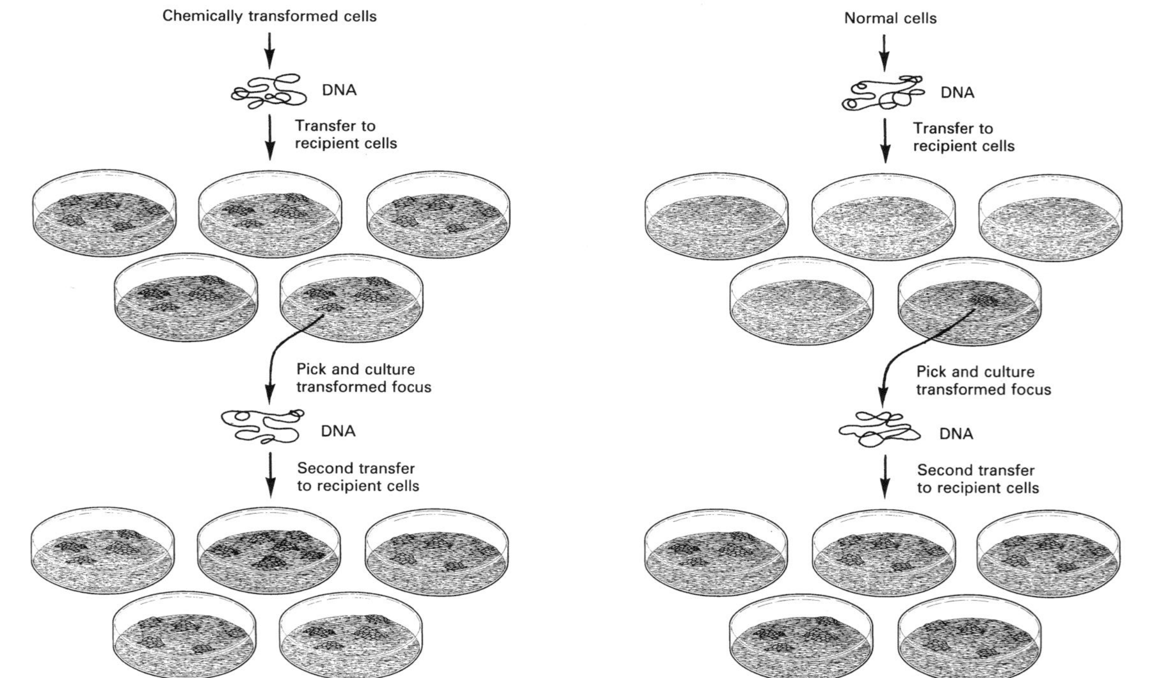

FIGURE 5.1
Detection of cellular oncogenes by gene transfer. Gene transfer assays can detect cellular oncogenes that are activated in transformed cells used as donors of DNA, as well as proto-oncogenes that become activated in the process of gene transfer. Thus, DNAs of some chemically transformed cell lines contain activated oncogenes that induce multiple colonies of transformed cells following transfer to nontransformed recipient mouse cells (left side). Normal proto-oncogenes become activated infrequently in gene transfer assays, resulting in the formation of rare transformed cell colonies (right side). In both cases, the transformed cells contain activated oncogenes that induce transformation with high frequencies in subsequent rounds of gene transfer.

that had apparently been activated at some stage of the carcinogen-induced transformation process; (2) normal cells contained proto-oncogenes that could be activated experimentally; and (3) cellular, in addition to viral, oncogenes could be detected by gene transfer. These experiments thus provided the first approach to studying alterations of specific cellular genes in tumors that had the potential of being causal contributors to the development of neoplasms.

Oncogenes of Human Tumors

In early 1981, both Weinberg's group and our own reported results of the initial application of the gene transfer approach to human tumors (Fig. 5.2). DNA extracted from a human bladder carcinoma cell line called EJ was found to induce transformation of NIH 3T3 cells with high efficiencies in

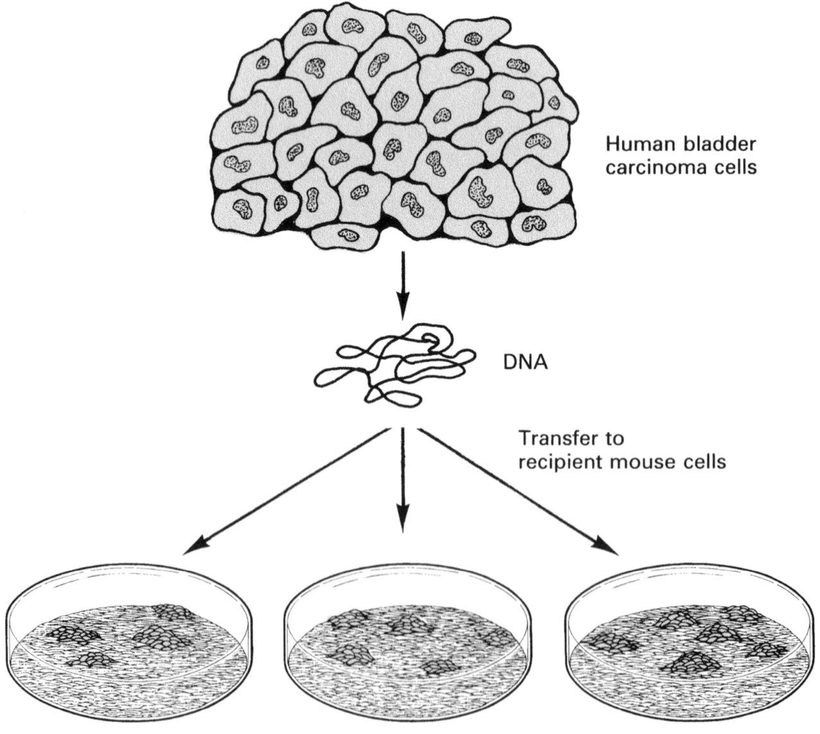

Figure 5.2
Detection of the first human oncogene. DNA extracted from the EJ human bladder carcinoma cell line induced high efficiency transformation of recipient mouse cells, indicating that this human tumor contained a biologically active oncogene.

gene transfer assays. This carcinoma therefore provided the first example of a biologically active human tumor oncogene.

Continuing studies have now identified active oncogenes in a large number of human neoplasms (and also rodent tumors) of a wide variety of cell types, including sarcomas, carcinomas of many different tissues, neuroblastomas, leukemias, and lymphomas. Overall, from about 10% to 20% of individual tumor DNAs induce transformation with high efficiencies. In contrast, normal DNAs (in some cases from the same individual) lack efficient transforming activity. A substantial fraction of human tumors therefore contain oncogenes, detectable by the gene transfer assay, that have been activated in the course of neoplasm development. It is consequently a reasonable assumption that such tumor oncogenes played a role in the pathogenesis of the neoplasms from which they have been identified.

Human Tumor Oncogenes and the *ras* Family

The gene transfer assay established the existence of biologically active human oncogenes. Could they be related to the other group of potentially transforming cellular genes discussed in chapter 4, the proto-oncogene homologs of retroviral oncogenes? In 1982, three research groups—our own, Weinberg's, and that of Mariano Barbacid and Stuart Aaronson—found that these two groups of genes overlapped. The first human oncogene identified from the EJ bladder carcinoma was a cellular homolog of the *ras*H oncogene of Harvey sarcoma virus (Fig. 5.3). In addition, an oncogene activated in a human lung carcinoma was identified as the human proto-oncogene corresponding to the Kirsten sarcoma virus oncogene, *ras*K. The *ras*H and *ras*K proto-oncogenes had thus been activated as oncogenes in two very different settings: (1) as part of the genome of acutely transforming viruses isolated from rats and (2) in human tumors that were not caused by retroviruses. These findings unified studies of the molecular biology of human cancer and transforming retroviruses. Years of previous work on the retroviral oncogenes appeared to become directly applicable to the study of cancer in human beings.

The *ras*H and *ras*K proto-oncogenes are two members of a gene family that encode similar protein products. Further analysis of cellular oncogenes identified by gene transfer from a wide variety of both human and rodent neoplasms established that many of the cellular oncogenes identified by this approach were homologs of either *ras*H or *ras*K. In addition, in 1983 a new human tumor oncogene was found to be related to but not identical with either *ras*H or *ras*K. This oncogene thus defined a third member of the cellular *ras* gene family. It was designated *ras*N because of its initial isolation from a neuroblastoma in Michael Wigler's laboratory, but it has also been detected as an active oncogene in many other types of neoplasms.

These three members of the *ras* gene family have been identified as active oncogenes in about 15% of many different types of human tumors, including sarcomas, carcinomas, neuroblastomas, leukemias, and lymphomas. It thus appears that *ras* oncogenes may take part in the development of a significant fraction of human neoplasms of diverse origins. Moreover, more sensitive methods (such as direct structural analysis using the polymerase chain reaction to amplify *ras* genes and detect activating mutations [see below]) have revealed

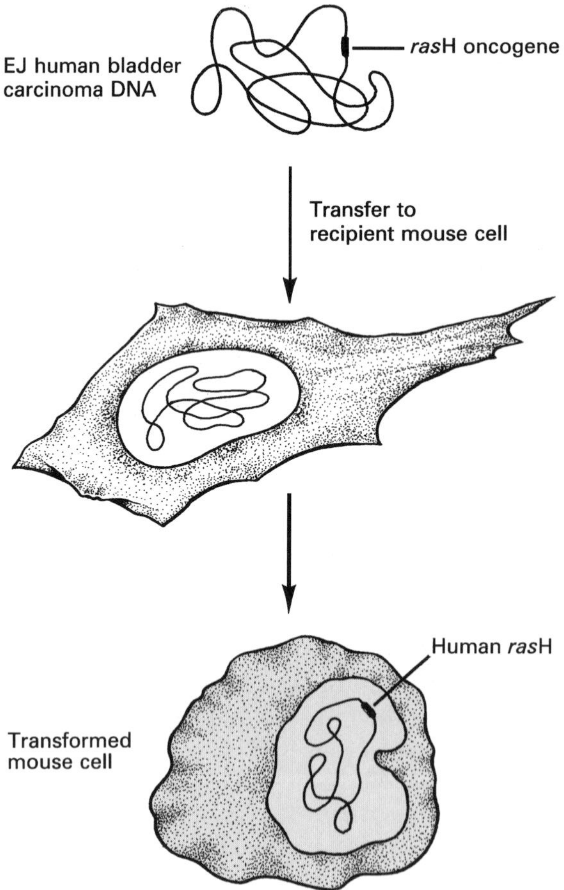

FIGURE 5.3
Identification of the EJ bladder carcinoma oncogene as *ras*H. Mouse cells transformed by EJ DNA contain a human *ras*H oncogene, which is reproducibly acquired from the human tumor DNA in the transformation process. The biologically active oncogene of this human neoplasm is therefore the cellular homolog of the *ras*H oncogene originally identified in Harvey sarcoma virus.

a much higher frequency of *ras* oncogene activation in some types of neoplasms, including acute myeloid leukemias (~25%), colon carcinomas (~50%), lung carcinomas (~25%), and pancreatic carcinomas (~90%). The activation of *ras* oncogenes in these neoplasms, and the potential roles of *ras* oncogenes in tumor pathogenesis, will be further discussed in chapter 11.

Mechanism of *ras* Oncogene Activation

The active *ras* oncogenes identified in tumor DNAs by gene transfer efficiently induce cell transformation, whereas the normal *ras* proto-oncogenes do not. What is the molecular basis for this difference in biological activity? By the end of 1982, three research groups (Weinberg's in collaboration with Edward Scolnick's laboratory, Wigler's, and Barbacid's) reported results of analysis of the nucleotide sequences of the normal human *ras*H proto-oncogene compared with the *ras*H oncogene isolated from the EJ human bladder carcinoma. The critical difference between the oncogene and the proto-oncogene was found to be a point mutation resulting in a single amino acid substitution in the Ras protein. The normal *ras*H gene encoded glycine at position 12: this amino acid was altered to valine in the EJ *ras*H oncogene product (Fig. 5.4). Experiments using chimeric genes, which represented recombinants between the normal and EJ genes, established that this single change was sufficient to convert the normal proto-oncogene into an actively transforming oncogene. Changing a single amino acid thus dramatically altered the biological activity of the *ras*H gene product, changing a normal protein into one that efficiently induced neoplastic transformation.

Subsequent studies of *ras* oncogenes in different tumors established that point mutations leading to single amino acid substitutions are a universal feature of *ras* activation. Many different amino acid substitutions at codon 12, not just valine, will activate *ras* transforming potential. Furthermore, *ras* oncogenes are also activated by a number of different amino acid substitutions at other

	1	2	3	4	5	6	7	8	9	10	11	12	13		188	189
Normal	Met	Thr	Glu	Tyr	Lys	Leu	Val	Val	Val	Gly	Ala	Gly	Gly		Leu	Ser
Human *ras*H	ATG	ACG	GAA	TAT	AAG	CTG	GTG	GTG	GTG	GGC	GCC	GGC	GGT	......	CTC	TCC
												↓				
Activated												GTC				
EJ *ras*H	Met	Thr	Glu	Tyr	Lys	Leu	Val	Val	Val	Gly	Ala	Val	Gly	......	Leu	Ser

FIGURE 5.4
Activation of the *ras*H oncogene by point mutation. A single nucleotide change, which alters codon 12 from GGC (glycine) to GTC (valine), is responsible for the potent transforming activity of the *ras*H oncogene in EJ bladder carcinoma DNA.

critical positions, particularly at codons 13 and 61 in many human tumors. In addition to these structural mutations, a point mutation in an intron of *ras*H has been found to increase transforming activity by affecting splicing to result in elevated expression of the *ras*H protein.

The biochemistry of the Ras proteins and the effect of activating mutations on their function will be discussed in detail in chapter 14. In brief, the *ras* gene products are plasma membrane-associated guanine nucleotide-binding proteins that take part in intracellular signal transduction. The general consequence of activating mutations in *ras* is the abrogation of normal control of Ras protein function, thereby converting a normally regulated cell protein into one that is constitutively active. It would appear that such deregulation of normal Ras function is responsible for the transforming activity of the mutated oncogene products.

Other Oncogenes Isolated by Gene Transfer

Although *ras* oncogenes have been the most frequently detected in gene transfer experiments, a variety of other cellular oncogenes also have been identified by this approach. Molecular clones of these oncogenes have been isolated in most cases by taking advantage of species-specific repetitive DNA sequences for use as hybridization probes, an approach first employed in Weinberg's laboratory to isolate the EJ *ras*H oncogene (Fig. 5.5). Almost all gene transfer experiments have used NIH 3T3 mouse cells as recipients because these cells are unusually efficient at taking up and integrating exogenous DNA. The problem of isolating a molecular clone of a human oncogene detected in such a gene transfer experiment can therefore be approached by isolating human DNA sequences from the transformed recipient mouse cells. Human DNA contains a family of short interspersed repetitive sequences (called *Alu* sequences) that are present on average once in every 5 kilobases of genomic DNA. A similar repetitive sequence family is present in the mouse genome, but the mouse and human sequences have diverged sufficiently that they do not hybridize to each other under appropriate conditions. Thus, human oncogene sequences have been isolated as molecular clones from transformed mouse cells with the use of *Alu* as a hybridization probe.

Most of the oncogenes that have been isolated and characterized in this way are novel genes that are distinct from the retroviral oncogenes. However, two retroviral proto-oncogenes in addition to *ras* have been detected as human oncogenes by gene transfer. These are the human homologs of *ros* (an oncogene first identified in an avian sarcoma virus) and c-*raf*-1 (an oncogene identified in both avian and murine viruses). The cellular oncogenes identified by gene transfer thus define a group of oncogenes that overlaps but is not identical with the retroviral oncogenes.

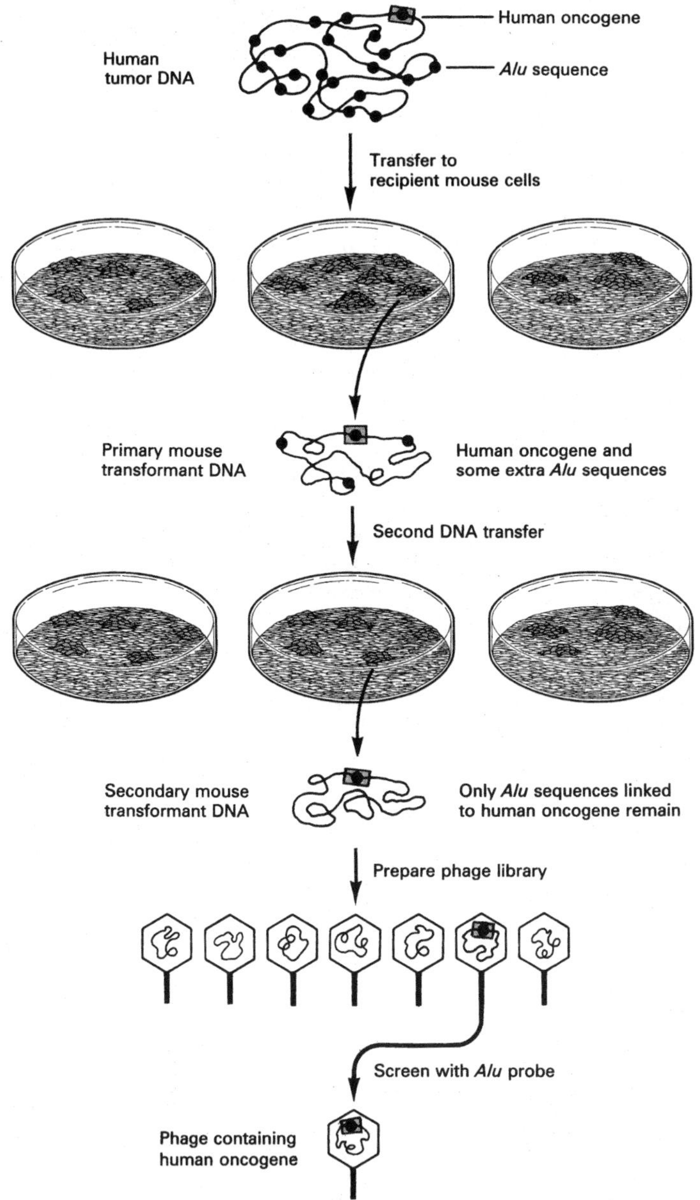

FIGURE 5.5

Molecular cloning of novel human oncogenes. Human DNA is used to transform recipient mouse cells. Transformed cells are isolated and used as donors of DNA in an additional cycle of gene transfer to dilute out human DNA sequences other than those that are closely linked to the human oncogene of interest. A phage library is then prepared from DNA of such secondary transformants and screened by hybridization with probe for the *Alu* human repetitive sequence family.

Characterization of the oncogenes identified by gene transfer has shown that they can be divided into two groups, as discussed earlier in this chapter (Fig. 5.1): (1) oncogenes that were activated in the neoplasms from which they were isolated and (2) normal proto-oncogenes that were activated in the process of gene transfer. The oncogenes that have clearly been activated in tumor cells (in addition to *ras*H, *ras*K, and *ras*N) include *neu* (isolated from rat neuroblastomas and gliomas), *met* (isolated from a chemically transformed human osteosarcoma cell line), *ret* (isolated from human thyroid carcinomas), and *trk* (isolated from a human colon carcinoma and thyroid carcinomas) (Table 5.1). Activation of *neu* occurs by a point mutation that is present in the tumors from which the oncogene was obtained. Activation of the *met*, *ret*, and *trk* oncogenes, in contrast, is by DNA rearrangements in the tumors resulting in deletion of the normal 5' coding sequences of the proto-oncogenes. The oncogenes are expressed as fusion proteins in which the normal amino-terminal domain has been replaced by coding sequences from a different cell gene. This is analogous to the activation of a number of retroviral oncogenes as fusion proteins with coding sequences derived from viral replicative genes.

In contrast, the oncogenes *ret*, *ros*, c-*raf*-1, B-*raf*, *mas*, *hst*, *fgf*-5 *dbl*, *tre*, *cot*, *ovc*, and *vav* were derived from normal human proto-oncogenes activated by DNA rearrangements that occurred in the process of gene transfer (Table 5.2). In each of these cases, the active oncogene isolated from transformed recipient cells

Table 5.1
Tumor Oncogenes Detected by Gene Transfer

Oncogene	Tumor	Activation Mechanism
*ras*H, *ras*K, and *ras*N	human and rodent carcinomas, sarcomas, neuroblastomas, leukemias, and lymphomas	point mutation
neu	rat neuroblastomas and glioblastomas	point mutation
met	chemically transformed human osteosarcoma cell line	recombinant fusion protein
ret	human thyroid carcinomas	recombinant fusion protein
trk	human colon carcinoma and thyroid carcinomas	recombinant fusion protein

TABLE 5.2
Human Oncogenes Activated During Gene Transfer Assays

Oncogene	Activation Mechanism
ret *ros* c-*raf-1* B-*raf* *dbl* *vav* *cot* *ovc* *tre*	recombinant fusion proteins
hst *fgf*-5 *mas*	aberrant gene expression

comprises two segments of donor DNA that were not linked in DNA of the human cells that served as donor in the initial gene transfer. These oncogenes were therefore formed by recombination between previously unlinked fragments of human DNA during integration of the donor DNA into the recipient cell genome. Activation of some proto-oncogenes by such recombination events is surprisingly frequent. For example, more than a dozen *raf* oncogenes have been independently activated in this manner in several different laboratories. It is further noteworthy that *ret*, which was initially isolated as an oncogene activated in the process of gene transfer, is also activated by DNA rearrangements in thyroid carcinomas, as discussed earlier.

The coding sequences of the activated *mas*, *hst*, and *fgf*-5 oncogenes are the same as those of the proto-oncogenes. Activation of these oncogenes thus appears to result not from a change in their protein products but from increased gene expression owing to recombination of the proto-oncogenes with abnormal regulatory sequences (Fig. 5.6, left side). Activation of the other nine oncogenes listed in Table 5.2, however, requires deletion of the proto-oncogene 5' coding sequences to yield a truncated oncogene that is expressed as a recombinant fusion protein (Fig. 5.6, right side). In at least some of these cases, discussed in detail in later chapters, these amino-terminal deletions appear to remove a regulatory domain of the normal proto-oncogene protein, thereby resulting in unregulated constitutive activity of the oncogene product.

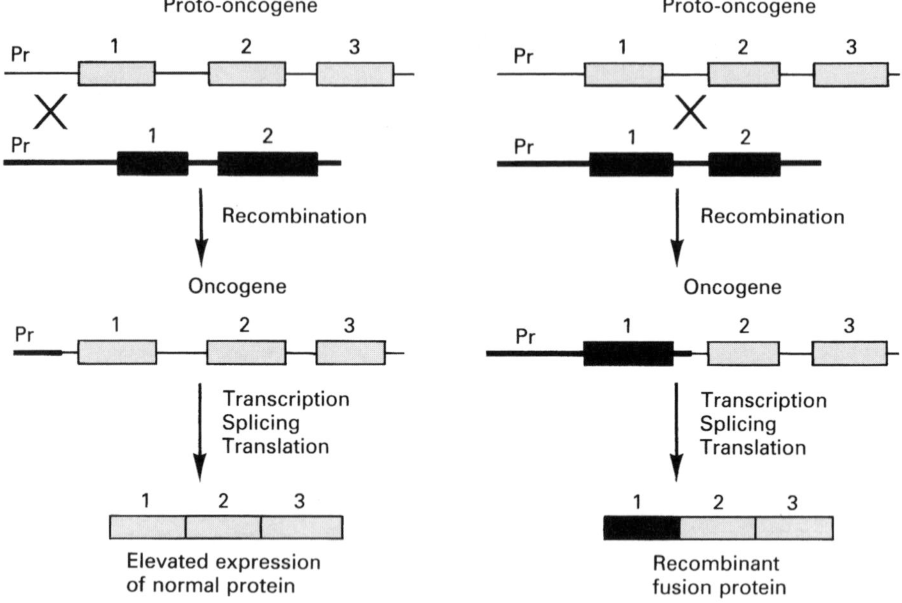

FIGURE 5.6
Oncogene activation by DNA rearrangements. Proto-oncogenes can be activated by recombination with other DNA sequences, either in neoplasm development or in the process of gene transfer. In some cases (left side), these recombination events result only in the replacement of the regulatory sequences (promoter, Pr) of the proto-oncogene with regulatory sequences derived from another gene: oncogene activation is then a result of abnormal expression of the normal proto-oncogene product. In other cases (right side), recombination events occur within coding regions, leading to the formation of recombinant fusion proteins.

Summary

Biologically active cellular oncogenes were first identified by the ability of tumor DNAs to induce transformation in gene transfer assays. Such experiments have led to the identification of more than a dozen distinct human oncogenes that are activated either by point mutations or DNA rearrangements in human neoplasms or by DNA rearrangements that occur in the process of gene transfer. Some of the cellular oncogenes identified by gene transfer are homologs of retroviral oncogenes (for example, the *ras* genes), whereas others are new oncogenes not previously found in retroviruses. In all cases, the biologically active cellular oncogenes are mutant forms of normal proto-oncogenes that differ in the regulation of their expression or in the structure and function of their gene products. The finding of such oncogenes activated by somatic mutations or DNA rearrangements in a significant fraction of human tumors clearly implicates cellular oncogene activation in the pathogenesis of human cancers.

References

Transformation by Cellular DNA

Avery, O.T., MacLeod, C.M., and McCarty, M. 1944. Studies on the chemical nature of the substance inducing transformation of pneumococcal types. *J. Exp. Med.* 79:137–158.

Cooper, G.M., Okenquist, S., and Silverman, L. 1980. Transforming activity of DNA of chemically transformed and normal cells. *Nature* 284:418–421.

Hill, M., and Hillova, J. 1972. Virus recovery in chicken cells tested with Rous sarcoma cell DNA. *Nature New Biol.* 237:35–39.

Shih, C., Shilo, B.-Z., Goldfarb, M.P., Dannenberg, A., and Weinberg, R.A. 1979. Passage of phenotypes of chemically transformed cells via transfection of DNA and chromatin. *Proc. Natl. Acad. Sci. USA* 76:5714–5718.

Wigler, M., Pellicer, A., Silverstein, S., and Axel, R. 1978. Biochemical transfer of single copy eucaryotic genes using total cellular DNA as donor. *Cell* 14:725–731.

Oncogenes of Human Tumors

Krontiris, T.G., and Cooper, G.M. 1981. Transforming activity of human tumor DNAs. *Proc. Natl. Acad. Sci. USA* 78:1181–1184.

Marshall, C.J., Hall, A., and Weiss, R.A. 1982. A transforming gene present in human sarcoma cell lines. *Nature* 299:171–173.

Murray, M.J., Shilo, B.-Z., Shih, C., Cowing, D., Hsu, H.W., and Weinberg, R.A. 1981. Three different human tumor cell lines contain different oncogenes. *Cell* 25:355–361.

Perucho, M., Goldfarb, M., Shimizu, K., Lama, C., Fogh, J., and Wigler, M. 1981. Human tumor derived cell lines contain common and different transforming genes. *Cell* 27:467–476.

Pulciani, S., Santos, E., Lauver, A.V., Long, L.K., Aaronson, S.A., and Barbacid, M. 1982. Oncogenes in solid human tumors. *Nature* 300:539–542.

Shih, C., Padhy, L.C., Murray, M., and Weinberg, R.A. 1981. Transforming genes of carcinomas and neuroblastomas introduced into mouse fibroblasts. *Nature* 290:261–264.

Human Tumor Oncogenes and the *ras* Family

Bos, J.L. 1989. *Ras* oncogenes in human cancer: a review. *Cancer Res.* 49:4682–4689.

Der, C.J., and Cooper, G.M. 1983. Altered gene products are associated with activation of cellular *ras*K genes in human lung and colon carcinomas. *Cell* 32:201–208.

Der, C.J., Krontiris, T.G., and Cooper, G.M. 1982. Transforming genes of human bladder and lung carcinoma cell lines are homologous to the *ras* genes of Harvey and Kirsten sarcoma viruses. *Proc. Natl. Acad. Sci. USA* 79:3637–3640.

Ellis, R.W., DeFeo, D., Shih, T.Y., Gonda, M.A., Young, H.A., Tsuchida, N., Lowy, D.R., and Scolnick, E.M. 1981. The p21 *src* genes of Harvey and Kirsten sarcoma viruses originate from divergent members of a family of normal vertebrate genes. *Nature* 292:506–511.

Eva, A., Tronick, S.R., Gol, R.A., Pierce, J.H., and Aaronson, S.A. 1983. Transforming genes of human hematopoietic tumors: frequent detection of *ras*-related oncogenes whose activation appears to be independent of tumor phenotype. *Proc. Natl. Acad. Sci. USA* 80:4926–4930.

Hall, A., Marshall, C.J., Spurr, N.K., and Weiss, R.A. 1983. Identification of the transforming gene in two human sarcoma cell lines as a new member of the *ras* gene family located on chromosome 1. *Nature* 303:396–400.

Murray, M.J., Cunningham, J.M., Parada, L.F., Dautry, F., Lebowitz, P., and Weinberg, R.A. 1983. The HL-60 transforming sequence: a *ras* oncogene coexisting with altered *myc* genes in hematopoietic tumors. *Cell* 33:749–757.

Parada, L.F., Tabin, C.J., Shih, C., and Weinberg, R.A. 1982. Human EJ bladder carcinoma oncogene is homologue of Harvey sarcoma virus *ras* gene. *Nature* 297:474–478.

Santos, E., Tronick, S.R., Aaronson, S.A., Pulciani, S., and Barbacid, M. 1982. T24 human bladder carcinoma oncogene is an activated form of the normal human homologue of BALB- and Harvey-MSV transforming genes. *Nature* 298:343–347.

Shimizu, K., Goldfarb, M., Perucho, M., and Wigler, M. 1983. Isolation and preliminary characterization of the transforming gene of a human neuroblastoma cell line. *Proc. Natl. Acad. Sci. USA* 80:383–387.

Shimizu, K., Goldfarb, M., Suard, Y., Perucho, M., Li, Y., Kamata, T., Feramisco, J., Stavnezer, E., Fogh, J., and Wigler, M. 1983. Three human transforming genes are related to the viral *ras* oncogenes. *Proc. Natl. Acad. Sci. USA* 80:2112–2116.

Mechanism of *ras* Oncogene Activation

Barbacid, M. 1987. *ras* genes. *Ann. Rev. Biochem.* 56:779–827.

Bos, J.L., Toksoz, D., Marshall, C.J., Verlaan-de Vries, M., Veeneman, G.H., van der Eb, A.J., van Boom, J.H., Janssen, J.W.G., and Steenvoorden, A.C.M. 1985. Amino acid substitutions at codon 13 of the N-*ras* oncogene in human acute myeloid leukemia. *Nature* 315:726–730.

Capon, D.J., Seeburg, P.H., McGrath, J.P., Hayflick, J.S., Edman, U., Levinson, A.D., and Goeddel, D.V. 1983. Activation of Ki-*ras*- 2 gene in human colon and lung carcinomas by two different point mutations. *Nature* 304:507–513.

Cohen, J.B., Broz, S.D., and Levinson, A.D. 1989. Expression of the H-*ras* proto-oncogene is controlled by alternative splicing. *Cell* 58:461–472.

Cohen, J.B., and Levinson, A.D. 1988. A point mutation in the last intron responsible for increased expression and transforming activity of the c-Ha-*ras* oncogene. *Nature* 334:119–124.

Der, C.J., Finkel, T., and Cooper, G.M. 1986. Biological and biochemical properties of human *ras*H genes mutated at codon 61. *Cell* 44:167–176.

Reddy, E.P., Reynolds, R.K., Santos, E., and Barbacid, M. 1982. A point mutation is responsible for the acquisition of transforming properties by the T24 human bladder carcinoma oncogene. *Nature* 300:149–152.

Seeburg, P.H., Colby, W.W., Hayflick, J.S., Capon, D.J., Goeddel, D.V., and Levinson, A.D. 1984. Biological properties of human c-Ha-*ras*1 genes mutated at codon 12. *Nature* 312:71–75.

Tabin, C.J., Bradley, S.M., Bargmann, C.I., Weinberg, R.A., Papageorge, A.G., Scolnick, E.M., Dhar, R., Lowy, D.R., and Chang, E.H. 1982. Mechanism of activation of a human oncogene. *Nature* 300:143–149.

Taparowsky, E., Suard, Y., Fasano, O., Shimizu, K., Goldfarb, M., and Wigler, M. 1982. Activation of the T24 bladder carcinoma transforming gene is linked to a single amino acid change. *Nature* 300:762–765.

Yuasa, Y., Srivastava, S.K., Dunn, C.Y., Rhim, J.S., Reddy, E.P., and Aaronson, S.A. 1983. Acquisition of transforming properties by alternative point mutations within c-*bas/has* human proto-oncogene. *Nature* 303:775–779.

Other Oncogenes Isolated by Gene Transfer

Bargmann, C.I., Hung, M.-C., and Weinberg, R.A. 1986. Multiple independent activations of the *neu* oncogene by a point mutation altering the transmembrane domain of p185. *Cell* 45:649–657.

Birchmeier, C., Birnbaum, D., Waitches, G., Fasano, O., and Wigler, M. 1986. Characterization of an activated human *ros* gene. *Mol. Cell. Biol.* 6:3109–3116.

Cooper, C.S., Park, M., Blair, D.G., Tainsky, M.A., Huebner, K., Croce, C.M., and Vande Woude, G.F. 1984. Molecular cloning of a new transforming gene from a chemically transformed human cell line. *Nature* 311:29–33.

Delli Bovi, P., Curatola, A.M., Kern, F.G., Greco, A., Ittmann, M., and Basilico, C. 1987. An oncogene isolated by transfection of Kaposi's sarcoma DNA encodes a growth factor that is a member of the FGF family. *Cell* 50:729–737.

Eva, A., and Aaronson, S.A. 1985. Isolation of a new human oncogene from a diffuse B-cell lymphoma. *Nature* 316:273–275.

Fukui, M., Yamamoto, T., Kawai, S., Maruo, K., and Toyoshima, K. 1985. Detection of a *raf*-related and two other transforming DNA sequences in human tumors maintained in nude mice. *Proc. Natl. Acad. Sci. USA* 82:5954–5958.

Grieco, M., Santoro, M., Berlingieri, M.T., Melillo, R.M., Donghi, R., Bongarzone, I., Pierotti, M.A., Della Porta, G., Fusco, A., and Vecchio, G. 1990. PTC is a novel rearranged form of the *ret* proto-oncogene and is frequently detected *in vivo* in human thyroid papillary carcinomas. *Cell* 60:557–563.

Halverson, D., Modi, W., Dean, M., Gelmann, E.P., Dunn, K.J., Clanton, D., Oskarsson, M., O'Brien, S.J., and Blair, D.G. 1990. An oncogenic chromosome 8-9 gene fusion isolated following transfection of human ovarian carcinoma cell line DNA. *Oncogene* 5:1085–1089.

Ikawa, S., Fukui, M., Ueyama, Y., Tamaoki, N., Yamamoto, T., and Toyoshima, K. 1988. B-*raf*, a new member of the *raf* family, is activated by DNA rearrangement. *Mol. Cell. Biol.* 8:2651–2654.

Katzav, S., Martin-Zanca, D., and Barbacid, M. 1989. *vav*, a novel human oncogene derived from a locus ubiquitously expressed in hematopoietic cells. *EMBO J.* 8:2283–2290.

Martin-Zanca, D., Hughes, S.H., and Barbacid, M. 1986. A human oncogene formed by the fusion of truncated tropomyosin and protein tyrosine kinase sequences. *Nature* 319:743–748.

Miyoshi, J., Higashi, T., Mukai, H., Ohuchi, T., and Kakunaga, T. 1991. Structure and transforming potential of the human *cot* oncogene encoding a putative protein kinase. *Mol. Cell. Biol.* 11:4088–4096.

Nakamura, T., Hillova, J., Mariage-Samson, R., and Hill, M. 1988. Molecular cloning of a novel oncogene generated by DNA recombination during transfection. *Oncogene Res.* 2:357–370.

Park, M., Dean, M., Cooper, C.S., Schmidt, M., O'Brien, S.J., Blair, D.G., and Vande Woude, G.F. 1986. Mechanism of *met* oncogene activation. *Cell* 45:895–904.

Schechter, A.L., Stern, D.F., Vaidyanathan, L., Decker, S.J., Drebin, J.A., Greene, M.I., and Weinberg, R.A. 1984. The *neu* oncogene: an *erb*B-related gene encoding a 185,000-Mr tumor antigen. *Nature* 312:513–516.

Shih, C., and Weinberg, R.A. 1982. Isolation of a transforming sequence from a human bladder carcinoma cell line. *Cell* 29:161–169.

Shimizu, K., Yoshimichi, N., Sekiguchi, M., Hokamura, K., and Tanaka, K. 1985. Molecular cloning of an activated human oncogene, homologous to v-*raf*, from primary stomach cancer. *Proc. Natl. Acad. Sci. USA* 82:5641–5645.

Stanton, V.P., Jr., and Cooper, G.M. 1987. Activation of human *raf* transforming genes by deletion of normal amino-terminal coding sequences. *Mol. Cell. Biol.* 7:1171–1179.

Taira, M., Yoshida, T., Miyagawa, K., Sakamoto, H., Terada, M., and Sugimura, T. 1987. cDNA sequence of human transforming gene *hst* and identification of the coding sequence required for transforming activity. *Proc. Natl. Acad. Sci. USA* 84:2980–2984.

Takahashi, M., Ritz, J., and Cooper, G.M. 1985. Activation of a novel human transforming gene, *ret*, by DNA rearrangement. *Cell* 42:581–588.

Takahashi, M., and Cooper, G.M. 1987. *ret* transforming gene encodes a fusion protein homologous to tyrosine kinases. *Mol. Cell. Biol.* 7:1378–1385.

Young, D., Waitches, G., Birchmeier, C., Fasano, O., and Wigler, M. 1986. Isolation and characterization of a new cellular oncogene encoding a protein with multiple potential transmembrane domains. *Cell* 45:711–719.

Zhan, X., Bates, B., Hu, X., and Goldfarb, M. 1988. The human FGF-5 oncogene encodes a novel protein related to fibroblast growth factors. *Mol. Cell. Biol.* 8:3487–3495.

❖ CHAPTER 6

Cellular Oncogene Targets for Retroviral Integration

Gene transfer assays identified cellular oncogenes by their biological activity: the ability to induce transformation of cultured cells. Alternative approaches, discussed in this and the following two chapters, have relied on the detection of specific genetic alterations in tumors as the primary criterion of cellular oncogene identification. As was the case with gene transfer experiments, the study of tumor-specific genetic alterations has uncovered a substantial number of cellular oncogenes that are implicated in neoplasm pathogenesis. The first insights leading to the identification of cellular oncogenes in this way came from studies of neoplasms induced by the weakly oncogenic retroviruses that do not contain viral oncogenes.

Tumor Induction by Retroviruses without Oncogenes

In chapter 3, the efficient transforming properties and rapid tumor induction characteristic of the acutely transforming viruses were ascribed to the activity of viral oncogenes. However, many retroviruses that do not carry oncogenes also induce tumors in infected animals. Viruses of this type include avian leukosis virus (ALV), mouse mammary tumor virus (MMTV), and leukemia viruses of mice, rats, cats, and primates (MuLV, FeLV, and GaLV). The genomes of these viruses contain only the *gag*, *pol*, and *env* genes that function in virus replication. How then is their oncogenicity to be explained?

The biology of tumor induction by retroviruses without oncogenes differs substantially from that characteristic of the acutely transforming viruses (Fig. 6.1). For example, the avian acutely transforming viruses RSV (*src* oncogene) or

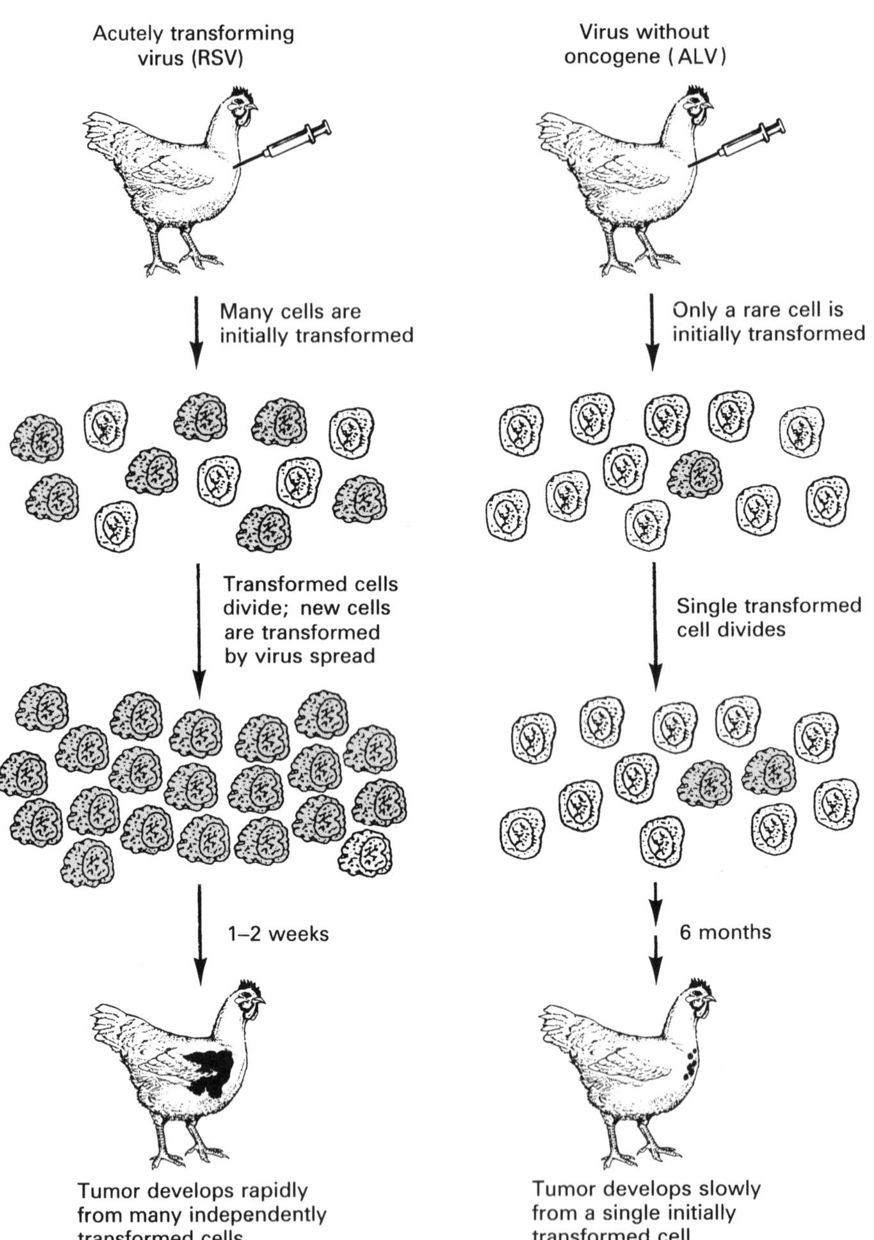

FIGURE 6.1
Tumor induction by retroviruses with and without oncogenes. Inoculation of a chicken with an acutely transforming virus (RSV, left) results in both infection and transformation of a large number of cells. Tumors then develop rapidly both from cell proliferation and from the transformation of additional cells by virus spread. Inoculation with ALV (right) also results in infection of a large number of cells, but only a rare cell is transformed. Tumors therefore develop slowly by clonal proliferation.

MC29 (*myc* oncogene) induce sarcomas or leukemias within from one to two weeks of inoculation of a susceptible chicken. In contrast, lymphomas induced by the related virus ALV develop only after latent periods of several months. Such long latent periods are characteristic of tumors induced by the retroviruses without oncogenes, as well as of tumors induced by chemical carcinogens and of most human neoplasms. The substantial lag time between virus infection and tumor development implies that only a rare cell infected with ALV becomes transformed and suggests that a series of additional changes must occur before the full neoplastic phenotype is attained.

The rarity of cell transformation by viruses without oncogenes is also revealed in the monoclonality of the tumors induced by these viruses. Whereas tumors induced by viruses such as RSV develop from a large number of independently transformed cells, the neoplasms induced by viruses such as ALV (like chemically induced and human neoplasms) develop from clonal expansion of a single transformed cell. This was initially established by analysis of the sites of proviral integration in ALV-induced lymphomas (Fig. 6.2). Analysis of the DNAs of such lymphomas indicated that the ALV proviruses were integrated at unique sites in the DNA of each neoplasm. Because proviral integration occurs at many sites in the genome of infected cells, the unique integration sites of proviral DNA in lymphomas indicated that the neoplasms arose from single infected cells—that is, they were monoclonal in origin, indicating a rare transformation event.

The Structure and Integration Sites of ALV Proviruses in Lymphomas

Examination of the ALV proviruses in chicken lymphomas revealed much more about the genesis of this neoplasm than just its monoclonality. First, many lymphomas contained only defective proviruses, from which most of the viral genome had been deleted. This result implied that lymphomagenesis does not require viral gene expression but must involve some other interaction of the virus with the host cell. Second, ALV proviruses were apparently integrated in the same region of cellular DNA in different independently arising neoplasms. Given the essentially random assortment of host cell sites available for viral DNA integration, the finding of proviruses integrated within the same region in multiple independent lymphomas suggested that development of the tumors posed a biological selection for proviral integration in this region. In particular, these results suggested the hypothesis that neoplasia was induced by integration of proviral DNA at a particular target locus in the host chromosome.

Taken together, these observations led to the formulation of a model for tumorigenesis by retroviruses without oncogenes that has become known as

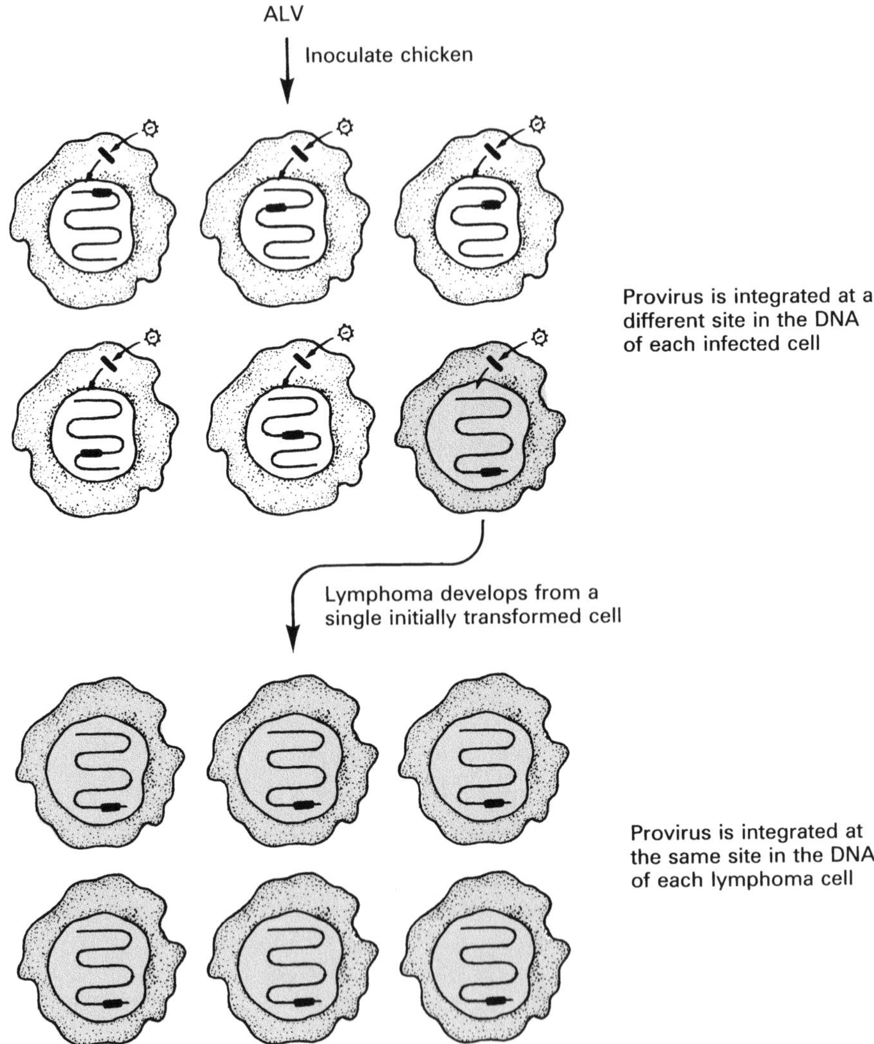

FIGURE 6.2
Clonality of ALV proviral integration sites in lymphomas. ALV proviral DNAs are integrated at different sites in different infected cells. However, individual lymphomas develop by clonal expansion of a single initially transformed cell. Therefore, each cell of a lymphoma contains a provirus integrated at the same site.

insertional mutagenesis. According to this model, integration of proviral DNA in a specific host cell target gene is a mutagenic event that initiates neoplasia. In principle, insertional mutagenesis could occur by either of two mechanisms. First, insertion of proviral DNA within a gene could result in its inactivation. Alternatively, integration of proviral DNA could activate gene expression. For example, proviral insertion could disrupt normal regulatory sequences and bring a target host gene under control of the viral LTRs. In both of these scenarios, the virus would induce neoplasia indirectly, as a result of mutations in host cell genes, rather than by expression of viral genes.

Activation of c-*myc* in Bursal Lymphomas

The critical evidence needed to substantiate the insertional mutagenesis hypothesis was identification of a target gene that, when altered by proviral integration, could lead to neoplasia. This was provided by the laboratories of Bill Hayward and Susan Astrin in 1981. These investigators set out to test the hypothesis that lymphomas resulted from activation of a cellular gene by adjacent proviral insertion, leading to enhanced expression of the target gene by transcription from the proviral LTR. Furthermore, Hayward and Astrin asked whether the target host genes were proto-oncogene homologs of already known retroviral oncogenes. Accordingly, they investigated both the organization and the expression of a series of candidate proto-oncogenes in ALV-induced chicken bursal lymphomas. The results of their initial experiments showed that the c-*myc* proto-oncogene was activated in more than 80% of these neoplasms (Fig. 6.3). Proviral DNAs, in particular functional LTRs, were integrated upstream of the lymphoma c-*myc* genes. Moreover, the activated c-*myc* genes in these lymphomas were expressed at high levels as fusion transcripts that were initiated at the viral 3' LTR promoter. Subsequent studies confirmed these observations and also demonstrated occasional activation of c-*myc* by downstream integration of LTRs or by upstream integration of LTRs in opposite transcriptional orientation to the c-*myc* gene. In these cases, the LTRs lead to elevated c-*myc* transcription by acting as enhancers (which increase transcription independent of position and orientation) rather than as promoters.

The identification of a cellular proto-oncogene as a highly reproducible target for activation in lymphomas elegantly substantiated insertional mutagenesis as a mechanism of neoplasm induction. The expected frequency of proviral integration at a particular target locus is roughly compatible with the frequency of development of rare monoclonal neoplasms, assuming that proviral integration occurs randomly in genomic DNA and that neoplasm development is a consequence of increased cell proliferation resulting from activation of the target gene. However, the long latent period required for development of these neoplasms suggests that additional events are required for the progression of

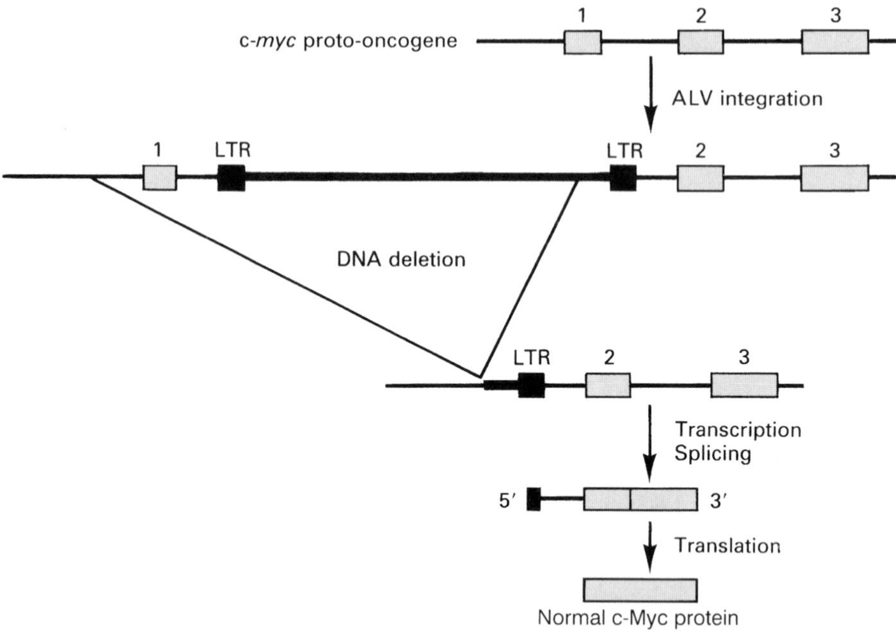

FIGURE 6.3
Activation of the c-*myc* proto-oncogene by ALV integration. The c-*myc* proto-oncogene consists of three exons (filled boxes), but only exons 2 and 3 contain protein-coding sequence. The most frequent site of ALV DNA integration in lymphomas is between exons 1 and 2. This is usually followed by deletion of the 5' portion of the provirus. Transcription from the 3' LTR then leads to enhanced expression of the normal c-*myc* protein.

initially transformed cells to frank neoplasia. The possible biological roles of oncogenes and tumor suppressor genes in such multistep pathways of neoplasm development will be discussed further in chapter 11.

Activation of Other Proto-oncogenes by Insertional Mutagenesis

The initial experiments of Hayward and Astrin with chicken bursal lymphomas stimulated a search for proto-oncogene activation in other virus-induced neoplasms (Table 6.1). Activation of c-*myc* was found not to be restricted to avian lymphomas: the c-*myc* proto-oncogene was also identified as a target of proviral insertion in T-cell lymphomas of both mice and cats. In these neoplasms, as in chicken bursal lymphomas, proviral insertion frequently results in c-*myc* activation. A related gene (N-*myc*), which represents another

Table 6.1
Oncogenes Activated by Retroviral Integration

Oncogene	Retrovirus	Neoplasm
c-*myc*	ALV	avian bursal lymphomas
	MuLV and FeLV	mouse and cat T-cell lymphomas
N-*myc*	MuLV	mouse T-cell lymphomas
*erb*B	ALV	avian erythroleukemias
*ras*H	ALV	avian nephroblastoma
*ras*K	MuLV	mouse bone marrow cell line
fos	ALV	avian nephroblastoma
myb	MuLV	mouse myeloid leukemias and plasmacytoid lymphosarcomas
	ALV	avian bursal lymphomas
fms	MuLV	mouse myeloid leukemias
mos	A particle	mouse plasmacytomas
wnt-1	MMTV	mouse mammary carcinomas
wnt-3	MMTV	mouse mammary carcinomas
int-2	MMTV	mouse mammary carcinomas
pim-1	MuLV	mouse T-cell lymphomas
lck	MuLV	mouse T-cell lymphomas
evi-1	MuLV	mouse myeloid leukemias
hox 2.4	A particle	mouse myelomonocytic leukemia
spi-1	SFFV	mouse erythroleukemias
fli-1	MuLV	mouse erythroleukemias
bmi-1	MuLV	mouse T-cell lymphomas
tpl-2	MuLV	rat T-cell lymphomas
nov	ALV	avian nephroblastoma
interleukin-2	GaLV	gibbon ape leukemia
interleukin-3	A particle	mouse myelomonocytic leukemia
GM-CSF	A particle	mouse myelomonocytic leukemia
CSF-1	MuLV	mouse monocyte tumor

member of the *myc* gene family, is also activated by proviral insertion in mouse T-cell lymphomas.

Although ALV primarily induces bursal lymphomas, it will also induce other types of neoplasms, including erythroleukemias and nephroblastomas. In both of these tumors, studies have revealed other proto-oncogene targets for insertional mutagenesis. In erythroleukemias, the *erb*B proto-oncogene is activated by insertion of proviral DNA. In this case, proviral integration reproducibly results in deletion of the amino-terminal coding portion of the proto-oncogene, as well as in abnormal gene expression. These truncations are similar to the amino-terminal deletion that occurred upon activation of the *erb*B oncogene in avian erythroblastosis virus and, as discussed in chapter 13, are critical to *erb*B oncogene activation.

Two members of the *ras* gene family also have been found to be activated by insertions of proviral DNA. In a chicken nephroblastoma, insertion of an ALV proviral LTR has been found to activate expression of the *ras*H proto-oncogene. Similarly, integration of a murine leukemia virus activates *ras*K expression in a mouse bone marrow cell line. In another chicken nephroblastoma, the *fos* proto-oncogene is activated.

Retroviral induction of myeloid leukemias and plasmacytoid lymphosarcomas in mice proceeds by activation of yet another proto-oncogene, *myb*. In these neoplasms, proviral insertion results in truncation of amino- and/or carboxy-terminal sequences of the *myb* proto-oncogene. Both amino- and carboxy-terminal deletions are seen together in the viral *myb* oncogene of avian myeloblastosis virus. Activation of *myb* has also been detected in some chicken B-cell lymphomas induced by ALV. In addition, the *fms* proto-oncogene is also activated by MuLV insertion in some mouse myeloid leukemias.

Finally, the *mos* proto-oncogene is activated in some mouse plasmacytomas by adjacent integration of endogenous retrovirus-related sequences (intracisternal A particles). The transposition of these endogenous proviruses in these neoplasms results in their insertion upstream of the *mos* proto-oncogene, leading to active expression of a gene that would normally be silent in this cell type.

Identification of New Cellular Oncogenes as Targets for Insertional Mutagenesis

The activation of known proto-oncogenes by proviral integration established insertional mutagenesis as a mechanism for neoplasm induction by viruses without oncogenes. A corollary of this mechanism is that proviral integration sites in tumors might identify new cellular oncogenes, in addition to the proto-oncogenes known from study of the acutely transforming viruses. According to this model, any cellular gene that was a reproducible target for proviral insertion in neoplasms would be a candidate cellular oncogene whose activation might contribute to tumorigenesis. Consistent with this expectation, a number of proviral integration sites that do not correspond to known proto-oncogenes have been detected in retrovirus-induced tumors, particularly mouse mammary carcinomas, leukemias, and lymphomas.

The activation of a candidate oncogene by proviral insertion not only provides a means of identifying such potential oncogenes, but also permits their isolation as molecular clones for subsequent characterization. The integrated viral DNA sequences provide a marker that can be used to isolate molecular clones containing flanking cell sequences, ultimately including the candidate cellular oncogene thought to be activated by the proviral insertions.

An example of this approach to isolation of a new oncogene is provided by *wnt*-1 (formerly called *int*-1), which is activated by integration of MMTV proviral DNA in mouse mammary carcinomas and was isolated in the laboratory of Harold Varmus in 1982 (Fig. 6.4). An initial molecular clone was obtained from

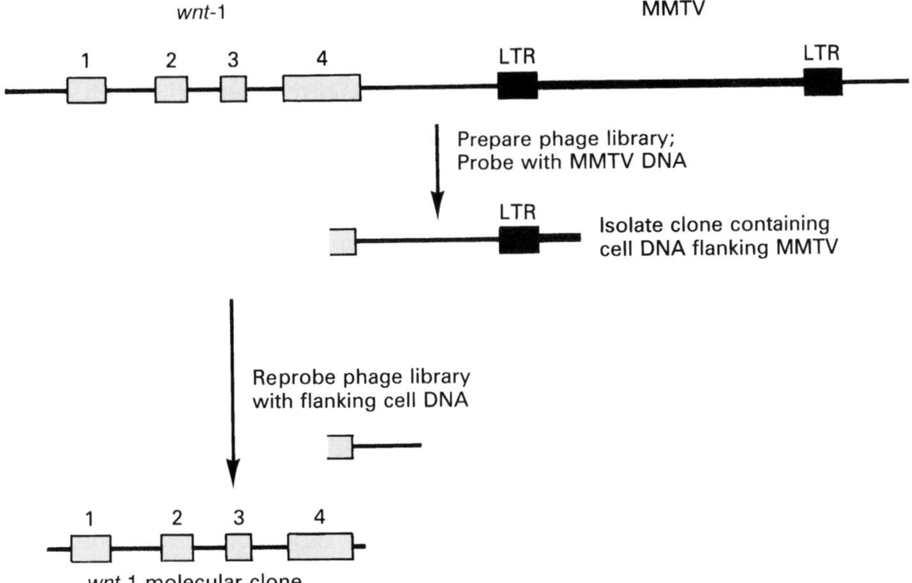

FIGURE 6.4
Molecular cloning of the *wnt*-1 proto-oncogene. MMTV proviral DNA is integrated downstream of *wnt*-1 in the mammary carcinoma illustrated. The MMTV LTR acts as an enhancer to elevate *wnt*-1 expression. The cloning strategy indicated is a simplification of that employed by Nusse and Varmus (*Cell* 31:99–109, 1982). An MMTV provirus together with its flanking cellular DNA was initially cloned from a mammary carcinoma phage library using MMTV DNA as probe. Specific probes for flanking cellular sequences were then used to isolate the entire *wnt*-1 gene, which consists of four exons.

DNA of a mammary carcinoma by the use of a fragment of MMTV DNA as hybridization probe. Flanking sequences of this initial clone were then used as probes to isolate a 25 kb segment of normal mouse DNA containing the region of interest. This genomic locus was found to contain a transcriptional unit that was activated in tumors by either upstream or downstream insertion of MMTV DNA. The *wnt*-1 gene thus appeared to be a cellular oncogene becaue of its frequent activation by insertional mutagenesis. The designation of *wnt*-1 as an oncogene was later substantiated directly when it was shown that introduction of this gene into cultured mammary epithelial cells induced abnormal growth characteristic of neoplastic transformation. This demonstration of the biological activity of *wnt*-1, which was initially identified by the indirect criteria of structural alteration in tumors, provided important further justification for this approach to oncogene isolation.

Several other candidate oncogenes have also been isolated as proviral integration sites (Table 6.1). These include additional genes from mammary carcinomas as well as a number of genes from murine leukemias and lymphomas. Characterization of the structure and biological activity of these genes, as for *wnt*-1, has served to establish their identity as cellular oncogenes.

Other targets for activation by insertional mutagenesis have been identified as genes encoding known growth factors. For example, the gibbon ape leukemia cell line MLA 144 constitutively produces the growth factor interleukin-2 (T-cell growth factor), which appears to be required for proliferation of these cells. Analysis of the interleukin-2 gene in MLA 144 cells revealed that its constitutive expression is a consequence of integration of proviral DNA of gibbon ape leukemia virus in the 3' untranslated region of the gene. Similar examples are provided by mouse myelomonocytic leukemia cell lines, in which insertion of endogenous intracisternal A particle proviruses has resulted in constitutive production of interleukin-3 (multipotential colony-stimulating factor) and GM-CSF (granulocyte-macrophage colony stimulating factor). Yet another case is provided by activation of the gene encoding colony-stimulating factor-1 (CSF-1) by proviral insertion in a mouse monocyte tumor.

Summary

Although the weakly oncogenic retroviruses do not contain viral oncogenes, they do induce tumors following long latent periods in infected animals. In many cases, neoplasm induction by these viruses involves activation of cellular oncogenes by adjacent integration of proviral DNA. More than twenty cellular oncogenes have been identified as targets for such insertional mutagenesis events. Some of these oncogenes have also been found in acutely transforming retroviruses or encode known growth factors, whereas others are novel oncogenes first identified as targets for retroviral integration. The numerous examples of oncogenes that are activated by insertional mutagenesis clearly establishes the validity of this approach to cellular oncogene identification. Moreover, these studies set a precedent for identifying cellular oncogenes that are activated in neoplasms by alterations in their structure and expression.

References

General Reference

Weiss, R., Teich, N., Varmus, H., and Coffin, J., eds. 1985. *Molecular biology of tumor viruses: RNA tumor viruses,* 2d ed. Cold Spring Harbor, New York.

Tumor Induction and Proviral Integration Sites

Cooper, M.D., Payne, L.N., Dent, P.B., Burmester, B.R., and Good, R.A. 1968. Pathogenesis of avian lymphoid leukosis. I. Histogenesis. *J. Natl. Cancer Inst.* 41:373–389.

Neel, B.G., Hayward, W.S., Robinson, H.L., Fang, J., and Astrin, S.M. 1981. Avian leukosis virus-induced tumors have common proviral integration sites and synthesize discrete new RNAs: oncogenesis by promoter insertion. *Cell* 23:323–334.

Neiman, P., Payne, L.N., and Weiss, R.A. 1980. Viral DNA in bursal lymphomas induced by avian leukosis viruses. *J. Virol.* 34:178–186.

Payne, G.S., Courtneidge, S.A., Crittenden, L.B., Fadly, A.M., Bishop, J.M., and Varmus, H.E. 1981. Analysis of avian leukosis virus DNA and RNA in bursal tumors: viral gene expression is not required for maintenance of the tumor state. *Cell* 23:311–322.

Ponten, J. 1964. The *in vivo* growth mechanism of avian Rous sarcoma. *Natl. Cancer Inst. Monograph* 17:131–45.

Activation of c-*myc* in Bursal Lymphomas

Fung, Y.-K.T., Crittenden, L.B., and Kung, H.-J. 1982. Orientation and position of avian leukosis virus DNA relative to the cellular oncogene c-*myc* in B-lymphoma tumors of highly susceptible $15I_5 \times 7_2$ chickens. *J. Virol.* 44:742–746.

Hayward, W.S., Neel, B.G., and Astrin, S.M. 1981. Activation of a cellular *onc* gene by promoter insertion in ALV-induced lymphoid leukosis. *Nature* 290:475–480.

Payne, G.S., Bishop, J.M., and Varmus, H.E. 1982. Multiple arrangements of viral DNA and an activated host oncogene in bursal lymphomas. *Nature* 295:209–214.

Activation of Other Proto-oncogenes by Insertional Mutagenesis

Canaani, E., Dreazen, O., Klar, A., Rechavi, G., Ram, D., Cohen, J.B., and Givol, D. 1983. Activation of the c-*mos* oncogene in a mouse plasmacytoma by insertion of an endogenous intracisternal A-particle genome. *Proc. Natl. Acad. Sci. USA* 80:7118–7122.

Collart, K.L., Aurigemma, R., Smith, R.E., Kawai, S., and Robinson, H.L. 1990. Infrequent involvement of c-*fos* in avian leukosis virus-induced nephroblastoma. *J. Virol.* 64:3541–3544.

Corcoran, L.M., Adams, J.M., Dunn, A.R., and Cory, S. 1984. Murine T lymphomas in which the cellular *myc* oncogene has been activated by retroviral insertion. *Cell* 37:113–122.

George, D.L., Glick, B., Trusko, S., and Freeman, N. 1986. Enhanced c-Ki-*ras* expression associated with Friend virus integration in a bone marrow-derived mouse cell line. *Proc. Natl. Acad. Sci. USA* 83:1651–1655.

Gisselbrecht, S., Fichelson, S., Sola, B., Bordereaux, D., Hampe, A., André, C., Galibert, F., and Tambourin, P. 1987. Frequent c-*fms* activation by proviral insertion in mouse myeloblastic leukaemias. *Nature* 329:259–261.

Kanter, M.R., Smith, R.E., and Hayward, W.S. 1988. Rapid induction of B-cell lymphomas: insertional activation of c-*myb* by avian leukosis virus. *J. Virol.* 62:1423–1432.

Levy, L.S., Gardner, M.B., and Casey, J.W. 1984. Isolation of a feline leukaemia provirus containing the oncogene *myc* from a feline lymphosarcoma. *Nature* 308:853–856.

Li, Y., Holland, C.A., Hartley, J.W., and Hopkins, N. 1984. Viral integrations near c-*myc* in 10–20% of MCF247-induced AKR lymphomas. *Proc. Natl. Acad. Sci. USA* 81:6801–6805.

Mullins, J.I., Brody, D.S., Binari, R.C., Jr., and Cotter, S.M. 1984. Viral transduction of c-*myc* in naturally occurring feline leukaemias. *Nature* 308:856–858.

Neil, J.C., Hughes, D., McFarlane, R., Wilkie, N.M., Onions, D.E., Lees, G., and Jarrett, O. 1984. Transduction and rearrangement of the *myc* gene by feline leukaemia virus in naturally occurring T- cell leukaemias. *Nature* 308:814–820.

Nilsen, T.W., Maroney, P.A., Goodwin, R.G., Rottman, F.M., Crittenden, L.B., Raines, M.A., and Kung, H.-J. 1985. c-*erb*B activation in ALV-induced erythroblastosis: novel RNA processing and promoter insertion result in expression of an amino-truncated EGF receptor. *Cell* 41:719–726.

Pizer, E., and Humphries, E.H. 1989. RAV-1 insertional mutagenesis: disruption of the c-*myb* locus and development of avian B-cell lymphomas. *J. Virol.* 63:1630–1640.

Shen-Ong, G.L.C., Morse, H.C., Potter, M., and Mushinski, J.F. 1986. Two modes of c-*myb* activation in virus-induced mouse myeloid tumors. *Mol. Cell. Biol.* 6:380–392.

Shen-Ong, G.L.C., Potter, M., Mushinski, J.F., Lavu, S., and Reddy, E.P. 1984. Activation of the c-*myb* locus by viral insertional mutagenesis in plasmacytoid lymphosarcomas. *Science* 226:1077–1080.

Steffen, D. 1984. Proviruses are adjacent to c-*myc* in some murine leukemia virus-induced lymphomas. *Proc. Natl. Acad. Sci. USA* 81:2097–2101.

Van Lohuizen, M., Breuer, M., and Berns, A. 1989. N-*myc* is frequently activated by proviral insertion in MuLV-induced T cell lymphomas. *EMBO J.* 8:133–136.

Westaway, D., Papkoff, J., Moscovici, C., and Varmus, H.E. 1986. Identification of a provirally activated c-Ha-*ras* oncogene in an avian nephroblastoma via a novel procedure: cDNA cloning of a chimaeric viral-host transcript. *EMBO J.* 5:301–309.

Identification of New Cellular Oncogenes as Targets for Insertional Mutagenesis

Baumbach, W.R., Colston, E.M., and Cole, M.D. 1988. Integration of the BALB/c ecotropic provirus into the colony-stimulating factor-1 growth factor locus in a *myc* retrovirus-induced murine monocyte tumor. *J. Virol.* 62:3151–3155.

Ben-David, Y., Giddens, E.B., Letwin, K., Bernstein, A. 1991. Erythroleukemia induction by Friend murine leukemia virus: insertional activation of a new member of the *ets* gene family, *Fli*-1, closely linked to c-*ets*-1. *Genes Dev.* 5:908–918.

Brown, A.M.C., Wildin, R.S., Prendergast, T.J., and Varmus, H.E. 1986. A retrovirus vector expressing the putative mammary oncogene *int*-1 causes partial transformation of a mammary epithelial cell line. *Cell* 46:1001–1009.

Chen, S.J., Holbrook, N.J., Mitchell, K.F., Vallone, C.A., Greengard, J.S., Crabtree, G.R., and Lin, Y. 1985. A viral long terminal repeat in the interleukin 2 gene of a cell line that constitutively produces interleukin 2. *Proc. Natl. Acad. Sci. USA* 82:7284–7288.

Cuypers, H.T., Selten, G., Quint, W., Zijlstra, M., Robanus-Maandag, E., Boelens, W., Van Wezenbeek, P., Melief, C., and Berns, A. 1984. Murine leukemia virus-induced T-cell lymphomagenesis: integration of proviruses in a distinct chromosomal region. *Cell* 37:141–150.

Duhrsen, U., Stahl, J., and Gough, N.M. 1990. *In vivo* transformation of factor-dependent hemopoietic cells: role of intracisternal A-particle transposition for growth factor gene activation. *EMBO J.* 9:1087–1096.

Haupt, Y., Alexander, W.S., Barri, G., Klinken, S.P., and Adams, J.M. 1991. Novel zinc finger gene implicated as *myc* collaborator by retrovirally accelerated lymphomagenesis in Eμ-*myc* transgenic mice. *Cell* 65:753–763.

Joliot, V., Martinerie, C., Dambrine, G., Plassiart, G., Brisac, M., Crochet, J., and Perbal, B. 1992. Proviral rearrangements and overexpression of a new cellular gene (*nov*) in myeloblastosis-associated virus type 1-induced nephroblastomas. *Mol. Cell. Biol.* 12:10–21.

Moreau-Gachelin, F., Tavitian, A., and Tambourin, P. 1988. *Spi*-1 is a putative oncogene in virally induced murine erythroleukemias. *Nature* 331:277–280.

Morishita, K., Parker, D.S., Mucenski, M.L., Jenkins, N.A., Copeland, N.G., and Ihle, J.N. 1988. Retroviral activation of a novel gene encoding a zinc finger protein in IL-3-dependent myeloid leukemia cell lines. *Cell* 54:831–840.

Nusse, R., and Varmus, H.E. 1982. Many tumors induced by the mouse mammary tumor virus contain a provirus integrated in the same region of the host genome. *Cell* 31:99–109.

Patriotis, C., Makris, A., Bear, S.E., and Tsichlis, P.N. 1993. Tumor progression locus 2 (*Tpl*-2) encodes a protein kinase involved in the progression of rodent T-cell lymphomas and in T-cell activation. *Proc. Natl. Acad. Sci. USA* 90:2251–2255.

Perkins, A., Kongsuwan, K., Visvader, J., Adams, J., and Cory, S. 1990. Homeobox gene expression plus autocrine growth factor production elicits myeloid leukemia. *Proc. Natl. Acad. Sci. USA* 87:8398–8402.

Peters, G., Brookes, S., Smith, R., and Dickson, C. 1983. Tumorigenesis by mouse mammary tumor virus: evidence for a common region for provirus integration in mammary tumors. *Cell* 33:369–377.

Van Lohuizen, M., Verbeek, S., Scheijen, B., Wientjens, E., van der Gulden, H., and Berns, A. 1991. Identification of cooperating oncogenes in Eμ-*myc* transgenic mice by provirus tagging. *Cell* 65:737–752.

Voronova, A.F., and Sefton, B.M. 1986. Expression of a new tyrosine protein kinase is stimulated by retrovirus promoter insertion. *Nature* 319:682–685.

Ymer, S., Tucker, W.Q.J., Sanderson, C.J., Hapel, A.J., Campbell, H.D., and Young, I.G. 1985. Constitutive synthesis of interleukin-3 by leukaemia cell line WEHI-3B is due to retroviral insertion near the gene. *Nature* 317:255–258.

Oncogenes and Chromosome Translocation

The success achieved in identifying cellular oncogenes as targets for proviral integration suggested the possibility that other sorts of specific genetic damage in neoplasms might similarly relate to oncogene activation. The potential importance of karyotypic abnormalities in neoplastic cells, including translocations, duplications, deletions, and loss of chromosomes, had been recognized since 1914. Did such abnormalities play a causal role in neoplastic transformation or were they secondary events due merely to the genetic instability of transformed cells? The first insight into the functional importance of chromosomal aberrations came from understanding their relation to oncogene activation.

Chromosome Abnormalities in Neoplasms

In many neoplasms, chromosomal abnormalities appeared erratic: there was no consistency in the aberrations observed in different tumors of the same type or even between different cells of the same individual neoplasm. In these cases, the observed karyotypic abnormalities do not appear likely to be primary contributors to the genesis of transformation.

In other instances, however, reproducible tumor-specific chromosome abnormalities have been identified. The initial finding was the work of Peter Nowell, reported in 1960. Nowell observed that human chronic myelogenous leukemia was consistently characterized by the presence of an abnormal small chromosome 22, which was called the *Philadelphia chromosome* after the city in which it was discovered. The same chromosomal abnormality was found in the leukemic cells of more than 90% of patients with chronic myelogenous leukemia; so the Philadelphia chromosome clearly represented a reproducible aberration that was closely associated with development of this neoplasm. Consistent chromosome abnormalities have since been identified in a number of other types of neoplastic disease. The reproducibility of such lesions suggested the possibility that they represent genetic alterations that impart a selective growth advantage to the neoplastic cells and thus play a causal role in tumor development—a hypothesis that was validated by a conjunction of the analysis of chromosome translocations with studies of cellular oncogenes.

Translocation of c-*myc* in Burkitt's Lymphomas and Plasmacytomas

One hypothesis for the role of chromosome translocations in neoplasia was that they might lead to DNA rearrangements that would result in activation of a cellular oncogene. The first experimental evidence was derived from studies of the c-*myc* oncogene, which had previously played a central role in elucidating cellular oncogene activation by nonacutely transforming retroviruses. At the end of 1982, studies from several different groups of investigators implicated translocation of c-*myc* in the genesis of Burkitt's lymphomas in humans and plasmacytomas in mice.

Burkitt's lymphomas and plasmacytomas are B-lymphocyte neoplasms that had previously been shown to contain specific chromosome translocations. Moreover, in both diseases, the characteristic translocations occurred at the chromosomal loci at which the immunoglobulin genes were located. In Burkitt's lymphomas, a region of chromosome 8 is translocated to either chromosome 14, chromosome 2, or chromosome 22. The sites of these translocations on chromosomes 14, 2, and 22 correspond to the immunoglobulin heavy-chain, κ light-chain, and λ light-chain genes, respectively. In mouse plasmacytomas, translocations occur between a distal region of chromosome 15 and either the immunoglobulin heavy-chain locus on chromosome 12 or the light-chain locus on chromosome 6. Because both Burkitt's lymphomas and plasmacytomas are neoplasms of B lymphocytes, which express immunoglobulin genes, translocation of a proto-oncogene into the immunoglobulin locus could readily be envisaged to result in proto-oncogene activation. This consideration led to the suggestion that the other partner in these translocations (the locus at the translocation breakpoint on human chromosome 8 and mouse chromosome 15) might be a proto-oncogene.

Based on the activation of c-*myc* by retroviral insertion in another B-cell neoplasm, chicken bursal lymphomas, several research groups investigated the possibility that c-*myc* was also activated in Burkitt's lymphomas and mouse plasmacytomas. The question was approached by determining the chromo-

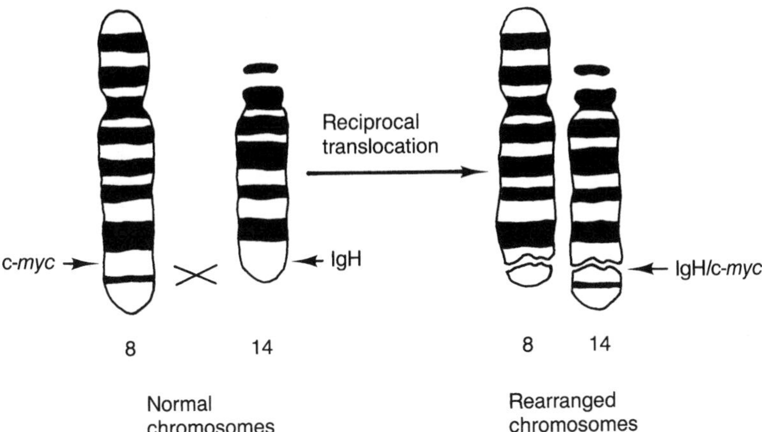

FIGURE 7.1
Translocation of c-*myc* from chromosome 8 to the immunoglobulin heavy-chain (IgH) locus on chromosome 14 in human Burkitt's lymphomas.

somal location of c-*myc*, by probing tumor DNAs directly for c-*myc* rearrangements, and by investigating the possible presence of c-*myc* sequences in molecular clones that had been derived from rearranged immunoglobulin genes. These related lines of investigation led to similar conclusions. The c-*myc* gene was located at the translocation breakpoints on human chromosome 8 and mouse chromosome 15 (Fig. 7.1). In both Burkitt's lymphomas and plasmacytomas, c-*myc* genes were rearranged by recombination with the immunoglobulin loci. Such translocations and c-*myc* rearrangements were a nearly universal feature of both neoplasms, implicating activation of the c-*myc* oncogene as a causal event in tumor development.

Molecular characterization of the structure and expression of the rearranged c-*myc* genes has led to the consensus that the translocations result in a loss of normal gene regulation, leading to constitutive c-*myc* expression. In plasmacytomas (Fig. 7.2), the translocations usually occur within the first intron of c-*myc* (the first exon is noncoding). In Burkitt's lymphomas, the translocations are more variable, sometimes occurring in either 5' or 3' flanking sequences rather than within the c-*myc* gene itself. The mechanisms by which these different rearrangements affect c- *myc* expression are not entirely understood: possibilities include the linkage of c-*myc* to immunoglobulin enhancer sequences, deletion of normal c-*myc* transcriptional regulatory sequences, and alterations in the stability of c-*myc* mRNA. In any event, the consequence of these rearrangements is abnormal constitutive expression of a c-*myc* coding sequence that is generally unaltered compared with its normal allele. The fact

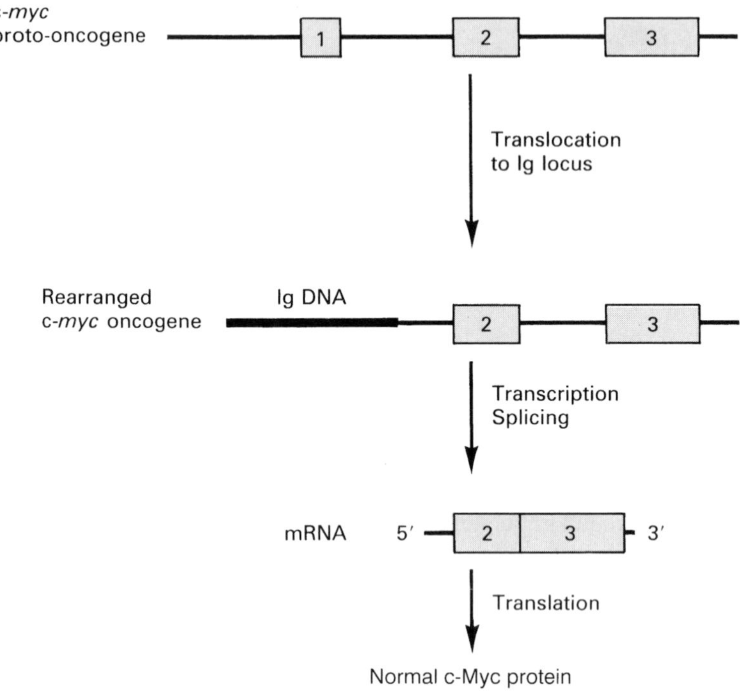

FIGURE 7.2
Activation of c-*myc* by translocation in mouse plasmacytomas. Translocation of c-*myc* to the immunoglobulin (Ig) locus most frequently results in deletion of noncoding exon 1 and 5' flanking sequences. Transcription of the rearranged gene is initiated from a cryptic promoter in the first intron, leading to constitutive expression of the normal c-*Myc* protein.

that such deregulated expression of a normal c-*myc* protein is sufficient to activate c-*myc* as an oncogene has been demonstrated directly by gene transfer experiments. Molecular clones of c-*myc* that result in constitutive expression of a normal Myc protein are able to induce transformation in a number of different cell culture systems.

Studies of c-*myc* rearrangement in these neoplasms thus established chromosome translocation as a mode of cellular oncogene activation. In addition, these results further intensified studies of c-*myc*, which had now been identified as an oncogene important in the development of a human lymphoma as well as several different animal neoplasms.

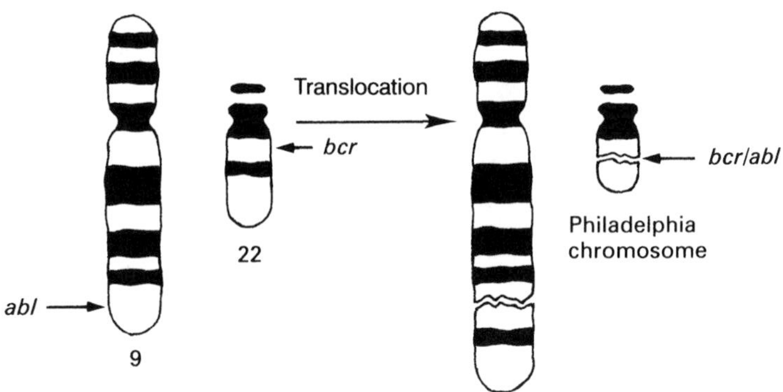

FIGURE 7.3
The Philadelphia chromosome. The *abl* proto-oncogene is translocated from chromosome 9 to the *bcr* locus on chromosome 22.

The Philadelphia Translocation and *abl*

The discovery of c-*myc* translocation heightened interest in the chromosomal locations of other proto-oncogenes. Perhaps additional proto-oncogenes could be identified at translocation breakpoints that would then prove to be activated by rearrangement. As noted before, the first consistent chromosomal abnormality identified in a neoplasm was the Philadelphia chromosome in chronic myelogenous leukemia. More than ten years after its initial description, the Philadelphia chromosome was found to be the result of a reciprocal translocation between chromosomes 9 and 22 (Fig. 7.3). In 1982, the *abl* proto-oncogene was mapped to chromosome 9 and was found to be translocated to the Philadelphia chromosome in chronic myelogenous leukemias. These findings clearly raised the prospect that *abl* was activated by this translocation. Substantiation of this hypothesis was forthcoming from molecular analysis of the effect of translocation on *abl* structure and function.

Elucidation of the mechanism of *abl* activation by the Philadelphia translocation came both from analysis of translocation breakpoints and from studies of *abl* expression. The *abl* proto-oncogene contains two alternative first exons (designated 1A and 1B), each with its own promoter, which are spliced to a common exon 2 to yield two different proto-oncogene transcripts. The Philadelphia translocation breakpoints on chromosome 9 occur over a range of 50 kb and can fall either upstream or downstream of *abl* exon 1A (Fig. 7.4). On chromosome 22, most of the breakpoints fall within a region of 6 kb, which has been termed the breakpoint cluster region (*bcr*). Molecular analysis established that *bcr* is a functional gene, which is disrupted near its middle by the transloca-

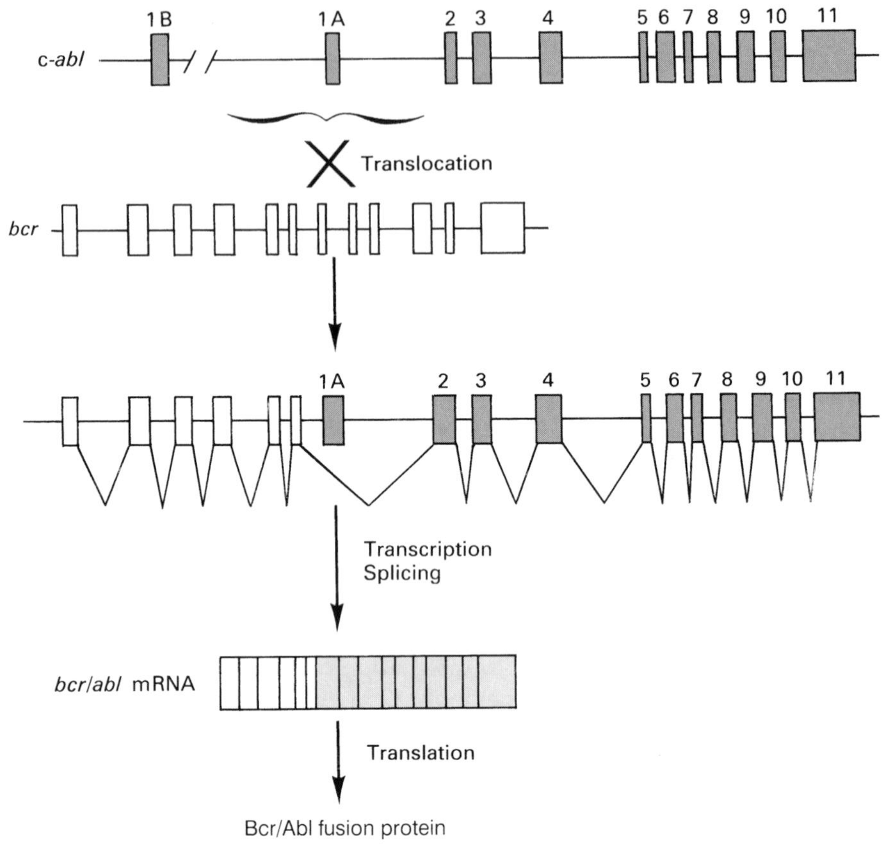

FIGURE 7.4
Activation of *abl* by translocation. The *abl* proto-oncogene is joined to the middle of the *bcr* gene. Transcription initiates at the *bcr* promoter and continues through *abl*. Splicing then generates a fused *bcr/abl* mRNA, from which *abl* exon 1A has been deleted. The *bcr/abl* mRNA is translated to yield a recombinant Bcr/Abl fusion protein.

tions. A translocation therefore creates a fusion gene, in which the 5' half of *bcr* is joined upstream of *abl* in the same transcriptional orientation.

Transcription of the *bcr/abl* fusion gene is initiated at the *bcr* promoter and continues through *abl*, yielding a long primary transcript containing both *bcr* and *abl* sequences. The *bcr* and *abl* coding sequences are then joined by RNA splicing, which results in formation of a fused *bcr/abl* mRNA of ~8.5 kb. The splice donor site of the last remaining *bcr* exon joins the first splice acceptor site in *abl*, which occurs at the start of *abl* exon 2, to form the junction of *bcr* and *abl* coding sequences. Because *abl* exon 1A does not contain a splice acceptor site, it is deleted from the spliced *bcr/abl* mRNA. This fused mRNA is then translated to yield a Bcr/Abl fusion protein in which the normal amino terminus of the *abl* proto-oncogene protein has been deleted and replaced with *bcr* amino-terminal coding sequences. A similar translocation activates the *abl* proto-oncogene in some acute lymphocytic leukemias, except that in these cases the translocation breakpoint falls farther upstream in the *bcr* gene.

The mechanism by which translocations activate *abl* thus appears to be fundamentally different from activation of c-*myc*: the *abl* translocations lead to formation of an altered protein, whereas c-*myc* translocations result in abnormal expression of the normal gene product. Interestingly, the Bcr/Abl fusion protein closely resembles the viral *abl* oncogene product, which is a fusion protein containing viral *gag* sequences at its amino terminus. In both of these fusion proteins, the normal amino terminus encoded by the *abl* proto-oncogene is deleted. Furthermore, the Bcr/Abl protein has been found to display enhanced tyrosine kinase activity, similar to that of the viral oncogene protein. Substantiation of the biological significance of this aberrant gene product has also been provided by the ability of the *bcr/abl* fusion gene to induce cell transformation. It thus appears that the Philadelphia chromosome translocation results in oncogene activation as a consequence of structural and functional changes in the Abl protein, rather than by an alteration in gene expression.

More Translocations—More Oncogenes

With chromosome translocation clearly established as a mode of oncogene activation, it became reasonable to ask whether those translocations that were reproducible features of neoplasm development would regularly be associated with activation of cellular oncogenes, potentially including new genes that had not previously been described. Activation of c-*myc* by translocation to the immunoglobulin locus has been reported in some human B-cell malignancies in addition to Burkitt's lymphomas. Moreover, the c-*myc* oncogene is also activated in occasional human T-cell neoplasms by translocation to the locus encoding one of the T-cell receptor subunits. Activation of c-*myc* in these lymphoid neoplasms thus appears to be a consequence of translocation of the oncogene into a new locus that is expressed in the target cell type. Such translocations may readily occur because the immunoglobulin and T-cell receptor loci normally undergo genetic rearrangements during maturation of B and T lymphocytes. The recombination systems responsible for these normal rearrangements may also increase the frequency of aberrant recombination events: if such events were to result in oncogene activation, the resultant cell would acquire enhanced growth potential.

On the basis of these considerations, one approach to identifying additional oncogenes that take part in the development of human neoplasms is molecular cloning and characterization of genes that are located near the breakpoints of characterized chromosome translocations. The rationale for this approach is that such studies would allow direct isolation of the candidate oncogene, whether or not it was a gene that had been previously described. The prototype neoplasm that has been analyzed in this way is follicular B-cell lymphoma, which is characterized by a translocation between chromosomes 14 and 18. The translocation site on chromosome 14 is the immunoglobulin heavy-chain locus, suggesting the hypothesis that the translocation might activate an oncogene from chromosome 18 (Fig. 7.5). Molecular cloning indicated that the breakpoints clustered in a small region of chromosome 18 DNA. Probes from this region were then used to establish the existence of an active transcriptional unit in this locus, thereby identifying a new candidate oncogene that was designated *bcl*-2. Cloning and further analysis of both normal and translocated *bcl*-2 transcripts have determined the sequence of the Bcl-2 protein, which has been identified in human cells. In addition, these studies have indicated that *bcl*-2 (like c-*myc*) is abnormally expressed as a result of translocation into the immunoglobulin heavy-chain locus. Such deregulated expression as a consequence of a consistent chromosome translocation clearly supports the suggestion that *bcl*-2 is a functional oncogene, as has been directly demonstrated by functional analysis.

Continuing molecular analysis of translocation breakpoints has identified a number of additional oncogenes that are activated in human neoplasms, particularly leukemias and lymphomas (Table 7.1). One example is the gene encoding the growth factor interleukin-3, which is translocated and abnormally expressed in some leukemias. Two other genes activated by translocations (*erg* and *fli*-1) are related to the *ets* oncogenes, first identified in avian retroviruses. In addition, translocations activate both the human homolog of the retroviral *rel* oncogene and

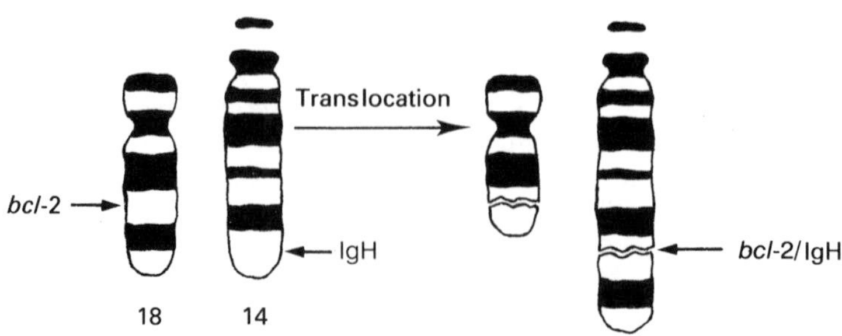

FIGURE 7.5
Translocation of *bcl*-2 from chromosome 18 to the immunoglobulin heavy-chain locus on chromosome 14 in human follicular B-cell lymphoma.

TABLE 7.1
Oncogenes Activated by Translocation

Oncogene	Neoplasm	Activation Mechanism
c-*myc*	Burkitt's lymphomas and other B- and T-cell neoplasms	aberrant expression
	mouse plasmacytomas	
bcl-2	follicular B-cell lymphomas	
bcl-3	chronic B-cell leukemia	
bcl-6	diffuse B-cell lymphomas	
*hox*11	acute T-cell leukemia	
IL-3	acute pre B-cell leukemia	
lyl-1	acute T-cell leukemia	
PRAD-1	chronic B-cell leukemia and parathyroid adenoma	
rhom-1	acute T-cell leukemia	
rhom-2	acute T-cell leukemia	
tal-1	acute T-cell leukemia	
tal-2	acute T-cell leukemia	
tan-1	acute T-cell leukemia	
bcr/abl	chronic myelogenous and acute lymphocytic leukemias	fusion proteins
dek/can	acute myeloid leukemia	
*E2A/pbx*1	acute pre B-cell leukemia	
PML/RAR	acute promyelocytic leukemia	
c16/*erg*[1]	myeloid leukemia	
rel/nrg	B-cell lymphoma	
CBFß/MYH11	acute myeloid leukemia	
ews/fli-1	Ewing sarcoma	
lyt-10/$C\alpha_1$	B-cell lymphoma	
hrx/enl	acute leukemias	
hrx/af4	acute leukemias	
aml1/mtg8	acute myeloid leukemias	
NPM/ALK	large-cell lymphomas	

[1]The fusion partner of *erg* on chromosome 16 (indicated c16) has not yet been identified.

a related member of the *rel* gene family (*lyt*-10). A variety of other novel oncogenes also have been isolated by this approach. Some of these genes are activated as a result of abnormal expression (like c-*myc* and *bcl*-2), whereas other translocations result in the formation of recombination fusion proteins, analogous to activation of the *bcr/abl* oncogene. It is noteworthy that most of the oncogenes activated by these translocations (including c-*myc* and the *ets*- and *rel*-related genes) encode transcription factors whose abnormal expression or function induces transformation as a result of aberrations in gene expression in neoplastic cells.

Summary

Chromosome translocations occur frequently in human neoplasms and constitute an important mechanism for oncogene activation. The activations of c-*myc* and *abl* by translocation are the most reproducible alterations of oncogenes in human neoplasia, occurring in virtually all Burkitt's lymphomas and chronic myelogenous leukemias, respectively. Moreover, molecular analysis of additional translocations, particularly in leukemias and lymphomas, has provided a powerful approach to identifying a number of novel oncogenes with important roles in the development of human cancers.

References

General References

Cleary, M.L. 1991. Oncogenic conversion of transcription factors by chromosomal translocations. *Cell* 66:619–622.

Rabbitts, T.H. 1991. Translocations, master genes, and differences between the origins of acute and chronic leukemias. *Cell* 67:641–644.

Solomon, E., Borrow, J., and Goddard, A.D. 1991. Chromosome aberrations and cancer. *Science* 254:1153–1160.

Chromosome Abnormalities in Neoplasms

Boveri, T. 1914. *Zur frage der erstehung maligner tumoren* (On the problem of the origin of malignant tumors). Gustav Fischer, Jena.

Heim, S., and Mitelman, F. 1987. *Cancer cytogenetics.* A.R. Liss, New York.

Nowell, P.C., and Hungerford, D.A. 1960. Chromosome studies on normal and leukemic human leukocytes. *J. Natl. Cancer Inst.* 25:85–109.

Translocation of *c-myc* in Burkitt's Lymphomas and Plasmacytomas

Adams, J., Gerondakis, S., Webb, E., Corcoran, L.M., and Cory, S. 1983. Cellular *myc* oncogene is altered by chromosomal translocation to the immunoglobulin locus in murine plasmacytomas and is rearranged similarly in human Burkitt lymphomas. *Proc. Natl. Acad. Sci. USA* 80:1982–1986.

Crews, S., Barth, R., Hood, L., Prehn, J., and Calame, K. 1982. Mouse c-*myc* oncogene is located on chromosome 15 and translocated to chromosome 12 in plasmacytomas. *Science* 218:1319–1321.

Dalla-Favera, R., Bregni, M., Erikson, J., Patterson, D., Gallo, R.C., and Croce, C.M. 1982. Human c-*myc onc* gene is located on the region of chromosome 8 that is translocated in Burkitt lymphoma cells. *Proc. Natl. Acad. Sci. USA* 79:7824–7827.

Keath, E.J., Caimi, P.G., and Cole, M.D. 1984. Fibroblast lines expressing activated c-*myc* oncogenes are tumorigenic in nude mice and syngeneic animals. *Cell* 39:339–348.

Leder, P., Battey, J., Lenoir, G., Moulding, C., Murphy, W., Potter, H., Stewart, T., and Taub, R. 1983. Translocations among antibody genes in human cancer. *Science* 222:765–771.

Marcu, K.B., Bossone, S.A., and Patel, A.J. 1992. *myc* function and regulation. *Ann. Rev. Biochem.* 61:809–860.

Marcu, K.B., Harris, L.J., Stanton, L.W., Erikson, J., Watt, R., and Croce, C.M. 1983. Transcriptionally active c-*myc* oncogene is contained within NIARD, a DNA sequence associated with chromosome translocation in B-cell neoplasia. *Proc. Natl. Acad. Sci. USA* 80:519–524.

Shen-Ong, G.L.C., Keath, E.J., Piccoli, S.P., and Cole, M.D. 1982. Novel *myc* oncogene RNA from abortive immunoglobulin-gene recombination in mouse plasmacytomas. *Cell* 31:443–452.

Taub, R., Kirsch, I., Morton, C., Lenoir, G., Swan, D., Tronick S., Aaronson, S., and Leder, P. 1982. Translocation of the c-*myc* gene into the immunoglobulin heavy chain locus in human Burkitt lymphoma and murine plasmacytoma cells. *Proc. Natl. Acad. Sci. USA* 79:7837–7841.

The Philadelphia Translocation and *abl*

Chan, L.C., Karhi, K.K., Rayter, S.I., Heisterkamp, N., Eridani, S., Powles, R., Lawler, S.D., Groffen, J., Foulkes, J.G., Greaves, M.F., and Wiedemann, L.M. 1987. A novel *abl* protein expressed in Philadelphia chromosome positive acute lymphoblastic leukaemia. *Nature* 325:635–637.

De Klein, A., van Kessel, A.G., Grosveld, G., Bartram, C.R., Hagemeijer, A., Bootsma, D., Spurr, N.K., Heisterkamp, N., Groffen, J., and Stephenson, J.R. 1982. A cellular oncogene is translocated to the Philadelphia chromosome in chronic myelocytic leukemia. *Nature* 300:765–767.

Gale, R.P., and Canaani, E. 1984. An 8-kilobase *abl* RNA transcript in chronic myelogenous leukemia. *Proc. Natl. Acad. Sci. USA* 81:5648–5652.

Gishizky, M.L., and Witte, O.N. 1992. Initiation of deregulated growth of multipotent progenitor cells by *bcr-abl in vitro. Science* 256:836–839.

Heisterkamp, N., Stam, K., Groffen, J., de Klein, A., and Grosveld, G. 1985. Structural organization of the *bcr* gene and its role in the Ph′ translocation. *Nature* 315:758–761.

Konopka, J.B., Watanabe, S.M., and Witte, O.N. 1984. An alteration of the human c-*abl* protein in K562 leukemia cells unmasks associated tyrosine kinase activity. *Cell* 37:1035–1042.

Kurzrock, R., Shtalrid, M., Romero, P., Kloetzer, W.S., Talpas, M., Trujillo, J.M., Blick, M., Beran, M., and Gutterman, J.U. 1987. A novel c-*abl* protein product in Philadelphia-positive acute lymphoblastic leukaemia. *Nature* 325:631–635.

Rowley, J.D. 1973. A new consistent chromosomal abnormality in chronic myelogenous leukemia identified by quinacrine fluorescence and Giemsa staining. *Nature* 243:290–293.

Shtivelman, E., Lifshitz, B., Gale, R.P., and Canaani, E. 1985. Fused transcript of *abl* and *bcr* genes in chronic myelogenous leukaemia. *Nature* 315:550–554.

Shtivelman, E., Lifshitz, B., Gale, R.P., Roe, B.A., and Canaani, E. 1986. Alternative splicing of RNAs transcribed from the human *abl* gene and from the *bcr-abl* fused gene. *Cell* 47:277–284.

More Translocations—More Oncogenes

Begley, C.G., Aplan, P.D., Davey, M.P., Nakahara, K., Tchorz, K., Kurtzberg, J., Hershfield, M.S., Haynes, B.F., Cohen, D.I., Waldmann, T.A., and Kirsch, I.R. 1989. Chromosomal translocation in a human leukemic stem-cell line disrupts the T-cell antigen receptor delta-chain diversity region and results in a previously unreported fusion transcript. *Proc. Natl. Acad. Sci. USA* 86:2031–2035.

Boehm, T., Foroni, L., Kaneko, Y., Perutz, M.F., and Rabbitts, T.H. 1991. The *rhombotin* family of cysteine-rich LIM-domain oncogenes: distinct members are involved in T-cell translocations to human chromosomes 11p15 and 11p13. *Proc. Natl. Acad. Sci. USA* 88:4367–4371.

Borrow, J., Goddard, A.D., Sheer, D., and Solomon, E. 1990. Molecular analysis of acute promyelocytic leukemia breakpoint cluster region on chromosome 17. *Science* 249:1577–1580.

Chen, Q., Cheng, J.T., Tasi, L.H., Schneider, N., Buchanan, G., Carroll, A., Crist, W., Ozanne, B., Siciliano, M.J., and Baer, R. 1990. The *tal* gene undergoes chromosome translocation in T cell leukemia and potentially encodes a helix-loop-helix protein. *EMBO J.* 9:415–424.

Cleary, M.L., Smith, S.D., and Sklar, J. 1986. Cloning and structural analysis of cDNAs for *bcl*-2 and a hybrid *bcl*- 2/immunoglobulin transcript resulting from the t(14;18) translocation. *Cell* 47:19–28.

de Thé, H., Chomienne, C., Lanotte, M., Degos, L., and Dejean, A. 1990. The t(15;17) translocation of acute promyelocytic leukaemia fuses the retinoic acid receptor α gene to a novel transcribed locus. *Nature* 347:558–561.

Ellisen, L.W., Bird, J., West, D.C., Soreng, A.L., Reynolds, T.C., Smith, S.D., and Sklar, J. 1991. *TAN*-1, the human homolog of the *Drosophila notch* gene, is broken by chromosomal translocations in T lymphoblastic neoplasms. *Cell* 66:649–661.

Erikson, J., Finger, L., Sun, L., Ar-Rushdi, A., Nishikura, K., Minowada, J., Finan, J., Emanuel, B.S., Nowell, P.C., and Croce, C.M. 1986. Deregulation of c-*myc* by translocation to the α locus of the T-cell receptor in T-cell leukemias. *Science* 232:884–886.

Gu, Y., Nakamura, T., Alder, H., Prasad, R., Canaani, O., Cimino, G., Croce, C.M., and Canaani, E. 1992. The t(4;11) chromosome translocation of human acute leukemias fuses the *ALL-1* gene, related to *Drosophila trithorax*, to the *AF-4* gene. *Cell* 71:701–708.

Hatano, M., Roberts, C.W., Minden, M., Crist, W.M., and Korsmeyer, S.J. 1991. Deregulation of a homeobox gene, *HOX11*, by the t(10;14) in T cell leukemia. *Science* 253:79–82.

Kamps, M.P., Murre, C., Sun, X., and Baltimore, D. 1990. A new homeobox gene contributes the DNA binding domain of the t(1;19) translocation protein in pre-B ALL. *Cell* 60:547–555.

Liu, P., Tarlé, S.A., Hajra, A., Claxton, D.F., Marlton, P., Freedman, M., Siciliano, M.J., and Collins, F.S. 1993. Fusion between transcription factor CBFß/PEBP2ß and a myosin heavy chain in acute myeloid leukemia. *Science* 261:1041–1044.

Lu, D., Thompson, J.D., Gorski, G.K., Rice, N.R., Mayer, M.G., and Yunis, J.J. 1991. Alterations of the *rel* locus in human lymphoma. *Oncogene* 6:1235–1241.

May, W.A., Gishizky, M.L., Lessnick, S.L., Lunsford, L.B., Lewis, B.C., Delattre, O., Zucman, J., Thomas, G., and Denny, C.T. 1993. Ewing sarcoma 11;22 translocation produces a chimeric transcription factor that requires the DNA-binding domain encoded by *FLI1* for transformation. *Proc. Natl. Acad. Sci. USA* 90:5752–5756.

Meeker, T.C., Hardy, D., Willman, C., Hogan, T., and Abrams, J. 1990. Activation of the interleukin-3 gene by chromosome translocation in acute lymphocytic leukemia with eosinophilia. *Blood* 76:285–289.

Melentin, J.D., Smith, S.D., and Cleary, M.L. 1989. *lyl*-1, a novel gene altered by chromosomal translocation in T cell leukemia, codes for a protein with a helix-loop-helix DNA binding motif. *Cell* 58:77–83.

Miyoshi, H., Kozu, T., Shimizu, K., Enomoto, K., Maseki, N., Kaneko, Y., Kamada, N., and Ohki, M. 1993. The t(8;21) translocation in acute myeloid leukemia results in production of an *AML1-MTG8* fusion transcript. *EMBO J.* 12:2715–2721.

Morris, S.W., Kirstein, M.N., Valentine, M.B., Kittmer, K.G., Shapiro, D.N., Saltman, D.L., and Look, A.T. 1994. Fusion of a kinase gene, *ALK*, to a nucleolar protein gene, *NPM*, in non-Hodgkin's lymphoma. *Science* 263:1281–1284.

Motokura, T., Bloom, T., Kim, H.G., Juppner, H., Ruderman, J.V., Kronenberg, H.M., and Arnold, A. 1991. A novel cyclin encoded by a *bcl*1-linked candidate oncogene. *Nature* 350:512–515.

Neri, A., Chang, C.-C., Lombardi, L., Salina, M., Carradini, P., Maiolo, A.T., Chaganti, R.S.K., and Dalla-Favera, R. 1991. B cell lymphoma-associated chromosomal translocation involves candidate oncogene *lyt*-10, homologous to NF-κB p50. *Cell* 67:1075–1087.

Nourse, J., Mellentin, J.D., Galili, N., Wilkinson, J., Stanbridge, E., Smith, S.D., and Cleary, M.L. 1990. Chromosomal translocation t(1;19) results in synthesis of a homeobox fusion mRNA that codes for a potential chimeric transcription factor. *Cell* 60:535–545.

Ohno, H., Takimoto, G., and McKeithan, T.W. 1990. The candidate proto-oncogene *bcl*-3 is related to genes implicated in cell lineage determination and cell cycle control. *Cell* 60:991–997.

Shimizu, K., Ichikawa, H., Tojo, A., Kaneko, Y., Maseki, N., Hayashi, Y., Ohira, M., Asano, S., and Ohki, M. 1993. An *ets*-related gene, *ERG*, is rearranged in human myeloid leukemia with t(16;21) chromosomal translocation. *Proc. Natl. Acad. Sci. USA* 90:10280–10284.

Tkachuk, D.C., Kohler, S., and Cleary, M.L. 1992. Involvement of a homolog of *Drosophila trithorax* by 11q23 chromosomal translocations in acute leukemias. *Cell* 71:691–700.

Tsujimoto, Y., Cossman, J., Jaffe, E., and Croce, C.M. 1985. Involvement of the *bcl*-2 gene in human follicular lymphoma. *Science* 228:1440–1443.

Von Lindern, M., Fornerod, M., van Baal, S., Jaegle, M., de Wit, T., Buijs, A., and Grosveld, G. 1992. The translocation (6;9), associated with a specific subtype of acute myeloid leukemia, results in the fusion of two genes, *dek* and *can*, and the expression of a chimeric, leukemia-specific *dek-can* mRNA. *Mol. Cell. Biol.* 12:1687–1697.

Xia, Y., Brown, L., Yang, C.Y.-C., Tsan, J.T., Siciliano, M.J., Espinosa, R., III, Le Beau, M.M., and Baer, R.J., 1991. *TAL*2, a helix-loop-helix gene activated by the (7;9)(q34;q32) translocation in human T-cell leukemia. *Proc. Natl. Acad. Sci. USA* 88:11416–11420.

Ye, B.H., Lista, F., Lo Coco, F., Knowles, D.M., Offit, K., Chaganti, R.S.K., and Dalla-Favera, R. 1993. Alterations of a zinc finger-encoding gene, *BCL-6,* in diffuse large-cell lymphoma. *Science* 262:747–750.

❖ CHAPTER 8

Amplification of Oncogenes in Tumors

We have considered a number of cases in which cellular oncogenes are activated by DNA rearrangements, including proviral insertions and chromosome translocations, that result in increased or deregulated gene expression. Another mechanism for increasing gene expression is DNA amplification, resulting in an increased number of gene copies per cell. In this case, elevated gene expression would be expected as a direct consequence of an increase in the numbers of templates available for transcription, rather than from changes in transcriptional regulatory sequences. As discussed in this chapter, DNA amplification appears to be a common mode of cellular oncogene activation in a wide variety of human neoplasms.

Gene Amplification and Chromosome Abnormalities

Gene amplification was first characterized as a common mechanism by which mammalian cells acquire resistance to metabolic inhibitors and is often seen in tumor cells that have become resistant to a chemotherapeutic agent. For example, methotrexate is a commonly used drug in cancer chemotherapy. It acts by inhibiting the enzyme dihydrofolate reductase, which plays a role in the biosynthesis of deoxyribonucleoside triphosphates and is therefore required for DNA replication. Resistance to methotrexate frequently develops by amplification of the dihydrofolate reductase gene. This leads to increased levels of the target enzyme, so that effective inhibition of dihydrofolate reductase by methotrexate is no longer achieved.

Studies of amplified genes in cells selected for resistance to metabolic inhibitors showed that DNA amplification is associated with two different types of karyotypic abnormalities (Fig. 8.1). In some cases, copies of amplified DNA are present as *double minute chromosomes,* which are small chromosome-like structures that lack centromeres. Alternatively, amplified DNA sequences are found within normal centromeric chromosomes. In this case, the amplified sequences can frequently be identified as abnormal areas of a chromosome—for example, regions that stain homogeneously rather than yielding a pattern of alternating light and dark bands characteristic of normal chromosomes. Double minutes and *homogeneous staining regions* appear to be interchangeable forms of amplified DNA. Because double minutes lack centromeres, they are not stably inherited at cell division—their maintenance requires continual selection for the amplified sequences. Integration of double minutes into a chromosome is thought to result in the formation of homogeneous staining regions, which represent a more stably maintained form of amplified DNA.

Double minutes and homogeneous staining regions are not encountered in cytological examination of normal mammalian cells, and DNA amplification does not appear to be a normal event in mammalian development. However, DNA amplification is at least 1,000 times as frequent in tumor cells as in normal cells. This genetic instability is likely to play an important role in tumor development under circumstances in which the amplified sequences confer a selective advantage, resulting in outgrowth of a cell population containing amplified DNA (for example, drug resistant variants). It is therefore noteworthy that double minutes and homogeneous staining regions are frequently observed in tumor cells, even when the neoplasms have not been treated with chemotherapeutic agents that might select for gene amplification in drug resistant variants. These findings indicate that gene amplification is a common event in neoplasm development, which presumably occurs because increased expression of the amplified genes gives a selective growth advantage to the tumor cells. Such considerations clearly engender the hypothesis that cellular oncogenes might be targets for such amplification events.

Amplification of c-*myc*

Studies of the c-*myc* oncogene formed the basis of our understanding of cellular oncogene activation by both retroviral insertional mutagenesis (chapter 6) and chromosome translocation (chapter 7). Similarly, c-*myc* was the first cellular oncogene found to be amplified in human neoplasms. In 1982, two research groups, the laboratories of Robert Gallo and Mark Groudine, reported that the c-*myc* gene was amplified about tenfold in the human promyelocytic leukemia cell line HL-60. Amplification of c-*myc* was detected not only in the tumor cells after growth in culture, but also in the primary uncultured leukemic cells

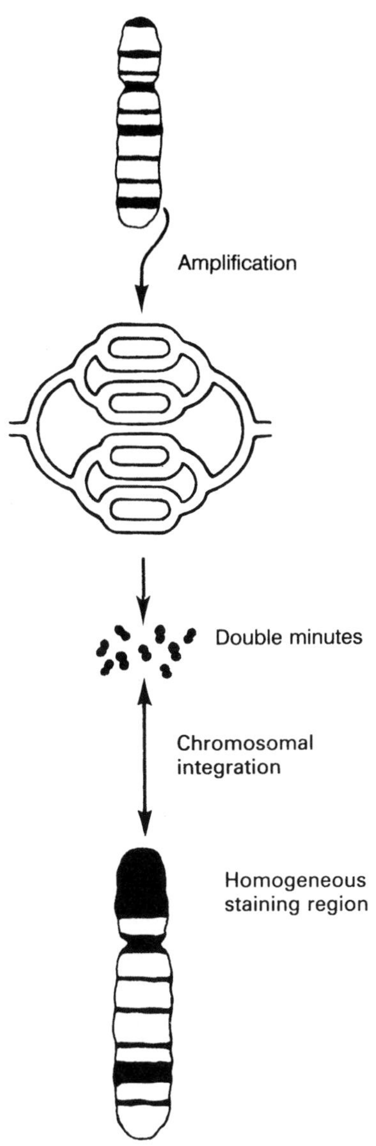

Figure 8.1
DNA amplification. A locus is amplified by repeated DNA replication. Recombination can generate tandem arrays of the amplified DNA, which can be excised from the chromosome to form double minutes. Double minutes can integrate into another chromosome and form a homogeneous staining region. Because chromosome integration is reversible, double minutes and homogeneous staining regions are interchangeable forms of amplified DNA.

from which the HL-60 cell line was derived. This amplification of c-*myc* DNA was associated with high levels of c-*myc* transcription in HL-60 cells. In addition, cytological studies and hybridization of c-*myc* probe to preparations of metaphase chromosomes indicated that the amplified c-*myc* sequences in HL-60 cells were associated with either double minutes or abnormal chromosome staining regions typical of DNA amplification.

Amplification and overexpression of c-*myc* are not restricted to HL-60 cells but have also been detected in a number of other tumors including breast, stomach, and lung carcinomas, malignant neuroendocrine cells derived from a colon carcinoma, neuroblastomas, and glioblastomas (Table 8.1). Amplification of c-*myc* is particularly common in breast cancers and small-cell lung carcino-

TABLE 8.1
Amplified Oncogenes of Human Tumors

Oncogene	Neoplasm
c-*myc*	leukemias breast, stomach, lung, and colon carcinomas neuroblastomas and glioblastomas
N-*myc*	neuroblastomas and retinoblastomas lung carcinomas
L-*myc*	lung carcinomas
*erb*B	glioblastomas, squamous cell carcinomas
*erb*B-2	breast, salivary gland, and ovarian carcinomas
int-2	breast and squamous cell carcinomas
hst	breast and squamous cell carcinomas
PRAD-1	breast and squamous cell carcinomas
abl	K562 chronic myelogenous leukemia cell line
myb	colon carcinoma, leukemias
ets-1	lymphoma
*ras*H	bladder carcinoma
*ras*K	lung, ovarian, and bladder carcinomas
*ras*N	breast carcinoma cell line
gli	glioblastomas
K-*sam*	stomach carcinomas
mdm-2	sarcomas

mas, occurring with frequencies of about 30%. Because c-*myc* has been encountered as an activated oncogene in a wide variety of different settings, its elevated expression as a correlate of gene amplification in a number of human neoplasms provides strong support for a role of DNA amplification in the activation of cellular oncogenes in the course of tumor development.

New Members of the *myc* Gene Family

DNA amplification, as evidenced cytologically by double minute chromosomes and homogeneous staining regions, is a particularly common feature of neuroblastomas. Consequently, this neoplasm appeared to serve as an especially attractive model for studies directed toward understanding the potential significance of amplified DNA to tumor development. In 1983, two research groups reported that the DNA sequences amplified in many neuroblastomas cross-hybridized with a c-*myc* probe. The amplified DNA, however, was distinct from the c-*myc* gene and was only readily detected with c-*myc* probe when enhanced hybridization resulted from the high number of amplified copies present in neuroblastoma DNAs. Because of its relationship to c-*myc* and its reproducible amplification in neuroblastomas, the sequence identified in these studies appeared to represent a new candidate oncogene, which was designated N-*myc*. As expected, the N-*myc* sequences were localized to double minutes and homogeneous staining regions and were transcribed at high levels in those neuroblastomas in which the gene was amplified.

Subsequent molecular cloning and characterization has verified that N-*myc* encodes a protein that is related to but distinct from the c-*myc* gene product (Fig. 8.2). In addition, gene transfer experiments have demonstrated the ability of N-*myc* to function in the induction of cell transformation. The N-*myc* gene is thus a new member of the *myc* oncogene family. Its amplification in neuroblastomas is not only a frequent event, but also highly correlated with the progression of these tumors to increasingly malignant stages associated with a poor clinical prognosis. As discussed further in chapter 11, these studies have provided some of the strongest available evidence for a relation between cellular oncogene activation and the clinical behavior of a human malignancy.

Amplification and expression of N-*myc* has also been detected in other neoplasms, including retinoblastomas and small-cell lung carcinomas (which also appear to be neoplasms of neuroendocrine origin) (Table 8.1). As noted earlier, amplification of c-*myc* is also a frequent event in lung cancers, particularly small-cell carcinomas. In addition, studies of these neoplasms revealed amplification of a third member of the *myc* gene family that, like N-*myc*, was initially detected as an amplified DNA sequence that cross-hybridized with c-*myc* probe. Because of its identification in small-cell lung carcinomas, this gene has been designated L-*myc*. Further characterization has shown that the protein

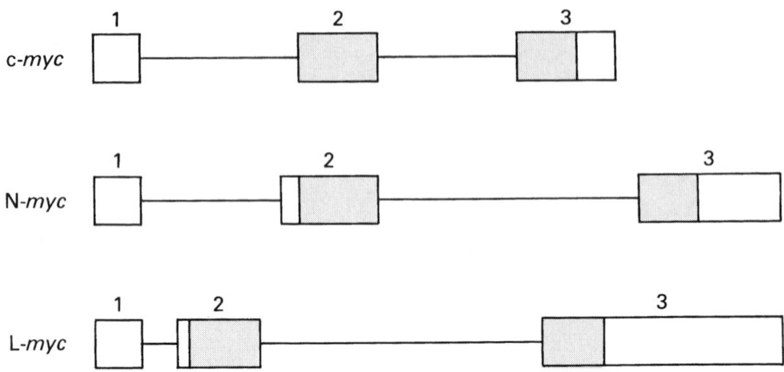

FIGURE 8.2
The *myc* gene family. The c-*myc*, N-*myc*, and L-*myc* genes each consist of three exons (depicted as boxes). In each case, protein coding sequence (filled boxes) is present only in exons 2 and 3. The three genes encode proteins containing from 360 to 460 amino acids with ~30% amino acid sequence identities.

encoded by L-*myc* is related to but distinct from both the c-*myc* and the N-*myc* gene products (Fig. 8.2). Any individual small-cell carcinoma appears to amplify and express only one of these three members of the *myc* gene family. Overall, however, amplification of either c-*myc*, N-*myc*, or L-*myc* is a frequent event in small-cell lung carcinomas, implicating enhanced expression of one of these three members of the *myc* gene family as an important event in pathogenesis of these neoplasms.

Amplification of *erb*B and *erb*B-2

Two other cellular oncogenes, *erb*B and *erb*B-2, also are frequently amplified in human neoplasms. The *erb*B oncogene was initially identified in avian erythroblastosis virus and is also activated as a cellular oncogene by insertional mutagenesis in chicken erythroleukemias (see chapter 6). As discussed in detail in chapter 13, the *erb*B oncogene is derived from the normal gene encoding the cell surface receptor for epidermal growth factor. Because expression of high levels of epidermal growth factor receptors had been reported in a number of human neoplasms, the possibility of *erb*B amplification in these tumors was investigated. These studies documented amplification and elevated transcription of *erb*B in a number of tumors, particularly in a high fraction of glioblastomas and squamous cell carcinomas (Table 8.1).

Amplification of *erb*B in these human neoplasms generally results in overexpression of the normal proto-oncogene product, which differs substantially from the protein encoded by the viral *erb*B oncogene. The normal amino and carboxy termini of the proto-oncogene product have both been deleted from the viral oncogene encoded protein. A similar deletion of the *erb*B proto-oncogene amino terminus, which encodes the extracellular ligand binding domain of the receptor, contributes to its activation by proviral insertional mutagenesis. The biological significance of amplification and enhanced expression of the normal *erb*B proto-oncogene was therefore not immediately evident by comparison with the transforming activity of its viral oncogene homolog. This contrasts with amplification of *myc* genes, because it had been directly demonstrated by gene transfer experiments that constitutive expression of a normal *myc* protein is sufficient to elicit cell transformation (see chapter 7). The consequence of overexpression of the normal *erb*B proto-oncogene was therefore investigated by gene transfer experiments, using a vector that led to expression of high levels of the normal *erb*B gene product (the epidermal growth factor receptor) in transfected cells. Analysis of these transfected cells indicated that their normal proliferative response to epidermal growth factor was enhanced and that, in the presence of epidermal growth factor, they displayed a transformed phenotype. These results thus provided direct support for the significance of amplification and overexpression of the normal *erb*B proto-oncogene in neoplastic transformation.

The human gene designated *erb*B-2 was initially isolated in 1985 by cross-hybridization with *erb*B probe. As was the case in studies that led to the isolation of the N-*myc* and L-*myc* genes, hybridization of *erb*B probe to DNA of a breast carcinoma cell line revealed amplification of a cross-hybridizing gene that was distinct from the cellular *erb*B oncogene. Amplification of the same *erb*B-related sequence in a human salivary gland carcinoma was independently reported at the same time by a second group of researchers. Molecular cloning and sequence analysis confirmed that the new gene, designated *erb*B-2, was related to but distinct from *erb*B, apparently encoding a cell surface receptor for a distinct ligand (Fig. 8.3). Interestingly, this analysis also revealed that *erb*B- 2 was the human homolog of the gene originally called *neu*, which had first been detected by gene transfer experiments as an activated oncogene in rat neuroblastomas (see chapter 5).

Activation of the transforming potential of *erb*B-2 (*neu*) in rat neuroblastomas is a consequence of a point mutation that yields a structurally altered gene product. As with *erb*B, it was therefore important to investigate the biological significance of amplification and overexpression of the normal *erb*B-2 protein. Gene transfer experiments with appropriate molecular constructs have, however, demonstrated that high-level expression of normal *erb*B-2 is indeed sufficient to induce cell transformation.

Amplification of *erb*B-2 is particularly common in human breast and ovarian cancers, having been detected in ~25% of these neoplasms. As further dis-

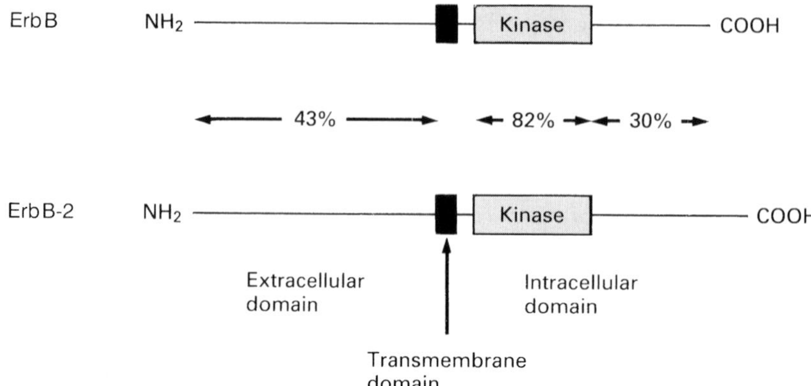

FIGURE 8.3
Relationship between the *erb*B and *erb*B-2 gene products. The *erb*B and *erb*B-2 proto-oncogene proteins consist of 1186 and 1260 amino acids, respectively. They are divided into extracellular, transmembrane, and intracellular (including tyrosine kinase) domains. The percentage of amino acid identities in different regions is indicated.

cussed in chapter 11, *erb*B-2 amplification is closely correlated with the clinical behavior of these neoplasms, such that tumors with *erb*B-2 amplification are more aggressive and associated with reduced patient survival.

Other Amplified Oncogenes

In addition to amplification of c-*myc* and *erb*B-2, amplification of three other oncogenes has been detected in about 20% of human breast carcinomas, as well as in several types of squamous cell carcinomas. These three genes (*hst*, *int*-2, and *PRAD*-1), which had previously been identified as oncogenes by other approaches, are all closely linked on chromosome 11 and are frequently amplified as a unit. The amplification of *int*-2 in human breast cancers is particularly interesting because *int*-2 was first identified as a cellular oncogene activated by promoter insertion in mouse mammary carcinomas (see chapter 6). However, neither *int*-2 nor *hst* is generally expressed in those tumors in which it is amplified, so it appears unlikely that either is the functional target of the chromosome 11 amplification. *PRAD*-1, on the other hand, is overexpressed in those breast and squamous cell carcinomas in which it has been amplified, suggesting that it is the key oncogene within this amplified unit of DNA.

A number of other known cellular oncogenes also have been found to be amplified, but amplification of these genes occurs only in occasional tumors or tumor cell lines, suggesting that they do not play a reproducible role in tumorigenesis. These genes include all three members of the *ras* gene family (*ras*H, *ras*K, and *ras*N), *abl*, *ets*-1, and *myb*. In some cases, the amplified oncogenes have also sustained additional lesions that may activate their transforming potential. For example, the chronic myelogenous leukemia cell line K562 contains an amplified *abl* gene that is also activated by the characteristic gene rearrangement associated with the Philadelphia chromosome translocation (see chapter 7). Similarly, amplification of rearranged c-*myc* oncogenes has been reported, as has amplification of *ras* oncogenes that have been activated by point mutations. In these cases, amplification may be a secondary event resulting in increased expression of an oncogene that has already been activated by DNA rearrangement or mutation.

Finally, three novel oncogenes (*gli*, K-*sam*, and *mdm*-2) have been identified by cloning and characterizing sequences that are amplified and overexpressed in tumors. As they have for oncogenes identified as the sites of chromosomal translocation, further studies of these amplified genes have substantiated their designation as oncogenes. In particular, *gli* and *mdm*-2 have been shown directly to induce cell transformation, and K-*sam* has been identified as a member of a family of genes encoding growth factor receptors. The *gli* gene is frequently amplified in glioblastomas, K-*sam* in stomach carcinomas, and *mdm*-2 in a variety of sarcomas, suggesting reproducible roles for these genes in tumor development.

Summary

Gene amplification is a frequent event that results in increased oncogene expression during the pathogenesis of a variety of human tumors. Amplification of more than a dozen oncogenes has been described in human neoplasms, including some oncogenes previously identified in retroviruses, by gene transfer assays, or as targets for retroviral insertional mutagenesis. Other oncogenes were identified for the first time as amplified sequences in human tumors. In some cases, for example N-*myc* in neuroblastomas and *erb*B-2 in breast carcinomas, oncogene amplification is highly reproducible and closely correlated with the progression of tumors to increasing malignancy. In other instances, amplification is sporadic and may be a secondary event that increases expression of oncogenes activated by other alterations.

References

General Reference

Alitalo, K., and Schwab, M. 1986. Oncogene amplification in tumor cells. *Adv. Cancer Res.* 47:235–281.

Gene Amplification and Chromosome Abnormalities

Alt, F.W., Kellems, R.E., Bertino, J.R., and Schimke, R.T. 1978. Selective multiplication of dihydrofolate reductase genes in methotrexate-resistant variants of cultured mouse cells. *J. Biol. Chem.* 253:1357–1370.

Barker, P.E. 1982. Double minutes in human tumor cells. *Cancer Genet. Cytogenet.* 5:81–94.

Biedler, J.L., and Spengler, B.A. 1976. Metaphase chromosome anomaly: association with drug resistance and cell specific products. *Science* 191:185–187.

Cowell, J.K. 1982. Double minutes and homogeneously staining regions: gene amplification in mammalian cells. *Ann. Rev. Genet.* 16:21–59.

Stark, G.R., and Wahl, G.M. 1984. Gene amplification. *Ann. Rev. Biochem.* 53:447–491.

Tlsty, T.D. 1990. Normal diploid human and rodent cells lack a detectable frequency of gene amplification. *Proc. Natl. Acad. Sci. USA* 87:3132–3136.

Wright, J.A., Smith, H.S., Watt, F.M., Hancock, M.C., Hudson, D.L., and Stark, G.R. 1990. DNA amplification is rare in normal human cells. *Proc. Natl. Acad. Sci. USA* 87:1791–1795.

Amplification of c-*myc*

Alitalo, K., Schwab, M., Lin, C.C., Varmus, H.E., and Bishop, J.M. 1983. Homogeneously staining chromosomal regions contain amplified copies of an abundantly expressed cellular oncogene (c-*myc*) in malignant neuroendocrine cells from a human colon carcinoma. *Proc. Natl. Acad. Sci. USA* 80:1707–1711.

Collins, S., and Groudine, M. 1982. Amplification of endogenous *myc*-related DNA sequences in a human myeloid leukaemia cell line. *Nature* 298:679–681.

Dalla-Favera, R., Wong-Staal, F., and Gallo, R.C. 1982. *onc* gene amplification in promyelocytic leukaemia cell line HL-60 and primary leukaemic cells of the same patient. *Nature* 299:61-63.

Escot, C., Theillet, C., Lidereau, R., Spyratos, F., Champeme, M.-H., Gest, J., and Callahan, R. 1986. Genetic alteration of the c-*myc* protooncogene (*MYC*) in human primary breast carcinomas. *Proc. Natl. Acad. Sci. USA* 83:4834–4838.

Little, C.D., Nau, M.M., Carney, D.N., Gazdar, A.F., and Minna, J.D. 1983. Amplification and expression of the c-*myc* oncogene in human lung cancer cell lines. *Nature* 306:194–196.

Nowell, P., Finan, J., Dalla-Favera, R., Gallo, R.C., ar-Rushdi, A., Romanczuk, H., Selden, J.R., Emanuel, B.S., Rovera, G., and Croce, C.M. 1983. Association of amplified oncogene c-*myc* with an abnormally banded chromosome 8 in a human leukaemia cell line. *Nature* 306:494–497.

New Members of the *myc* Gene Family

Brodeur, G.M., Seeger, R.C., Schwab, M., Varmus, H.E., and Bishop, J.M. 1984. Amplification of N-*myc* in untreated human neuroblastomas correlates with advanced disease stage. *Science* 224:1121–1124.

Kohl, N.E., Kanda, N., Schreck, R.R., Bruns, G., Latt, S.A., Gilbert, F., and Alt, F.W. 1983. Transposition and amplification of oncogene-related sequences in human neuroblastomas. *Cell* 36:359–367.

Marcu, K.B., Bossone, S.A., and Patel, A.J. 1992. *myc* function and regulation. *Ann. Rev. Biochem.* 61:809–860.

Nau, M.M., Brooks, B.J., Battey, J., Sausville, E., Gazdar, A.F., Kirsch, I.R., McBride, O.W., Bertness, V., Hollis, G.F., and Minna, J.D. 1985. L-*myc*, a new *myc*-related gene amplified and expressed in human small cell lung cancer. *Nature* 318:69–73.

Nau, M.M., Brooks, B.J., Jr., Carney, D.N., Gazdar, A.F., Battey, J.F., Sausville, E.A., and Minna, J.D. 1986. Human small cell lung cancers show amplification and expression of the N-*myc* gene. *Proc. Natl. Acad. Sci. USA* 83:1092–1096.

Schwab, M. 1988. The *myc*-box oncogenes. In *The Oncogene handbook*, ed. Reddy, E.P., Skalka, A.M., and Curran, T. Elsevier, Amsterdam. pp. 381–391.

Schwab, M., Alitalo, K., Klempnauer, K.H., Varmus, H.E., Bishop, J.M., Gilbert, F., Brodeur, G., Goldstein, M., and Trent, J. 1983. Amplified DNA with limited homology to *myc* cellular oncogene is shared by human neuroblastoma cell lines and a neuroblastoma tumour. *Nature* 305:245–248.

Schwab, M., Varmus, H.E., and Bishop, J.M. 1985. Human N-*myc* gene contributes to neoplastic transformation of mammalian cells in culture. *Nature* 316:160–162.

Wong, A.J., Ruppert, J.M., Eggleston, J., Hamilton, S.R., Baylin, S.B., and Vogelstein, B. 1986. Gene amplification of c-*myc* and N-*myc* in small cell carcinoma of the lung. *Science* 233:461–464.

Amplification of *erb*B and *erb*B-2

Coussens, L., Yang-Feng, T.L., Liao, Y.-C., Chen, E., Gray, A., McGrath, J., Seeburg, P.H., Libermann, T.A., Schlessinger, J., Francke, U., Levinson, A., and Ullrich, A. 1985. Tyrosine kinase receptor with extensive homology to EGF receptor shares chromosomal location with *neu* oncogene. *Science* 230:1132–1139.

Di Fiore, P.P., Pierce, J.H., Fleming, T.P., Hazan, R., Ullrich, A., King, C.R., Schlessinger, J., and Aaronson, S.A. 1987. Overexpression of the human EGF receptor confers an EGF-dependent transformed phenotype to NIH 3T3 cells. *Cell* 51:1063–1077.

Di Fiore, P.P., Pierce, J.H., Kraus, M.H., Segatto, O., King, C.R., and Aaronson, S.A. 1987. *erb*B-2 is a potent oncogene when overexpressed in NIH/3T3 cells. *Science* 237:178-182.

King, C.R., Kraus, M.H., and Aaronson, S.A. 1985. Amplification of a novel v-*erb*B-related gene in a human mammary carcinoma. *Science* 229:974-976.

Libermann, T.A., Nusbaum, H.R., Razon, N., Kris, R., Lax, I., Soreq, H., Whittle, N., Waterfield, M.D., Ullrich, A., and Schlessinger, J. 1985. Amplification, enhanced expression and possible rearrangement of EGF receptor gene in primary human brain tumours of glial origin. *Nature* 313:144–147.

Merlino, G.T., Xu, Y.H., Ishii, S., Clark, A.J., Semba, K., Toyoshima, K., Yamamoto, T., and Pastan, I. 1984. Amplification and enhanced expression of the epidermal growth factor receptor gene in A431 human carcinoma cells. *Science* 224:417–419.

Semba, K., Kamata, N., Toyoshima, K., and Yamamoto, T. 1985. A v-*erb*B-related proto-oncogene, c-*erb*B-2, is distinct from the c- *erb*B-1/epidermal growth factor-receptor gene and is amplified in a human salivary gland adenocarcinoma. *Proc. Natl. Acad. Sci. USA* 82:6497–6501.

Slamon, D.J., Godolphin, W., Jones, L.A., Holt, J.A., Wong, S.G., Keith, D.E., Levin, W.J., Stuart, S.G., Udove, J., Ullrich, A., and Press, M.F. 1989. Studies of the *HER-2/neu* proto-oncogene in human breast and ovarian cancer. *Science* 244:707–712.

Ullrich, A., Coussens, L., Hayflick, J.S., Dull, T.J., Gray, A., Tam, A.W., Lee, J., Yarden, Y., Libermann, T.A., Schlessinger, J., Downward, J., Mayes, E.L.V., Whittle, N., Waterfield, M.D., and Seeburg, P.H. 1984. Human epidermal growth factor receptor cDNA sequence and aberrant expression of the amplified gene in A431 epidermoid carcinoma cells. *Nature* 309:418–425.

Yamamoto, T., Kamata, N., Kawano, H., Shimizu, S., Kuroki, T., Toyoshima, K., Rikimaru, K., Nomura, N., Ishizaki, R., Pastan, I., Gamou, S., and Shimizu, N. 1986. High incidence of amplification of the epidermal growth factor receptor gene in human squamous carcinoma cell lines. *Cancer Res.* 46:414–416.

Other Amplified Oncogenes

Ali, I.U., Merlo, G., Callahan, R., and Lidereau, R. 1989. The amplification unit on chromosome 11q13 in aggressive primary human breast tumors entails the *bcl*-1, *int*-2 and *hst* loci. *Oncogene* 4:89–92.

Alitalo, K., Winquist, R., Lin, C.C., de la Chapelle, A., Schwab, M., and Bishop, J.M. 1984. Aberrant expression of an amplified c-*myb* oncogene in two cell lines from a colon carcinoma. *Proc. Natl. Acad. Sci. USA* 81:4534–4538.

Buckley, M.F., Sweeney, K.J., Hamilton, J.A., Sini, R.L., Manning, D.L., Nicholson, R.I., de Fazio, A., Watts, C.K., Musgrove, E.A., and Sutherland, R.L. 1993. Expression and amplification of cyclin genes in human breast cancer. *Oncogene* 8:2127–2133.

Fakharzadeh, S.S., Trusko, S.P., and George, D.L. 1991. Tumorigenic potential associated with enhanced expression of a gene that is amplified in a mouse tumor cell line. *EMBO J.* 10:1565–1569.

Hattori, Y., Odagiri, H., Nakatani, H., Miyagawa, K., Naito, K., Sakamoto, H., Katoh, O., Yoshida, T., Sugimura, T., and Terada, M. 1990. K-*sam*, an amplified gene in stomach cancer, is a member of the heparin-binding growth factor receptor genes. *Proc. Natl. Acad. Sci. USA* 87:5983–5987.

Jiang, W., Kahn, S.M., Tomita, N., Zhang, Y.-J., Lu, S.-H., and Weinstein, I.B. 1992. Amplification and expression of the human cyclin D gene in esophageal cancer. *Cancer Res.* 52:2980–2983.

Kinzler, K.W., Bigner, S.H., Bigner, D.D., Trent, J.M., Law, M.L., O'Brien, S.J., Wong, A.J., and Vogelstein, B. 1987. Identification of an amplified, highly expressed gene in a human glioma. *Science* 236:70–73.

Lammie, G.A., Fantl, V., Smith, R., Schuuring, E., Brookes, S., Michalides, R., Dickson, C., Arnold, A., and Peters, G. 1991. D11S287, a putative oncogene on chromosome 11q13, is amplified and expressed in squamous cell and mammary carcinomas and linked to *BCL*-1. *Oncogene* 6:439–444.

Oliner, J.D., Kinzler, K.W., Meltzer, P.S., George, D.L., and Vogelstein, B. 1992. Amplification of a gene encoding a p53-associated protein in human sarcomas. *Nature* 358:80–83.

Pulciani, S., Santos, E., Long, L.K., Sorrentino, V., and Barbacid, M. 1985. *ras* gene amplification and malignant transformation. *Mol. Cell. Biol.* 5:2836–2841.

Rovigatti, U., Watson, D.K., and Yunis, J.J. 1986. Amplification and rearrangement of Hu-*ets*-1 in leukemia and lymphoma with involvement of 11q23. *Science* 232:398–400.

Ruppert, J.M., Vogelstein, B., and Kinzler, K.W. 1991. The zinc finger protein GLI transforms primary cells in cooperation with adenovirus E1A. *Mol. Cell. Biol.* 11:1724–1728.

Schwab, M., Alitalo, K., Varmus, H.E., Bishop, J.M., and George, D. 1983. A cellular oncogene (c-Ki-*ras*) is amplified, overexpressed, and located within karyotypic abnormalities in mouse adrenocortical tumour cells. *Nature* 303:497–501.

Selden, J.R., Emanuel, B.S., Wang, E., Cannizzaro, L., Palumbo, A., Erikson, J., Nowell, P.C., Rovera, G., and Croce, C.M. 1983. Amplified C-λ and c-*abl* genes are on the same marker chromosome in K562 leukemia cells. *Proc. Natl. Acad. Sci. USA* 80:7289–7292.

❖ CHAPTER 9

Retinoblastoma and the Discovery of Tumor Suppressor Genes

Cellular oncogenes were ultimately defined by their ability to induce neoplastic transformation upon introduction into an appropriate target cell. These genes act in a dominant positive regulatory fashion in that the introduction of a functional oncogene into a nontransformed recipient cell results in development of a new phenotype—transformation—in the presence of two copies of the normal allele. Moreover, oncogene activation is a consequence of mutations or gene rearrangements that increase gene expression or increase the activity of the oncogene encoded proteins.

In contrast with the dominant oncogenes, a distinct class of genes that function as negative regulators of neoplastic disease has been revealed by studies of somatic cell hybrids and of some inherited human cancers. For this class of genes, loss of function leads to tumor development. Although molecular analysis of these genes is less advanced than studies of the dominant oncogenes, it is clear that loss of function of a number of cellular genes plays a critical role in the pathogenesis of many human neoplasms. Such genes have been referred to as cancer susceptibility genes, antioncogenes, recessive cancer genes, recessive oncogenes, and (as used here) *tumor suppressor genes.*

Somatic Cell Hybridization

The first approach to genetic analysis of neoplastic cells, initiated in 1969 by Henry Harris and his colleagues, was to fuse normal cells with tumor cells and analyze the properties of the resulting hybrids. Extensive studies of this type have now established that most hybrids between normal and

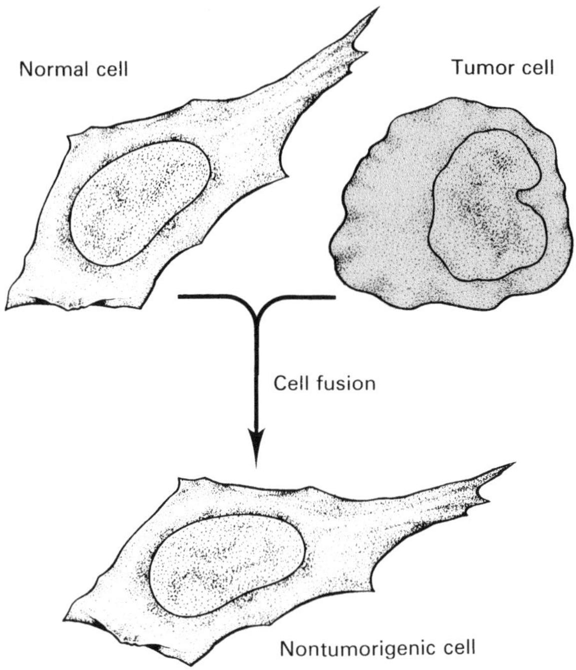

FIGURE 9.1
Somatic cell hybridization between neoplastic and normal cells. Fusion of malignant and normal cells usually yields hybrids that are nontumorigenic.

malignant cells are no longer tumorigenic (Fig. 9.1). Such suppression of tumorigenicity by fusion of a tumor cell with a normal cell implies that normal cells contain one or more genes that act as negative regulators of the neoplastic phenotype: the tumor suppressor genes. Interestingly, however, many nontumorigenic hybrids between normal and malignant cells still display some phenotypic properties of transformed cells *in vitro*, including loss of density-dependent inhibition of cell growth and loss of anchorage dependence. Thus, it appears that suppression of the tumorigenic phenotype in hybrids constitutes only partial, rather than complete, reversion of cell transformation. As discussed in more detail in chapter 11, this is consistent with the notion that both activation of dominant oncogenes and loss of tumor suppressor gene function independently contribute to development of the fully malignant phenotype of many tumors.

The interpretation that the nontumorigenic phenotype of hybrids between normal and malignant cells results from the action of specific normal cell genes is strengthened by the fact that such hybrids frequently revert to the tumorigenic phenotype following loss of specific chromosomes of the normal parent. For example, hybrids between normal human fibroblasts and HeLa cells (a human cervical carcinoma cell line) are nontumorigenic but revert to the tumorigenic phenotype following loss of normal chromosome 11. These findings suggest that tumorigenicity of the malignant parent (HeLa cells) resulted, at least in part, from loss of function of a tumor suppressor gene located on chromosome 11. Introduction of the wild-type allele by fusion with a normal cell would then restore function of the gene and suppress tumorigenicity. If the relevant normal chromosome 11 was subsequently lost, the hybrid would revert to the tumorigenic phenotype. Similar studies of hybrids between normal cells and other tumor cell lines have identified additional human chromosomes that carry tumor suppressor genes that may have been inactivated in other human neoplasms.

Somatic cell hybridization experiments have thus provided a clear indication that tumorigenicity is associated with loss of function of critical regulatory genes in malignant cells. It remained, however, for such tumor suppressor genes to be defined at the molecular level by a different approach—namely, the analysis of inherited human cancers.

Inherited and Sporadic Retinoblastoma

The disease that has provided the prototype for understanding the role of tumor suppressor genes in human tumors is retinoblastoma, an eye tumor that usually develops in children by age three. Provided that the disease is detected early, retinoblastoma can be successfully treated and many patients survive to reproductive age. This has allowed two forms of the disease to be distinguished: inherited and sporadic. Individuals with the inherited form of the disease transmit the retinoblastoma susceptibility gene to half of their offspring, consistent with Mendelian transmission of a single autosomal dominant locus (Fig. 9.2). Importantly, patients with inherited retinoblastoma frequently develop multiple tumors in both eyes, whereas patients who develop retinoblastoma without a family history of the disease (noninherited or sporadic retinoblastoma) develop only a single tumor in one eye. In addition, children with the sporadic form of the disease are generally older when they develop tumors than children with inherited disease.

Although susceptibility to retinoblastoma is transmitted in a genetically dominant manner, inheritance of the susceptibility gene does not by itself suffice to transform a normal retinal cell into a tumor cell. Thus, patients with inherited retinoblastoma develop only a few focal tumor growths over a background of several million normal retinal cells. Even though all retinal cells have inherited the susceptibility gene, most are unaffected and function normally: only a small fraction actually become neoplastic. It therefore appears that, at the cellular level, inheritance of the retinoblastoma susceptibility gene

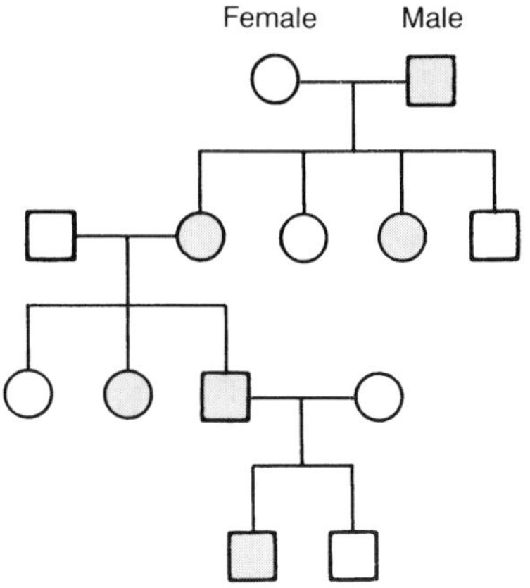

FIGURE 9.2
A family pedigree illustrating the inheritance of retinoblastoma. Affected individuals are depicted by filled symbols. In inherited retinoblastoma, the retinoblastoma susceptibility gene is transmitted to ~50% of offspring. In contrast, sporadic retinoblastoma occurs without a family history of the disease.

is not sufficient for neoplastic transformation and that additional alterations must occur before the tumor phenotype is expressed.

Based on statistical analysis of the comparative frequency and age of development of inherited and sporadic forms of retinoblastoma, Alfred Knudson in 1971 proposed that retinoblastoma is caused by two mutations (Fig. 9.3). In inherited disease, the first mutation is present in the germ line and the second mutation occurs somatically, resulting in tumor development. The probability of the second mutation is sufficiently high that tumors will nearly always develop among the more than 10^6 cells in each retina. In sporadic disease, both required mutations are somatic and only a very rare retinal cell in which two independent mutations have occurred will become neoplastic.

This two-mutation model accounted for the dominant inheritance of susceptibility to retinoblastoma in spite of the fact that the susceptibility gene did not function as a single dominant determinant of transformation at the cellular level. However, the model did not address the nature of the gene or genes involved. One possibility that would be compatible with Knudson's two-step model is that the first mutation activated a dominant oncogene and that a second mutation, perhaps activating a second oncogene, was required for development of neoplasms. Multiple events in tumor progression were alluded to earlier and several model systems (for example, transgenic mice carrying activated oncogenes) in which oncogene activation appears to be one step in a multistep pathway of tumor development will be discussed in chapter 11. However, this is not the case for retinoblastoma. Rather, the two mutations required for retinoblastoma development have proved to be the loss of both functional copies of the retinoblastoma susceptibility gene that would be present on complementary chromosomes in a normal diploid cell. These studies have thus

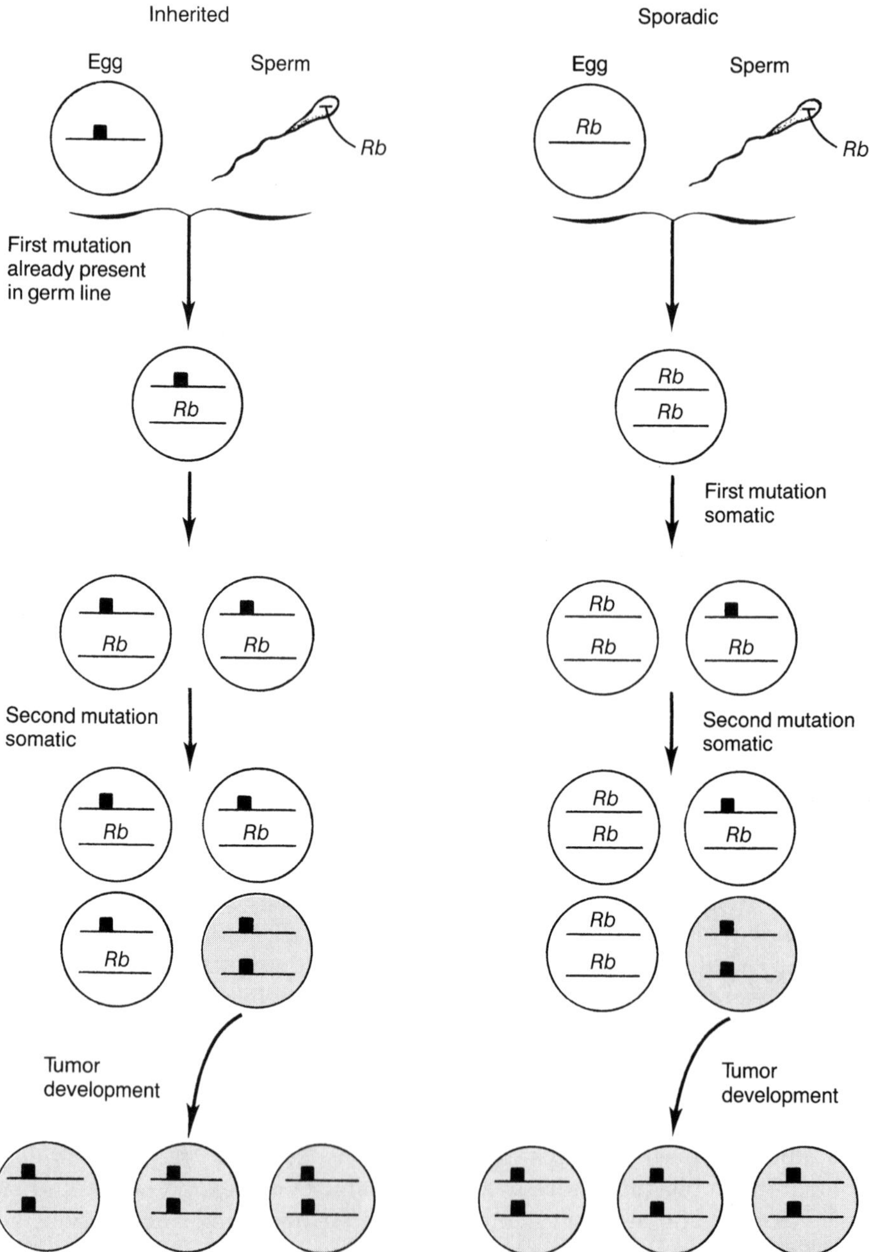

Figure 9.3
The two-mutation model of retinoblastoma development. In hereditary retinoblastoma, the first mutation is transmitted by the germ line of an affected parent. The second mutation occurs somatically in a retinal cell, leading to development of the tumor. In sporadic retinoblastoma, development of a tumor requires two somatic mutations.

revealed that loss rather than activation of retinoblastoma gene function promotes neoplasia; hence, the designation of the retinoblastoma susceptibility gene as a tumor suppressor gene.

Mapping the Retinoblastoma Gene

The first insight into the character of the retinoblastoma gene came from observations of chromosome morphology. These studies indicated that visible deletions in one copy of chromosome 13 could be detected in normal blood cells of about 5% of patients with the inherited form of the disease (Fig. 9.4). The extent of these deletions was variable in different patients. However, one particular chromosomal band, 13q14, was always lost. These observations suggested that deletion of a gene or genes localized to chromosome 13q14 was the first event required for development of at least some inherited retinoblastomas.

Karyotypic analysis also identified a locus on chromosome 13q14 as being involved in the pathogenesis of sporadic retinoblastoma. Chromosome 13 deletions involving band 13q14 were found in the tumors of about 25% of patients with the noninherited form of the disease. In these cases, normal blood cells of the same patients contained two normal chromosome 13 homologs, indicating that deletion of band 13q14 had occurred as a somatic mutation in the course of tumor development. Thus, deletions of the same chromosomal locus appeared to be involved in at least some cases of both inherited and sporadic retinoblastoma.

Although deletions of chromosome 13q14 seemed to consistently occur in some cases of retinoblastoma, detectable karyotypic abnormalities were not found in the majority of cases. However, a deletion large enough to be recognized at the level of chromosome morphology would have to extend over hundreds of kilobases of DNA. Clearly, loss of gene function could occur as a consequence of smaller deletions (or point mutations) that would not be detectable by karyotypic

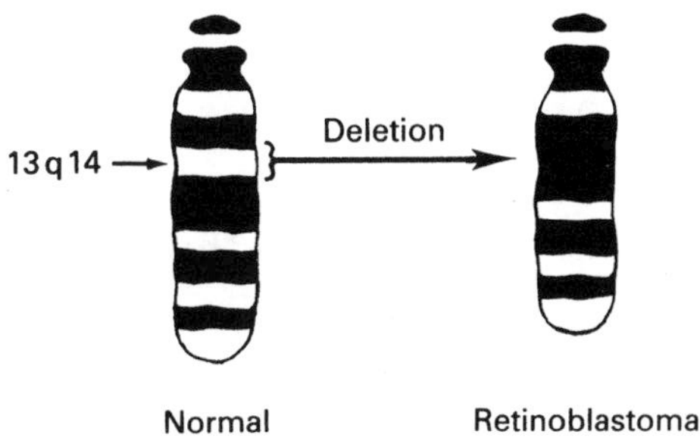

FIGURE 9.4
Chromosome deletions in retinoblastoma. Deletions of chromosome 13 band q14 are detected in both inherited and sporadic retinoblastoma.

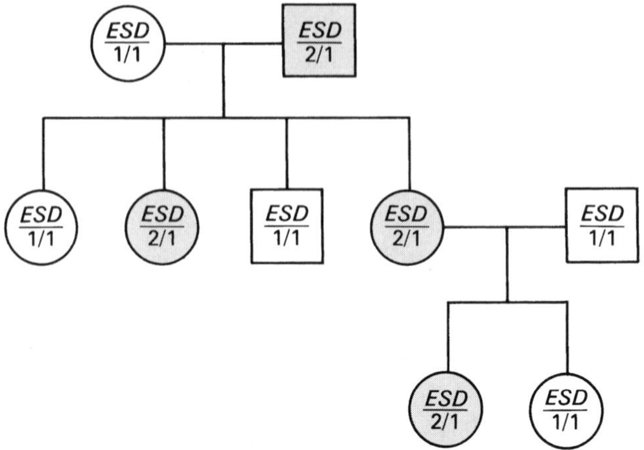

FIGURE 9.5
Genetic linkage of retinoblastoma susceptibility and esterase D. An illustrative family pedigree is shown. Individuals affected with retinoblastoma are indicated by filled symbols, and genes encoding esterase D isozymes 1 and 2 are indicated as *ESD*-1 and *ESD*-2. In this example, the retinoblastoma susceptibility gene cosegregrates with *ESD*-2.

analysis. In order to investigate the potential contribution of such mutations to the majority of retinoblastomas, which did not evidence gross chromosomal deletions, a fine structure genetic approach to defining the retinoblastoma locus was needed. This was accomplished by analyzing the linkage of the retinoblastoma susceptibility gene to another genetic marker, the gene encoding esterase D, which had independently been mapped to the same chromosomal locus. Two different alleles of this gene could be distinguished because they encoded electrophoretically separable forms (isozymes) of esterase D. Thus, heterozygosity at the esterase D locus provided a linked marker gene that could be followed in families with inherited retinoblastoma. Such studies demonstrated close linkage between esterase D and retinoblastoma development in families without recognizable chromosome deletions (Fig. 9.5). The children in these families who developed retinoblastoma always inherited the gene for one esterase D isozyme: children who did not develop retinoblastoma inherited the other. These findings indicated that susceptibility to retinoblastoma was inherited through a gene on chromosome 13, even in cases without detectable chromosome deletions.

The Retinoblastoma Gene is Recessive at the Cellular Level

The mapping of the retinoblastoma susceptibility gene to a chromosomal locus, which was deleted in some cases of both inherited and sporadic disease, suggested the possibility that the critical event in development of this neoplasm was loss of a functional gene. If this were the case, the two mutations required for retinoblastoma development might represent inactivation of both copies of the wild-type gene, present on two chromosomes of a normal cell. The first mutation—whether germ line or somatic—would result in a cell with one mutant and one normal allele. The second mutation would inactivate the remaining normal copy of the gene, leading to complete loss of gene function. Inactivation of the second wild-type allele in a cell already containing one mutant allele could occur either by an independent somatic mutation or by loss of the chromosome region containing the remaining normal locus (Fig. 9.6). The latter class of events, which could occur either by mitotic recombination or by chromosome loss with or without duplication of the remaining chromosome, would result in tumors that would be homozygous for the retinoblastoma gene and for other linked markers on chromosome 13. Two dif-

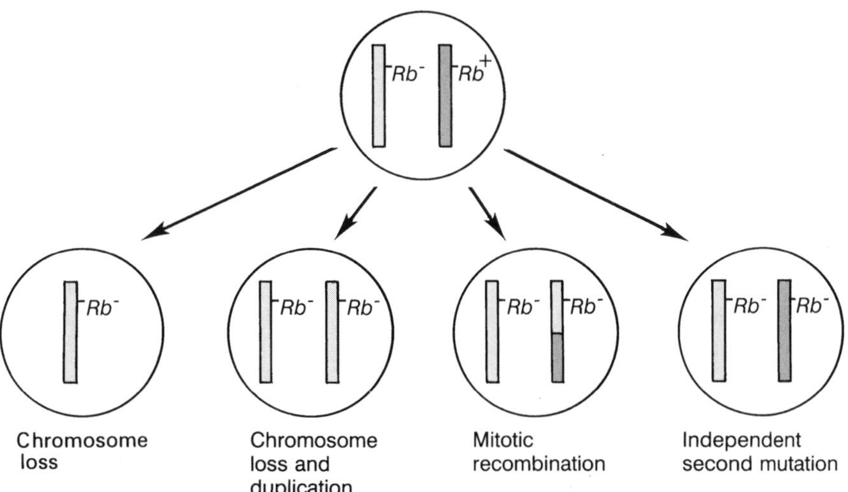

FIGURE 9.6
Mechanisms for inactivating the second copy of the retinoblastoma susceptibility gene. The first mutation leading to retinoblastoma development inactivates one allele (designated Rb^-) by a recessive mutation. Phenotypic expression of this recessive mutation requires inactivation of the remaining wild-type allele (designated Rb^+). This can occur by chromosome loss with or without duplication of the remaining chromosome, by mitotic recombination, or by an independent second mutation. The first two mechanisms lead to loss of heterozygosity for all markers on chromosome 13, and the third mechanism (mitotic recombination) leads to loss of heterozygosity for those markers linked to the retinoblastoma locus.

ferent experimental approaches verified this predicted homozygosity in tumors and thus provided direct evidence that neoplasia results from loss of retinoblastoma gene function.

The first approach again used esterase D as a linked genetic marker by comparing esterase D genes in normal somatic cells with retinoblastomas of the same patients (Fig. 9.7). Such studies identified a number of cases in which normal cells were heterozygous for esterase D, whereas only one of the two inherited alleles was present in the neoplasms. Thus, retinoblastomas were frequently reduced to homozygosity at the esterase D locus and, by implication, for the linked retinoblastoma gene. Moreover, in hereditary cases, the esterase D allele that was lost in the tumors was inherited from the normal unaffected parent. Thus, the mutant retinoblastoma gene was retained in the neoplasms, whereas the wild-type allele was lost. The development of homozygosity for a mutant retinoblastoma gene thus appeared to represent a second mutation that unmasked the mutant gene and led to tumor development. The need for homozygosity to detect a phenotype—in this case, cell transformation—validated the hypothesis that loss of retinoblastoma gene function resulted in tumorigenesis.

The esterase D studies were confirmed and extended by using restriction fragment-length polymorphisms (RFLPs) as additional markers to analyze chromosomal DNA sequences that were linked to the retinoblastoma gene. The use of RFLPs employs inherited differences in restriction endonuclease cleavage sites as genetic markers. Such markers are identified by testing cloned DNA probes in blot hybridization analysis against restriction endonuclease digested DNAs of several different individuals. A probe that hybridizes to fragments of varying sizes in DNAs of different individuals identifies an RFLP, which provides a polymorphic genetic marker defined by both a probe and a particular restriction endonuclease. Such markers are easily identified because they are not dependent on finding functional differences between expressed proteins—for example, isozymes of esterase D. Consequently, a large number of RFLPs have now been mapped to specific regions of human chromosomes and are used as markers in genetic linkage analysis.

The application of RFLPs to analysis of retinoblastomas was first reported by Webster Cavanee, Ray White, and their collaborators in 1983 (Fig. 9.8). These researchers used RFLPs derived from chromosome 13 to compare the DNAs of normal blood cells and retinoblastomas from a series of patients. In several cases, they identified individuals who were heterozygous for an RFLP in their blood cells but homozygous for the same RFLP in their tumors. It was also found in some cases of inherited retinoblastoma that the copy of chromosome 13 from the affected parent (and thus bearing the retinoblastoma mutation) was always the copy retained in the tumors.

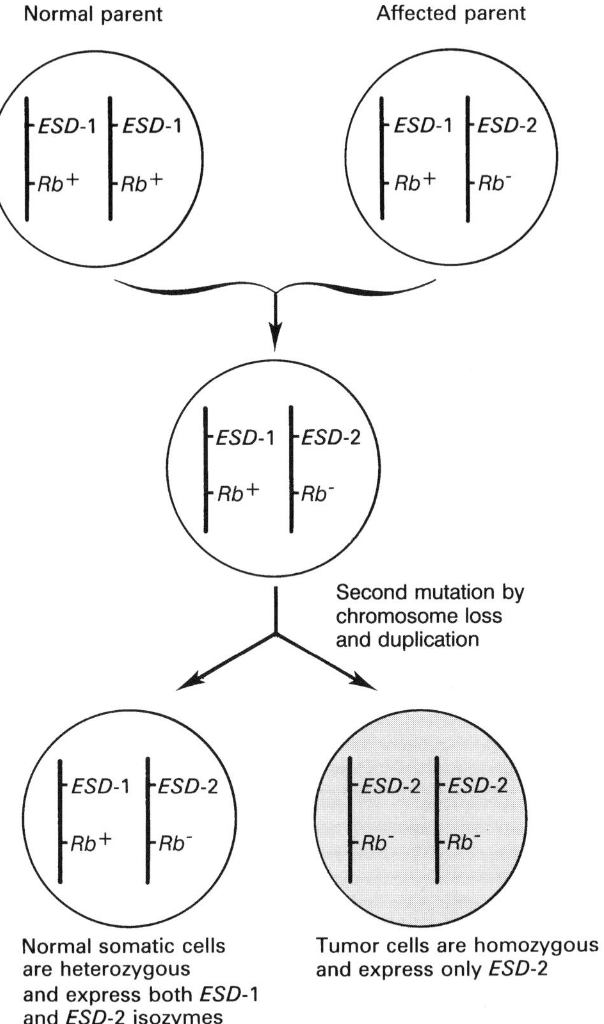

FIGURE 9.7
Loss of heterozygosity at the esterase D locus in retinoblastomas. In the case illustrated, an *Rb* mutation linked to *ESD*-2 is inherited from an affected parent. The normal parent is homozygous for *ESD*-1; so the genotype of the affected offspring is *ESD*-1/*ESD*-2. Normal somatic cells are therefore heterozygous at the *ESD* locus and express both esterase D-1 and esterase D-2 isozymes. In this example, the second mutation leading to retinoblastoma development occurs by chromosome loss and duplication of the remaining chromosome 13. The tumor cells are therefore homozygous for *ESD*-2 and express only esterase D-2.

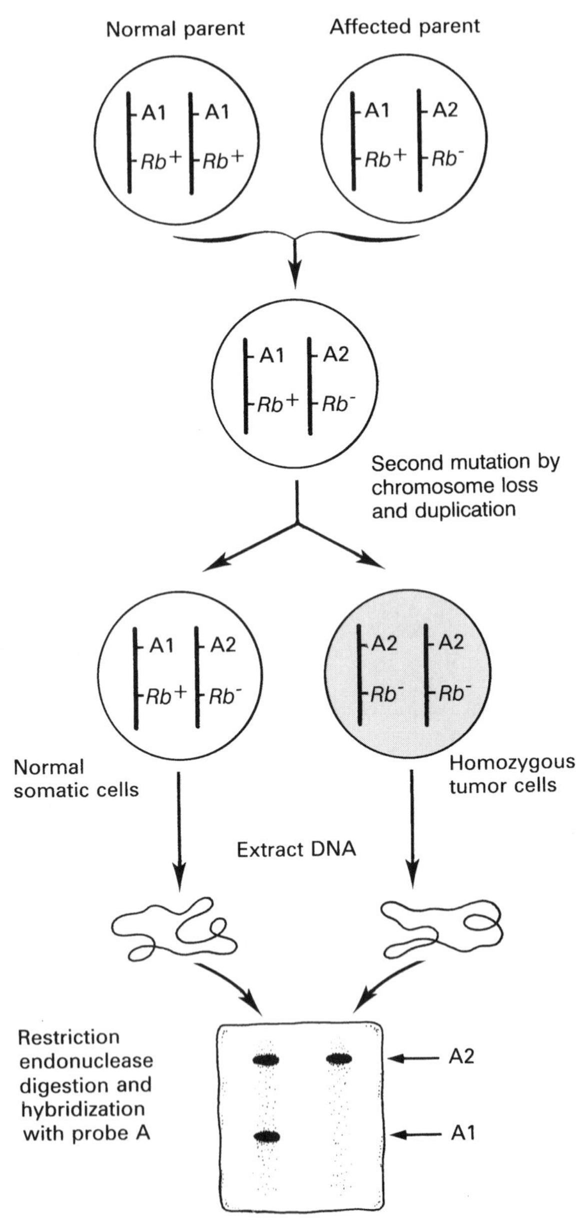

Figure 9.8
Loss of heterozygosity for chromosome 13 RFLPs in retinoblastomas. A restriction fragment-length polymorphism (RFLP) distinguishes alleles 1 and 2 at locus A. An *Rb* mutation linked to marker A2 is inherited, and the affected individual is heterozygous at the A locus. Normal somatic cells therefore contain both A1 and A2 markers, which can be identified by blot hybridization analysis of cell DNA. In contrast, retinoblastoma cells in this example have become homozygous for the A2 marker as a consequence of chromosome loss and duplication.

A further step in this analysis came from studies of Thaddeus Dryja and his colleagues in 1986. By screening retinoblastomas with additional probes derived from the 13q14 region, they identified tumors with homozygous deletions of at least 25 kb of DNA. Neoplasm development was therefore unambiguously associated with deletion and complete loss of retinoblastoma gene function.

Taken together, these genetic studies clearly established that retinoblastoma development was due to the inactivation or loss of both copies of the retinoblastoma susceptibility gene. These experiments thus define a new class of genes whose loss of function (rather than whose activation) can lead to neoplastic transformation.

Molecular Cloning of the *Rb* Gene

Although the foregoing genetic experiments elegantly defined a gene for retinoblastoma susceptibility, the application of genetics alone could not elucidate the nature or function of the gene product. However, the chromosomal localization of the retinoblastoma gene (*Rb*), and the finding of homozygous deletions in some tumors, provided a plausible approach to the isolation of the gene as a molecular clone, a step that would begin to extend the genetic studies to the molecular and biochemical levels of understanding.

Cloning the *Rb* gene was undertaken by three independent groups—a collaboration between the laboratories of Thaddeus Dryja and Robert Weinberg; the laboratory of Wen-Hwa Lee; and the laboratory of Yuen-Kai Fung and William Benedict—who reported isolation of the gene in late 1986 and early 1987. The general strategy used by all of the research groups was the same: existing probes to the 13q14 region were used to isolate a series of genomic clones of the surrounding DNA. These clones were then used as probes for hybridization to RNAs of normal retinal cells and retinoblastomas. Each group identified candidate probes that hybridized to a normal retinal transcript of ~4.7 kb that was not detected in RNAs of at least some retinoblastomas. The expression of this 4.7 kb RNA in normal retinal cells (and a variety of other cell types) but its absence in some retinoblastomas indicated that it was a plausible candidate for the transcript of the retinoblastoma gene. Consequently, cDNA clones of the candidate transcript were isolated and used for further characterization of gene structure and expression. These studies detected frequent deletions of the candidate gene in retinoblastomas. Importantly, in some cases, the boundaries of the deletions were contained entirely within the candidate gene—these internal deletions strongly suggested that the critical target was the cloned gene itself rather than a flanking gene. In addition, many retinoblastomas that did not display deletions of the candidate *Rb* gene were nevertheless abnormal in its transcription, either lacking detectable levels of RNA or expressing abnormally sized transcripts. Inactivation of the gene at the level of transcription could

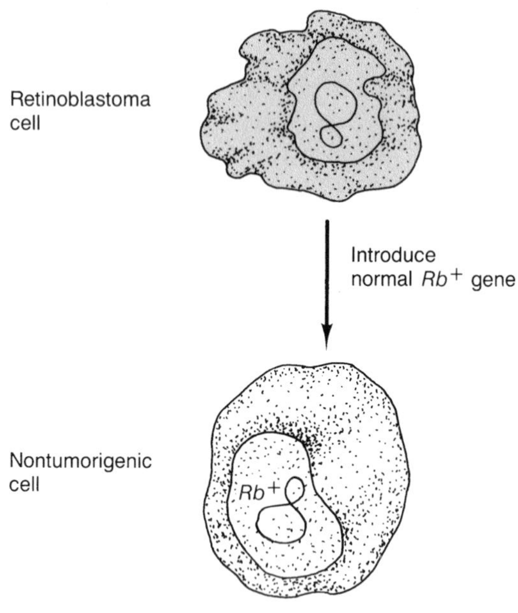

FIGURE 9.9
Suppression of tumorigenicity by *Rb*. Introduction of a normal, functional copy of the *Rb* gene suppresses the tumorigenicity of a retinoblastoma cell line in which the endogenous *Rb* gene had been inactivated.

therefore be detected in tumors without gross deletions, presumably as a consequence of small deletions or point mutations. Finally, point mutations were detected within the candidate gene even in those tumors expressing normal levels of the 4.7 kb RNA.

The highly frequent and reproducible mutational inactivation of the cloned gene in retinoblastomas thus provided strong circumstantial evidence that the retinoblastoma susceptibility gene had been isolated. Direct evidence supporting this assignment was then provided by gene transfer experiments, in which it has been demonstrated that introduction of a functional normal *Rb* gene suppresses the tumorigenicity of tumor cell lines in which the endogenous *Rb* gene had been deleted (Fig. 9.9). Thus, the cloned retinoblastoma gene displays the predicted biological activity—suppression of the malignant phenotype.

The *Rb* gene extends over approximately 200 kb of DNA and is composed of 27 exons (Fig. 9.10). Nucleotide sequencing of the *Rb* cDNA has identified a predicted protein product of 928 amino acids, which has been studied with the use of antibodies raised against the predicted amino acid sequence. As discussed in subsequent chapters, the *Rb* gene product is a nuclear protein that acts to regulate gene expression as cells progress through the cell cycle.

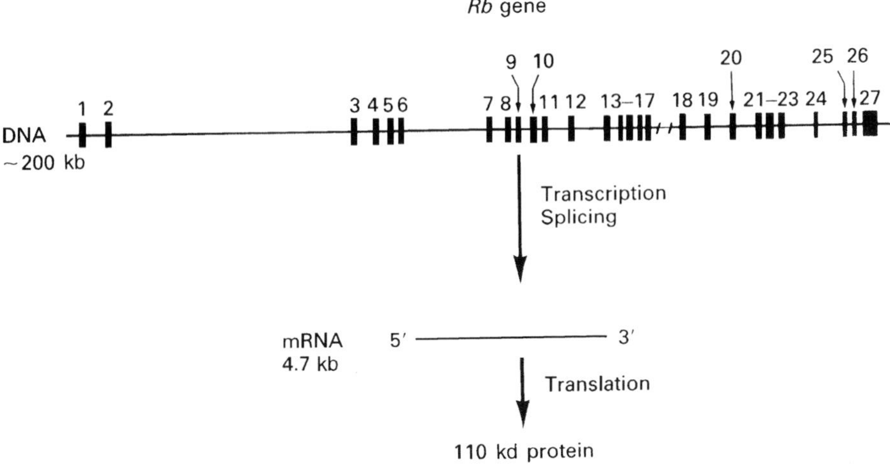

FIGURE 9.10
Organization and expression of the retinoblastoma gene. The retinoblastoma gene (*Rb*) encompasses ~200 kb and consists of 27 exons. The transcript is processed to a 4.7 kb mRNA and translated to yield a protein of ~110 kd (928 amino acids).

Role of the *Rb* Gene in Other Neoplasms

Although the *Rb* gene was initially identified by inherited susceptibility to retinoblastoma, the loss of function of this gene also contributes to the development of several other tumors. Surviving patients with inherited retinoblastoma develop second neoplasms with much higher frequencies than either normal individuals or survivors of sporadic retinoblastoma. Nearly half of the secondary neoplasms in inherited retinoblastoma patients are osteosarcomas, suggesting that loss of *Rb* function also predisposes to osteosarcoma development. Molecular analyses have supported this notion. Both inherited and sporadic osteosarcomas are frequently homozygous for chromosome 13 RFLPs, parallel to the situation in retinoblastoma. In addition, the *Rb* gene itself is deleted or otherwise mutated in a high fraction of osteosarcomas. Thus, a single tumor suppressor gene predisposes to the development of both of these inherited neoplasms.

Inherited *Rb* mutations are rare and retinoblastoma occurs with a low frequency, affecting only about 1 in 20,000 children. However, mutations of the *Rb* tumor suppressor gene also contribute to some of the common malignancies of adults, including lung, breast, and bladder cancers. In these tumors, which are more than 1,000 times as frequent as retinoblastoma, the inactivating mutations

FIGURE 9.11
Interaction of the Rb protein with SV40 T antigen. SV40 large T binds to and inactivates Rb in virus-infected cells. Transformation by SV40 thus results, in part, from inactivation of the *Rb* tumor suppressor gene product.

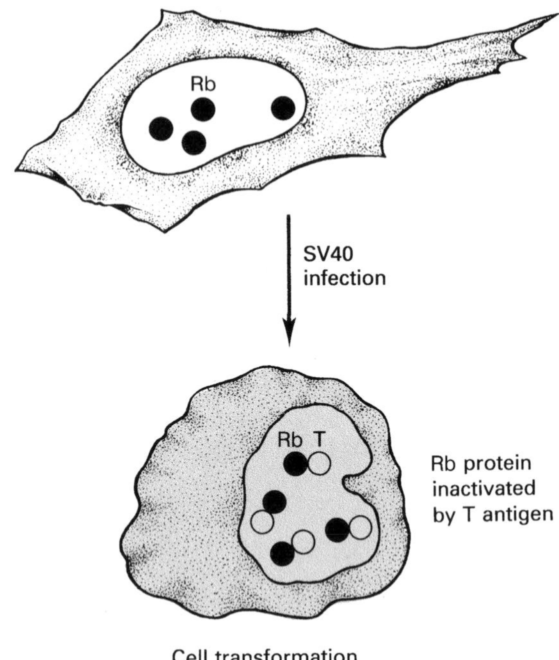

of *Rb* occur somatically rather than being inherited. In fact, these tumors are not common among survivors of inherited retinoblastoma, suggesting that mutations of genes other than *Rb* limit their development. However, because *Rb* mutations are found in a significant fraction of these malignancies, the role of the *Rb* tumor suppressor gene is not limited to the rare childhood cancer from which it was isolated—it is also implicated in some of the most commonly occurring tumors of adults.

As discussed in chapter 2, the *Rb* gene product also plays a central role in tumors induced by several DNA tumor viruses, particularly SV40, adenovirus, and human papillomaviruses. The transforming proteins of these viruses (T antigen, E1A, and E7) all act, at least in part, by binding to the Rb protein and inactivating its function (Fig. 9.11). The inactivation of the Rb protein by binding to the oncogene proteins of these DNA tumor viruses has the same consequences as mutational inactivation of the *Rb* gene—uncontrolled cell proliferation and tumor development. Moreover, the interaction of these oncogene proteins with Rb provides a good illustration of the relation between oncogene and tumor suppressor gene functions, with an oncogene protein inducing cell transformation by inactivating the product of a tumor suppressor gene.

Summary

A broad generic distinction can be made between the concepts of oncogenes and tumor suppressor genes. For the oncogenes, enhanced expression or function leads to transformation whereas, for the tumor suppressor genes, transformation is the result of loss of function. The *Rb* gene has provided the prototype of this group of genes, which function as negative regulators of cell proliferation and normally act to inhibit tumor development. Although first identified in a rare inherited childhood cancer, *Rb* is also inactivated by somatic mutations in several common adult malignancies, including lung cancers. Moreover, as discussed in the next chapter, the *Rb* gene has proved to be only the first of a growing family of tumor suppressor genes that play important roles in the development of a wide variety of human cancers.

References

General References

Hansen, M.F., and Cavenee, W.K. 1988. Retinoblastoma and the progression of tumor genetics. *Trends Genet.* 4:125–128.

Knudson, A.G. 1993. Antioncogenes and human cancer. *Proc. Natl. Acad. Sci. USA* 90:10914–10921.

Stanbridge, E.J. 1990. Human tumor suppressor genes. *Ann. Rev. Genet.* 24:615–657.

Weinberg, R.A. 1991. Tumor suppressor genes. *Science* 254:1138–1146.

Somatic Cell Hybridization

Harris, H. 1988. The analysis of malignancy by cell fusion: the position in 1988. *Cancer Res.* 48:3302–3306.

Harris, H., Miller, O.J., Klein, G., Worst, P., and Tachibana, T. 1969. Supression of malignancy by cell fusion. *Nature* 223:363–368.

Stanbridge, E.J. 1976. Suppression of malignancy in human cells. *Nature* 260:17–20.

Stanbridge, E.J., Flandermeyer, R.R., Daniels, D.W., and Nelson-Rees, W.A. 1981. Specific chromosome loss associated with the expression of tumorigenicity in human cell hybrids. *Somatic Cell Mol. Genet.* 7:699–712.

Inherited and Sporadic Retinoblastoma

Knudson, A.G., Jr. 1971. Mutation and cancer: statistical study of retinoblastoma. *Proc. Natl. Acad. Sci. USA* 68:820–823.

Mapping the Retinoblastoma Gene

Francke, U., and Kung, F. 1976. Sporadic bilateral retinoblastoma and $13q^-$ chromosomal deletion. *Med. Ped. Oncol.* 2:379–385.

Knudson, A.G., Meadows, A.T., Nichols, W.W., and Hill, R. 1976. Chromosomal deletion and retinoblastoma. *N. Engl. J. Med.* 295:1120–1123.

Sparkes, R.S., Murphree, A.L., Lingua, R.W., Sparkes, M.C., Field, L.L., Funderburk, S.J., and Benedict, W.F. 1983. Gene for hereditary retinoblastoma assigned to human chromosome 13 by linkage to esterase D. *Science* 219:971–973.

Yunis, J.J., and Ramsay, N. 1978. Retinoblastoma and subband deletion of chromosome 13. *Am. J. Dis. Child.* 132:161–163.

The Retinoblastoma Gene is Recessive at the Cellular Level

Benedict, W.F., Murphree, A.L., Banerjee, A., Spina, C.A., Sparkes, M.C., and Sparkes, R.S. 1983. Patient with 13 chromosome deletion: evidence that the retinoblastoma gene is a recessive cancer gene. *Science* 219:973–975.

Cavenee, W.K., Dryja, T.P., Phillips, R.A., Benedict, W.F., Godbout, R., Gallie, B.L., Murphree, A.L., Strong, L.C., and White, R.L. 1983. Expression of recessive alleles by chromosomal mechanisms in retinoblastoma. *Nature* 305:779–784.

Cavenee, W.K., Hansen, M.F., Nordenskjold, M., Kock, E., Maumenee, I., Squire, J.A., Phillips, R.A., and Gallie, B.L. 1985. Genetic origin of mutations predisposing to retinoblastoma. *Science* 228:501–503.

Dryja, T.P., Rapaport, J.M., Joyce, J.M., and Petersen, R.A. 1986. Molecular detection of deletions involving band q14 of chromosome 13 in retinoblastomas. *Proc. Natl. Acad. Sci. USA* 83:7391–7394.

Godbout, R., Dryja, T.P., Squire, J., Gallie, B.L., and Phillips, R.A. 1983. Somatic inactivation of genes on chromosome 13 is a common event in retinoblastoma. *Nature* 304:451–453.

Molecular Cloning of the *Rb* Gene

Bookstein, R., Shew, J.-Y., Chen, P.-L., Scully, P., and Lee, W.- H. 1990. Suppression of tumorigenicity of human prostate carcinoma cells by replacing a mutated *RB* gene. *Science* 247:712–715.

Dunn, J.M., Phillips, R.A., Becker, A.J., and Gallie, B.L. 1988. Identification of germline and somatic mutations affecting the retinoblastoma gene. *Science* 241:1797–1800.

Friend, S.H., Bernards, R., Rogelj, S., Weinberg, R.A., Rapaport, J.M., Albert, D.M., and Dryja, T.P. 1986. A human DNA segment with properties of the gene that predisposes to retinoblastoma and osteosarcoma. *Nature* 323:643–646.

Fung, Y.-K. T., Murphree, A.L., T'Ang, A., Qian, J., Hinrichs, S.H., and Benedict, W.F. 1987. Structural evidence for the authenticity of the human retinoblastoma gene. *Science* 236:1657–1661.

Hong, F.D., Huang, H.-J. S., To, H., Young, L.-J. S., Oro, A., Bookstein, R., Lee, E. Y.-H. P., and Lee, W.-H. 1989. Structure of the human retinoblastoma gene. *Proc. Natl. Acad. Sci. USA* 86:5502–5506.

Horowitz, J.M., Yandell, D.W., Park, S.-H., Canning, S., Whyte, P., Buchkovich, K., Harlow, E., Weinberg, R.A., and Dryja, T.P. 1989. Point mutational inactivation of the retinoblastoma antioncogene. *Science* 243:937–940.

Huang, H.-J. S., Yee, J.-K., Shew, J.-Y., Chen, P.-L., Bookstein, R., Friedmann, T., Lee, E. Y.-H. P., and Lee, W.-H. 1988. Suppression of the neoplastic phenotype by replacement of the *RB* gene in human cancer cells. *Science* 242:1563–1566.

Lee, W.-H., Bookstein, R., Hong, F., Young, L.-J., Shew, J.-Y., and Lee, E.Y.-H. P. 1987. Human retinoblastoma susceptibility gene: cloning, identification, and sequence. *Science* 235:1394–1399.

Takahashi, R., Hashimoto, T., Xu, H.-J., Hu, S.-X., Matsui, T., Miki, T., Bigo-Marshall, H., Aaronson, S.A., and Benedict, W.F. 1991. The retinoblastoma gene functions as a growth and tumor suppressor in human bladder carcinoma cells. *Proc. Natl. Acad. Sci. USA* 88:5257–5261.

Role of the *Rb* Gene in Other Neoplasms

Bookstein, R., Rio, P., Madreperla, S.A., Hong, F., Allred, C., Grizzle, W.E., and Lee, W.-H. 1990. Promoter deletion and loss of retinoblastoma gene expression in human prostate carcinoma. *Proc. Natl. Acad. Sci. USA* 87:7762–7766.

DeCaprio, J.A., Ludlow, J.W., Figge, J., Shew, J.-Y., Huang, C.- M., Lee, W.-H., Marsilio, E., Paucha, E., and Livingston, D.M. 1988. SV40 large tumor antigen forms a specific complex with the product of the retinoblastoma susceptibility gene. *Cell* 54:275–283.

Dyson, N., Howley, P.M., Munger, K., and Harlow, E. 1989. The human papilloma virus-16 E7 oncoprotein is able to bind to the retinoblastoma gene product. *Science* 243:934–937.

Friend, S.H., Horowitz, J.M., Gerber, M.R., Wang, X.-F., Bogenmann, E., Li, F.P., and Weinberg, R.A. 1987. Deletions of a DNA sequence in retinoblastomas and mesenchymal tumors: organization of the sequence and its encoded protein. *Proc. Natl. Acad. Sci. USA* 84:9059–9063.

Hansen, M.F., Koufos, A., Gallie, B.L., Phillips, R.A., Fodstad, O., Brogger, A., Gedde-Dahl, T., and Cavenee, W.K. 1985. Osteosarcoma and retinoblastoma: a shared chromosomal mechanism revealing recessive predisposition. *Proc. Natl. Acad. Sci. USA* 82:6216–6220.

Harbour, J.W., Lai, S.-L., Whang-Peng, J., Gazdar, A.F., Minna, J.D., and Kaye, F.J. 1988. Abnormalities in structure and expression of the human retinoblastoma gene in SCLC. *Science* 241:353–357.

Horowitz, J.M., Park, S.-H., Begenmann, E., Cheng, J.-C., Yandell, D.W., Kaye, F.J., Minna, J.D., Dryja, T.P., and Weinberg, R.A. 1990. Frequent inactivation of the retinoblastoma anti-oncogene is restricted to a subset of human tumor cells. *Proc. Natl. Acad. Sci. USA* 87:2775–2779.

Ishikawa, J., Xu, H.-J., Hu, S.-X., Yandell, D.W., Maeda, S., Kamidono, S., Benedict, W.F., and Takahashi, R. 1991. Inactivation of the retinoblastoma gene in human bladder and renal cell carcinomas. *Cancer Res.* 51:5736–5743.

Lee, E. Y.-H. P., To, H., Shew, J-Y., Bookstein, R., Scully, P., and Lee, W.-H. 1988. Inactivation of the retinoblastoma susceptibility gene in human breast cancers. *Science* 241:218–221.

Reissmann, P.T., Koga, H., Takahashi, R., Figlin, R.A., Holmes, E.C., Piantadosi, S., Cordon-Cardo, C., Slamon, D.J., and the Lung Cancer Study Group. 1993. Inactivation of the retinoblastoma susceptibility gene in non-small-cell lung cancer. *Oncogene* 8:1913–1919.

T'Ang, A., Varley, J.M., Chakraborty, S., Murphree, A.L., and Fung, Y.-K. T. 1988. Structural rearrangement of the retinoblastoma gene in human breast carcinoma. *Science* 242:263–266.

Whyte, P., Buchkovich, K.J., Horowitz, J.M., Friend, S.H., Raybuck, M., Weinberg, R.A., and Harlow, E. 1988. Association between an oncogene and an anti-oncogene: the adenovirus E1A proteins bind to the retinoblastoma gene product. *Nature* 334:124–129.

❖ CHAPTER 10

Tumor Suppressor Genes in Human Neoplasms

The isolation of *Rb* paved the way for the identification and molecular cloning of a growing list of tumor suppressor genes. Some of these genes, like *Rb,* were initially identified by genetic analysis of rare inherited forms of cancer. Others were identified as genes that are frequently inactivated by somatic mutations or deletions in more common cancers of adults. In either event, further studies have indicated that most of these additional tumor suppressor genes (like *Rb*) take part in the development of both inherited and noninherited tumors. The number of these genes continues to expand rapidly, and it is clear that the inactivation of tumor suppressor genes plays as important a role in tumor development as the activation of oncogenes.

The *p53* Gene

The second tumor suppressor gene to be characterized was *p53,* which has turned out to be the most common target of genetic alterations in human tumors. Interestingly, however, studies of the *p53* gene have had a long and complex history. In fact, *p53* was initially thought to be an oncogene: its true identity as a tumor suppressor gene was not appreciated until ten years after its initial discovery.

The *p53* gene product was first recognized in 1979 as a protein of 53 kd (hence, the designation p53) that was complexed to SV40 T antigen in SV40-transformed cells. The levels of p53 were found to be low in normal cells, but were substantially increased in SV40-transformed cells as well as in a variety of other transformed cell lines. It thus appeared that elevated p53 levels correlated with transformation, consistent with the possibility that *p53* was an oncogene. These observations were followed by the isolation of molecular clones of *p53*, and it was found that overexpression of these cloned *p53* genes was capable of inducing cell transformation in gene transfer assays. The biological activity of the *p53* clones used in these initial experiments thus provided direct support for the classification of *p53* as an oncogene.

This interpretation, however, was incorrect. Further studies showed that the *p53* clones initially shown to induce transformation were mutant rather than wild-type forms of the gene. When the normal *p53* gene was isolated, it was found to lack transforming activity. Moreover, the groups of both Arnold Levine and Moshe Oren demonstrated in 1989 that overexpression of normal *p53* in fact acted to inhibit rather than induce transformation—behavior expected for a tumor suppressor gene, not an oncogene. These observations suggested an alternative interpretation for the transforming activity of mutant *p53* genes. Namely, the ability of wild-type *p53* to suppress transformation suggested that the transforming activity of the mutant *p53* genes could be explained by their acting as dominant inhibitory mutants that interfered with the action of the normal p53 protein (Fig. 10.1). This interpretation was supported by the fact that p53 proteins form oligomeric complexes, and it appears that complexes formed between normal and mutant polypeptides are inactive. The presence of an excess of mutationally altered inactive p53 polypeptides within a cell therefore blocks normal p53 function, and it is this interference with normal p53 activity that results in cell transformation. As discussed in chapter 2, the binding of SV40 T antigen also inactivates p53, so its action is analogous to that of the dominant inhibitory proteins encoded by mutant *p53* genes.

The interpretation of *p53* as a tumor suppressor gene is supported by the inactivation of *p53* in a wide range of neoplasms (Table 10.1). These include murine retrovirus-induced erythroleukemias, in which *p53* is a target for proviral insertions resulting in its inactivation. Inactivation of the *p53* tumor suppressor gene in human tumors was first documented by Bert Vogelstein and his colleagues in 1989. The *p53* gene is located at the chromosomal locus 17p13, which is frequently deleted in colon and rectum carcinomas. Analysis of these tumors demonstrated that one copy of the *p53* gene was lost as a result of the 17p deletion. Importantly, the remaining copies almost always contained mutations resulting in amino acid substitutions in the p53 protein. It therefore appeared that the 17p deletions served to render the tumor cells homozygous for mutant *p53* genes, analogous to deletions of the *Rb* locus in retinoblastomas. Subsequent studies have shown that deletions and mutations of *p53* are common in a wide spectrum of additional types of human tumors, including lung, breast, esophageal, liver, bladder, and ovarian carcinomas, brain tumors, sarcomas, lymphomas, and leukemias. It is further noteworthy that *p53* mutations in liver tumors have been shown to be induced directly by a chemical carcino-

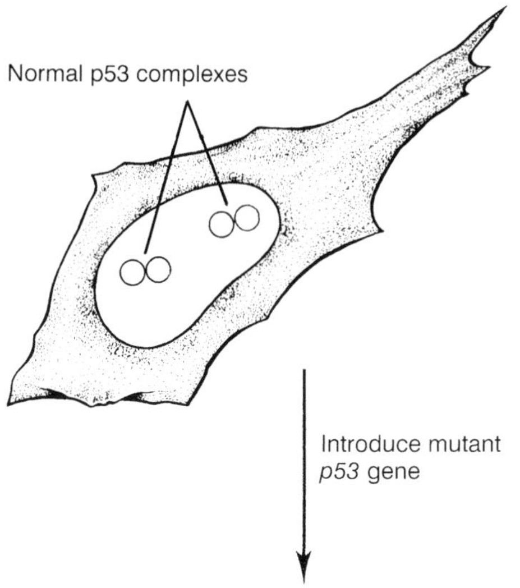

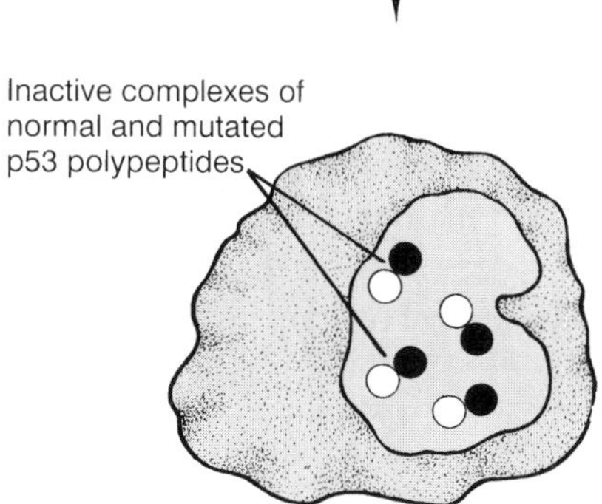

Figure 10.1
Transformation by dominant inhibitory *p53* mutants. The normal p53 protein functions as an oligomeric complex. Introduction of a mutant *p53* gene results in production of excess altered p53 polypeptides, which form complexes with the normal protein. These complexes of normal and mutated polypeptides are inactive, resulting in a loss of normal p53 function that leads to cell transformation.

gen, as discussed further in chapter 11. Overall, mutations of *p53* appear to contribute to approximately 50% of all human cancers, making *p53* the single most common target for genetic alterations leading to human tumor development.

Formal demonstration of the activity of *p53* as a tumor suppressor gene has also been provided by gene transfer experiments in which normal *p53* genes have been introduced into tumor cell lines in which the endogenous *p53* genes were deleted or mutated. As in similar studies with the *Rb* gene, these experiments demonstrated that wild-type *p53* was capable of suppressing cell growth and tumorigenicity.

Not only is *p53* frequently inactivated by somatic mutations, but germline mutations of *p53* were subsequently found to be responsible for rare inherited cancers associated with the *Li-Fraumeni cancer family syndrome.* Families with

TABLE 10.1
Cloned Tumor Suppressor Genes

Gene	Chromosomal Location	Inherited Cancer	Sporadic Cancer
Rb	13q14	retinoblastoma	retinoblastoma, sarcomas, bladder, breast, esophageal, and lung carcinomas
p53	17p13	Li-Fraumeni cancer family syndrome	bladder, breast, colorectal, esophageal, liver, lung, and ovarian carcinomas, brain tumors, sarcomas, lymphomas, and leukemias
DCC	18q21	——	colorectal carcinomas
APC	5q21	familial adenomatous polyposis	colorectal, stomach, and pancreatic carcinomas
WT1	11p13	Wilms tumor	Wilms tumor
NF1	17q11	neurofibromatosis type 1	colon carcinoma and astrocytoma
NF2	22q12	neurofibromatosis type 2	schwannoma and meningioma
VHL	3p25	von Hippel-Lindau syndrome	renal cell carcinomas
MTS1	9p21	melanoma	melanoma, brain tumors, leukemias, sarcomas, bladder, breast, kidney, lung, and ovarian carcinomas

this syndrome are characterized by the inheritance of a spectrum of tumors, including breast carcinomas, sarcomas, brain tumors, leukemias, and adrenocortical carcinomas. As in inherited retinoblastoma, tumor susceptibility is transmitted as a dominant Mendelian characteristic, so that approximately 50% of family members are affected (Fig. 10.2). The inactivation of *p53* in a similar broad spectrum of noninherited cancers suggested that the *p53* gene might also be responsible for this inherited cancer susceptibility—a prediction that was substantiated by the finding of germline *p53* mutations in Li-Fraumeni cancer families. Subsequent studies have also identified germline mutations of *p53* in some patients with a history of multiple tumors but without a family history characteristic of the Li-Fraumeni syndrome. Thus, germ line as well as somatic *p53* mutations play important roles in human cancer.

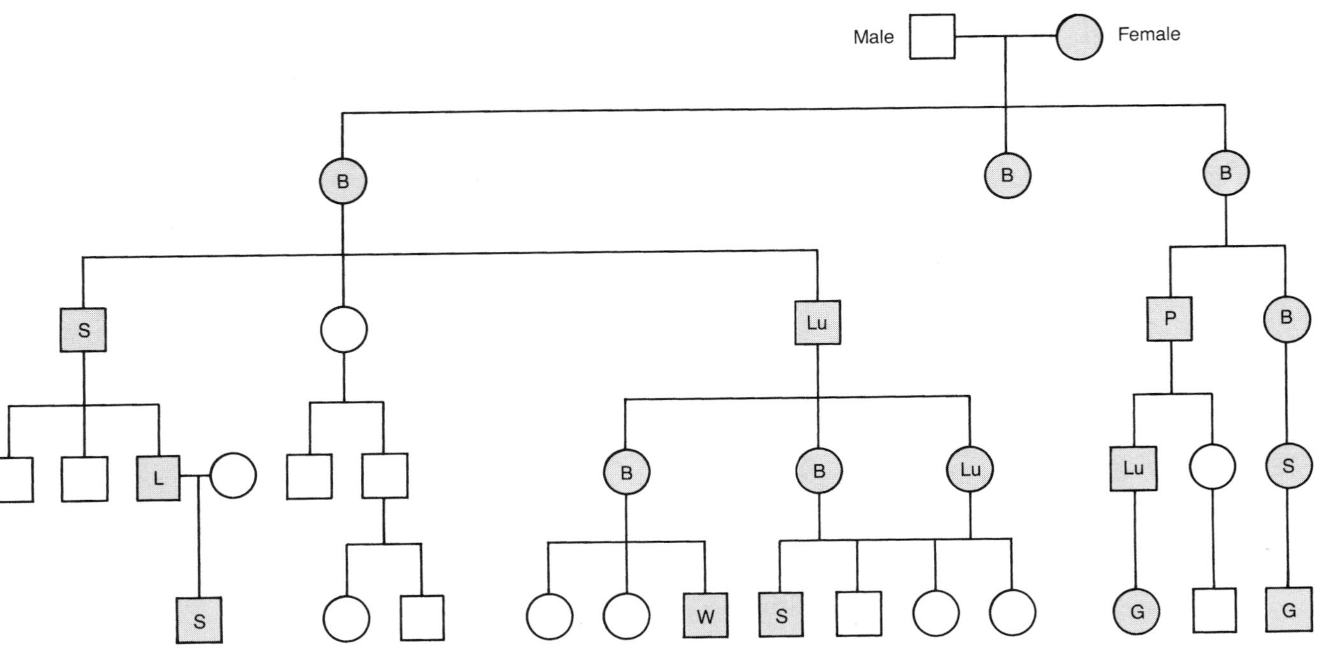

FIGURE 10.2
Pedigree of a family with Li-Fraumeni cancer family syndrome. Individuals with cancer are depicted by filled symbols. A blank symbol signifies cancer of unknown origin. The letter B signifies breast cancer; G, glioblastoma; L, leukemia; Lu, lung cancer; P, pancreatic cancer; S, sarcoma; and W, Wilms tumor. (From F.P. Li and J.F. Fraumeni, *J. Amer. Med. Assoc.* 247:2692–2694, 1982.)

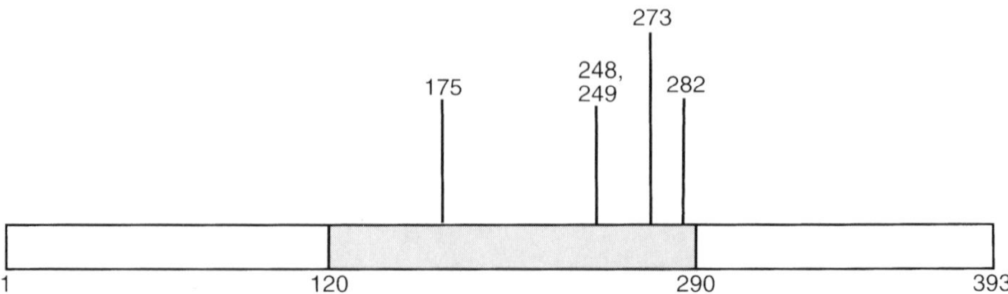

FIGURE 10.3
Sites of mutation in *p53*. The human *p53* gene encodes a protein of 393 amino acids. Nearly all of the mutations in human tumors fall between codons 120 and 290 (filled area) and about half alter codons 175, 248, 249, 273, or 282.

The human *p53* gene consists of eleven exons, spanning approximately 20 kb of DNA, and encodes a protein of 393 amino acids. The mutations of *p53* identified in human tumors cluster in four regions of the gene, which are highly conserved in *p53* genes of other vertebrate species (Fig. 10.3). Nearly all of these mutations fall between codons 120 and 290 and about 50% alter one of five codons in the *p53* gene: codons 175, 248, 249, 273, and 282. The phenotypic consequence of these mutations is inactivation of the p53 protein, which is now known to function as a transcription factor (discussed in chapter 16). As predicted from the ability of mutant *p53* genes to induce transformation, the mutationally altered proteins have been shown to act as dominant inhibitors of normal p53 function. It is further noteworthy that the mutations that inactivate p53 function in tumors are associated with increased stability of the p53 protein. This increased stability of the mutant proteins may facilitate their action as dominant inhibitors, as well as accounting for the apparent overexpression of p53 that was observed in early studies of a variety of tumor cell lines.

In addition to mutational inactivation of the *p53* gene, the p53 protein can be inactivated by association with the products of other oncogenes. As already noted, these include the oncogene proteins of DNA tumor viruses—SV40 large T antigen, the adenovirus E1B protein, and the E6 protein of human papillomaviruses. In addition, the cellular oncogene *mdm*-2 encodes a protein that functions by binding to and inhibiting the activity of p53. As noted in chapter 8, the *mdm*-2 gene is frequently amplified and overexpressed in human sarcomas. In those sarcomas containing amplified *mdm*-2 genes, the *p53* genes are not mutated and normal p53 protein is expressed. However, normal p53 function is effectively inhibited by its association with overexpressed Mdm-2 protein. The *p53* tumor suppressor gene is thus a common target not only for inactivation by mutations but also, at the protein level, for inhibition by both viral and cellular oncogene products.

Identification of Additional Tumor Suppressor Genes in Colorectal Carcinomas

Colon and rectum (colorectal) carcinomas have provided a particularly useful source of material for studies of the role of genetic alterations in development of a common human tumor. As discussed earlier, these tumors are characterized by frequent deletions of chromosome 17p, and it was the molecular analysis of these deletions that pointed to the involvement of *p53* in human cancer. Additional deletions of chromosomes 18q and 5q also are common in colorectal carcinomas, and molecular analysis of these deletions has similarly led to the identification and molecular cloning of new tumor suppressor genes (Table 10.1).

The first of these genes, *DCC*, was identified by Bert Vogelstein and colleagues as a result of cloning approximately 370 kb of DNA from the region of 18q chromosome deletions. Analysis of this region led to the identification of coding segments that were conserved between humans and rodents. These exon segments were then used as probes to isolate a cDNA clone, which defined the *DCC* (*d*eleted in *c*olorectal *c*arcinomas) gene. Consistent with its designation as a tumor suppressor gene, *DCC* is transcribed in normal colon epithelial cells (and other cell types), but its expression is reduced or absent in colorectal carcinomas. In addition, *DCC* is the target of homozygous deletions and additional somatic mutations in colorectal carcinomas, as expected for a tumor suppressor gene taking part in the development of these neoplasms. However, the predicted tumor suppressive activity of *DCC* in gene transfer assays has not yet been demonstrated; so its formal classification as a tumor suppressor gene awaits direct evidence of its biological activity. In addition, germ line mutations of *DCC* have not been implicated in inherited cancers, in contrast with the germ line mutations of other tumor suppressor genes.

In contrast with the lack of involvement of *DCC* in inherited cancers, the chromosome 5q region was not only known to be frequently deleted in colorectal carcinomas, but also identified as the site of a gene responsible for a rare inherited form of colon cancer called *familial adenomatous polyposis*. Individuals with this disease, which affects about 1 in 10,000 people, develop hundreds or thousands of colon polyps within the first twenty years of life. Unless treated, some of these polyps invariably develop into colon carcinomas. Familial adenomatous polyposis (like retinoblastoma) is inherited as a dominant Mendelian characteristic, which was mapped to chromosome 5q21 by genetic linkage analysis. The gene responsible (called *APC* for *a*denomatous *p*olyposis *c*oli) was then isolated by two groups of researchers in 1991. The *APC* gene contains germline mutations in familial adenomatous polyposis patients, and it is also inactivated by somatic mutations in the majority of sporadic colorectal carcinomas. *APC* is a large gene, encoding a protein of approximately 2,840 amino acids. Interestingly, both inherited and somatic mutations almost always result in the production of truncated

APC proteins, detection of which may provide a useful assay for diagnosis of individuals carrying the *APC* gene. Although the expected biological activity of *APC* has not been demonstrated in gene transfer experiments, its involvement in both inherited and sporadic colon carcinomas provides strong support for its characterization as a tumor suppressor gene. Moreover, an inbred mouse strain with an inherited predisposition to the development of multiple intestinal neoplasms has similarly been found to carry a mutation in *APC*. Mutations of *APC* have also been reported in stomach and pancreatic cancers, suggesting the possibility that it may play a broader role in human disease.

An additional gene located in the 5q21 region also is mutated in some sporadic colorectal carcinomas. It is possible that this gene (designated *MCC* for *m*utated in *c*olorectal *c*ancer) is a second tumor suppressor gene at this locus. However, because the *MCC* gene is unaffected in familial adenomatous polyposis patients, it is alternatively possible that the mutations observed within this gene do not play a role in tumor development. Further evidence is thus required to determine whether *MCC* in fact is a tumor suppressor gene.

It is noteworthy that familial adenomatous polyposis is only one type of inherited colon cancer, which is responsible for less than 1% of total colon cancer incidence. A more common inherited form of the disease is *hereditary nonpolyposis colorectal cancer* (*HNPCC*), which accounts for about 15% of all colorectal cancers. The gene responsible for most cases of HNPCC has been recently mapped to chromosome 2p and isolated as a molecular clone. Unexpectedly, however, this gene is not a tumor suppressor gene. Rather, it was noted that colon cancers arising in HNPCC patients, as well as some sporadic tumors, were characterized by genetic instabilities (in particular, destabilization of repeated DNA sequences) suggestive of a defect in mismatch repair of DNA. This prompted two research groups to examine human homologs of *E. coli* and yeast mismatch repair genes as candidates for the gene responsible for HNPCC. These studies identified a human mismatch repair gene (*MSH2*) that mapped to the relevant region of chromosome 2 and was mutated in HNPCC patients. A second DNA repair gene (*MLH1*), located on chromosome 3, has similarly been found to be responsible for inheritance of other cases of HNPCC. Mutations of the *MSH2* and *MLH1* genes thus appear to lead indirectly to tumor development, which occurs as a consequence of a high mutation frequency resulting from defective DNA repair.

Tumor Suppressor Genes in Other Inherited Cancers

Most of the tumor suppressor genes have been isolated as a result of genetic analysis of rare inherited cancers. These studies directly followed the prototype analysis of the *Rb* gene in retinoblastoma, as discussed in chapter 9.

The extension of this approach to additional inherited cancers has resulted in the molecular cloning of several new tumor suppressor genes (Table 10.1).

The most extensively studied inherited cancer, after retinoblastoma, was Wilms tumor—a childhood kidney cancer with an incidence of about 1 in 10,000. Analogous to children with inherited retinoblastoma, children with inherited Wilms tumor usually develop multiple neoplasms in both kidneys. Cytogenetic studies localized the gene responsible to chromosome 11p, and RFLP analysis demonstrated that Wilms tumors frequently developed homozygosity for markers on this chromosome. These genetic studies paralleled those of retinoblastoma, suggesting that the Wilms tumor gene was a tumor suppressor. Based on these findings, two groups of researchers proceeded to isolate a candidate Wilms tumor gene, designated *WT1*, from the 11p13 region.

Consistent with the hypothesis that *WT1* was a tumor suppressor gene, it was found to be transcribed in normal embryonic kidney but not in Wilms tumors. Moreover, Wilms tumors containing internal deletions and point mutations of *WT1* have been identified. In addition, the activity of *WT1* as a tumor suppressor has been directly demonstrated by gene transfer experiments in which a normal *WT1* gene was found to suppress growth of a Wilms tumor cell line. Conversely, a dominant negative mutant of *WT1* has been described that, like similar mutants of *p53*, can induce transformation by interfering with function of the normal WT1 protein. Taken together, these results clearly support the characterization of *WT1* as a tumor suppressor gene that contributes to Wilms tumor development.

It is noteworthy, however, that *WT1* may not be the only gene on chromosome 11 that is responsible for Wilms tumors. Some Wilms tumors are associated with deletions at 11p15 rather than 11p13, suggesting that a second tumor suppressor gene may reside at this locus. Consistent with this possibility, a modified chromosome 11 with a deletion of 11p13 but containing an intact 11p15 has been found to suppress the tumorigenicity of a Wilms tumor cell line. The presumed tumor suppressor gene in this region awaits molecular cloning and characterization.

Neurofibromatosis describes two distinct genetic disorders, both of which are dominantly inherited and predispose to tumors of the nervous system. Neurofibromatosis type 1, or von Recklinghausen's disease, is one of the most common human genetic disorders, with an incidence of about 1 in 3,000. The gene responsible was mapped to chromosome 17q11 and isolated by molecular cloning of this region of DNA. The *NF1* gene was then identified as the site of germ line mutations in neurofibromatosis patients that, as expected for a tumor suppressor gene, disrupt *NF1*. Somatic mutations of *NF1* have also been described in some sporadic tumors, including a colon carcinoma and an astrocytoma, and the possible involvement of *NF1* in a broader array of human cancers remains to be assessed. Neurofibromatosis type 2, which has an incidence of about 1 in 40,000, is caused by inactivation of a different tumor suppressor gene (*NF2*) on chromosome 22q12. The *NF2* gene is not only mutated in the germ line of patients with inherited neurofibromatosis, but also frequently inactivated by somatic mutations in sporadic tumors characteristic of the neurofibromatosis type 2 syndrome, such as schwannomas and meningiomas.

Another cloned tumor suppressor gene was identified as the gene responsible for the von Hippel-Lindau family cancer syndrome. This syndrome has an incidence of about 1 in 35,000 and is associated with hemangioblastoma, renal cell carcinoma, and pheochromocytoma. The *VHL* gene was localized by linkage analysis to chromosome 3p25 and then cloned. In addition to being the target of germ line mutations in von Hippel-Lindau syndrome patients, *VHL* is also frequently inactivated by somatic mutations in sporadic renal cell carcinomas.

A tumor suppressor gene mapping to chromosome 9p21, a region responsible for inherited melanomas and frequently deleted in sporadic melanomas and other tumors, also has been recently isolated. Mutations or deletions of this gene have been found in cell lines derived from melanomas and a variety of other tumor types, including brain tumors, leukemias, sarcomas, and bladder, breast, lung, kidney, and ovarian carcinomas. Because of its apparent involvement in a wide variety of tumors, the gene has been designated *MTS1* (for *M*ultiple *T*umor *S*uppressor *1*). In addition to its frequent involvement in diverse neoplasms, *MTS1* is of considerable interest because it functions (like *p53*) to regulate progression through the cell cycle (see chapter 18).

Candidates for Additional Tumor Suppressors

Several additional candidate tumor suppressor loci have been identified by genetic analysis but not yet isolated as molecular clones (Table 10.2). As noted earlier, one such candidate is a gene on chromosome 11p15 that appears to play a role in some Wilms tumors. Another candidate tumor suppressor gene on chromosome 9, mapping at q31, is implicated in Gorlin syndrome, which predisposes to basal cell skin carcinomas, medulloblastomas, and other types of neoplasms. A gene responsible for familial breast cancer (designated *BRCA1*) also has been mapped to chromosome 17q21. Molecular cloning of *BRCA1* is being actively pursued, particularly because it may also be involved in a significant fraction of common noninherited breast tumors, as well as in ovarian carcinomas. Also of notable interest for its involvement in a common human tumor is a candidate tumor suppressor gene on chromosome 3p21, which appears to be frequently deleted in lung cancers. Other putative tumor suppressor genes may be responsible for inheritance of neuroblastoma

TABLE 10.2
Candidate Tumor Suppressor Loci

Chromosomal Locus	Neoplasm
1p36	neuroblastoma
3p21	lung cancer
9q31	Gorlin syndrome: basal cell skin carcinoma and other cancers
11p15	Wilms tumor
11q13	multiple endocrine neoplasia-1
17q21	breast cancer

(chromosome 1p36) and multiple endocrine neoplasia type 1 (chromosome 11q13).

It thus appears that the list of tumor suppressor genes will continue to expand, although it is important to recognize that tumor suppressor genes are not the only type of gene responsible for the dominant inheritance of cancer susceptibility. As noted earlier in this chapter, the most common form of inherited colon cancer (hereditary nonpolyposis colon cancer) is due to mutation of genes encoding DNA repair enzymes, not to tumor suppressors. Another example is multiple endocrine neoplasia type 2, which has recently been found to result from inheritance of an activated *ret* oncogene rather than (as had been expected) a mutated tumor suppressor gene. Molecular and functional characterization of the genes responsible for cancer inheritance is thus needed to determine whether they are new tumor suppressor genes or other types of genes that predispose to tumor development.

Summary

A variety of tumor suppressor genes have been isolated both from genetic analysis of inherited cancers and from molecular characterization of regions of chromosome deletions in noninherited tumors. In most cases, further studies have shown that germ line mutations of these genes are responsible for one or more types of inherited cancers, whereas somatic mutations of the same genes are found in more common sporadic forms of cancer. The *p53* gene is particularly notable in that it is mutated in approximately 50% of all cancers, making *p53* the single most common target for genetic damage in human tumors. The list of cloned tumor suppressor genes continues to grow, and it is clear that these genes are as important as the oncogenes in tumor development.

References

General References

Knudson, A.G. 1993. Antioncogenes and human cancer. *Proc. Natl. Acad. Sci. USA* 90:10914–10921.

Levine, A.J. 1993. The tumor suppressor genes. *Ann. Rev. Biochem.* 62:623–651.

Marshall, C.J. 1991. Tumor suppressor genes. *Cell* 64:313–326.

Stanbridge, E.J. 1990. Human tumor suppressor genes. *Ann. Rev. Genet.* 24:615–657.

Weinberg, R.A. 1991. Tumor suppressor genes. *Science* 254:1138–1146.

The *p53* Gene

Baker, S.J., Fearon, E.R., Nigro, J.M., Hamilton, S.R., Preisinger, A.C., Jessup, J.M., van Tuinen, P., Ledbetter, D.H., Barker, D.F., Nakamura, Y., White, R., and Vogelstein, B. 1989. Chromosome 17 deletions and *p53* gene mutations in colorectal carcinomas. *Science* 244:217–221.

Baker, S.J., Markowitz, S., Fearon, E.R., Willson, J.K.V., and Vogelstein, B. 1990. Suppression of human colorectal carcinoma cell growth by wild-type *p53*. *Science* 249:912–915.

Ben-David, Y., Prideaux, V.R., Chow, V., Benchimol, S., and Bernstein, A. 1988. Inactivation of the *p53* oncogene by internal deletion or retroviral integration in erythroleukemic cell lines induced by Friend leukemia virus. *Oncogene* 3:179–185.

Chen, P.-L., Chen, Y.-M., Bookstein, R., and Lee, W.-H. 1990. Genetic mechanisms of tumor suppression by the human *p53* gene. *Science* 250:1576–1580.

Eliyahu, D., Michalovitz, D., Eliyahu, S., Pinhasi-Kimhi, O., and Oren, M. 1989. Wild-type p53 can inhibit oncogene-mediated focus formation. *Proc. Natl. Acad. Sci. USA* 86:8763–8767.

Finlay, C.A., Hinds, P.W., and Levine, A.J. 1989. The *p53* proto-oncogene can act as a suppressor of transformation. *Cell* 57:1083–1093.

Frebourg, T., Kassel, J., Lam, K.T., Gryka, M.A., Barbier, N., Andersen, T.I., Borresen, A.-L., and Friend, S.H. 1992. Germ-line mutations of the *p53* tumor suppressor gene in patients with high risk for cancer inactivate the p53 protein. *Proc. Natl. Acad. Sci. USA* 89:6413–6417.

Hollstein, M., Sidransky, D., Vogelstein, B., and Harris, C.C. 1991. *p53* mutations in human cancers. *Science* 253:49–53.

Lane, D.P., and Crawford, L.V. 1979. T antigen is bound to a host protein in SV40-transformed cells. *Nature* 278:261–263.

Levine, A.J., Momand, J., and Finlay, C.A. 1991. The *p53* tumour suppressor gene. *Nature* 351:453–456.

Linzer, D.I.H., and Levine, A.J. 1979. Characterization of a 54K dalton cellular SV40 tumor antigen present in SV40-transformed cells and uninfected embryonal carcinoma cells. *Cell* 17:43–52.

Malkin, D., Li, F.-P., Strong, L.C., Fraumeni, J.F., Jr., Nelson, C.E., Kim, D.H., Kassel, J., Gryka, M.A., Bischoff, F.Z., Tainsky, M.A., and Friend, S.H. 1990. Germ line *p53* mutations in a familial syndrome of breast cancer, sarcomas, and other neoplasms. *Science* 250:1233–1238.

Momand, J., Zambetti, G.P., Olson, D., George, D.L., and Levine, A.J. 1992. The *mdm-2* oncogene product forms a complex with the p53 protein and inhibits p53-mediated transactivation. *Cell* 69:1237–1245.

Nigro, J.M., Baker, S.J., Preisinger, A.C., Jessup, J.M., Hostetter, R., Cleary, K., Bigner, S.H., Davidson, N., Baylin, S., Devilee, P., Glover, T., Collins, F.S., Weston, A., Modali, R., Harris, C.C., and Vogelstein, B. 1989. Mutations in the *p53* gene occur in diverse human tumour types. *Nature* 342:705–708.

Oliner, J.D., Kinzler, K.W., Meltzer, P.S., George, D.L., and Vogelstein, B. 1992. Amplification of a gene encoding a p53-associated protein in human sarcomas. *Nature* 358:80–83.

Sarnow, P., Ho, Y.S., Williams, J., and Levine, A.J. 1982. Adenovirus E1B-58kd tumor antigen and SV40 large tumor antigen are physically associated with the same 54 kd cellular protein in transformed cells. *Cell* 28:387–394.

Scheffner, M., Werness, B.A., Huibregtse, J.M., Levine, A.J., and Howley, P.M. 1990. The E6 oncoprotein encoded by human papillomavirus types 16 and 18 promotes the degradation of p53. *Cell* 63:1129–1136.

Sidransky, D., von Eschenbach, A., Tsai, Y.C., Jones, P., Summerhayes, I., Marshall, F., Paul, M., Green, P., Hamilton, S.R., Frost, P., and Vogelstein, B. 1991. Identification of *p53* gene mutations in bladder cancers and urine samples. *Science* 252:706–709.

Srivastava, S. Zou, Z., Pirollo, K., Blattner, W., and Chang, E.H. 1990. Germ-line transmission of a mutated *p53* gene in a cancer-prone family with Li-Fraumeni syndrome. *Nature* 348:747–749.

Takahashi, T., Nau, M.M., Chiba, I., Birrer, M.J., Rosenberg, R.K., Vinocour, M., Levitt, M., Pass, H., Gazdar, A.F., and Minna, J.D. 1989. *p53*: a frequent target for genetic abnormalities in lung cancer. *Science* 246:491–494.

Identification of Additional Tumor Suppressor Genes in Colorectal Carcinomas

Bronner, C.E., Baker, S.M., Morrison, P.T., Warren, G., Smith, L.G., Lescoe, M.K., Kane, M., Earabino, C., Lipford, J., Lindblom, A., Tannergard, P., Bollag, R.J., Godwin, A.R., Ward, D.C., Nordenskjold, M., Fishel, R., Kolodner, R., and Liskay, R.M. 1994. Mutation in the DNA mismatch repair gene homologue *hMLH1* is associated with hereditary nonpolyposis colon cancer. *Nature* 368:258–261.

Fearon, E.R., Cho, K.R., Nigro, J.M., Kern, S.E., Simons, J.W., Ruppert, J.M., Hamilton, S.R., Preisinger, A.C., Thomas, G., Kinzler, K.W., and Vogelstein, B. 1990. Identification of a chromosome 18q gene that is altered in colorectal cancers. *Science* 247:49–56.

Fishel, R., Lescoe, M.K., Rao, M.R.S., Copeland, N.G., Jenkins, N.A., Garber, J., Kane, M., and Kolodner, R. 1993. The human mutator gene homolog *MSH2* and its association with hereditary nonpolyposis colon cancer. *Cell* 75:1027–1038.

Groden, J., Thliveris, A., Samowitz, W., Carlson, M., Gelbert, L., Albertsen, H., Joslyn, G., Stevens, J., Spirio, L., Robertson, M., Sargeant, L., Krapcho, K., Wolff, E., Burt, R., Hughes, J.P., Warrington, J., McPherson, J., Wasmuth, J., Le Paslier, D., Abderrahim, H., Cohen, D., Leppert, M., and White, R. 1991. Identification and characterization of the familial adenomatous polyposis coli gene. *Cell* 66:589–600.

Horii, A., Nakatsuru, S., Miyoshi, Y., Ichii, S., Nagase, H., Ando, H., Yanagisawa, A., Tsuchiya, E., Kato, Y., and Nakamura, Y. 1992. Frequent somatic mutations of the *APC* gene in human pancreatic cancer. *Cancer Res.* 52:6696–6698.

Horii, A., Nakatsuru, S., Miyoshi, Y., Ichii, S., Nagase, H., Kato, Y., Yanagisawa, A., and Nakamura, Y. 1992. The *APC* gene, responsible for familial adenomatous polyposis, in mutated in human gastric cancer. *Cancer Res.* 52:3231–3233.

Joslyn, G., Carlson, M., Thliveris, A., Albertsen, H., Gelbert, L., Samowitz, W., Groden, J., Stevens, J., Spirio, L., Robertson, M., Sargeant, L., Krapcho, K., Wolff, E., Burt, R., Hughes, J.P., Warrington, J., McPherson, J., Wasmuth, J., Le Paslier, D., Abderrahim, H., Cohen, D., Leppert, M., and White, R. 1991. Identification of deletion mutations and three new genes at the familial polyposis locus. *Cell* 66:601–613.

Kinzler, K.W., Nilbert, M.C., Su, L.-K., Vogelstein, B., Bryan, T.M., Levy, D.B., Smith, K.J., Preisinger, A.C., Hedge, P., McKechnie, D., Finniear, R., Markham, A., Groffen, J., Boguski, M.S., Altschul, S.F., Horii, A., Ando, H., Miyoshi, Y., Miki, Y., Nishisho, I., and Nakamura, Y. 1990. Identification of FAP locus genes from chromosome 5q21. *Science* 253:661–665.

Kinzler, K.W., Nilbert, M.C., Vogelstein, B., Bryan, T.M., Levy, D.B., Smith, K.J., Preisinger, A.C., Hamilton, S.R., Hedge, P., Markham, A., Carlson, M., Joslyn, G., Groden, J., White, R., Miki, Y., Miyoshi, Y., Nishisho, I., and Nakamura, Y. 1991. Identification of a gene located at chromosome 5q21 that is mutated in colorectal cancers. *Science* 251:1366–1370.

Leach, F.S., Nicolaides, N.C., Papadopoulos, N., Liu, B., Jen, J., Parsons, R., Peitomaki, P., Sistonen, P., Aaltonen, L.A., Nystrom- Lahti, M., Guan, X.-Y., Zhang, J., Meltzer, P.S., Yu, J.-W., Kao, F.-T., Chen, D.J., Cerosaletti, K.M., Fournier, R.E.K., Todd, S., Lewis, T., Leach, R.J., Naylor, S.L., Weissenbach, J., Mecklin, J.-P., Jarvinen, H., Petersen, G.M., Hamilton, S.R., Green, J., Jass, J., Watson, P., Lynch, H.T., Trent, J.M., de la Chapelle, A., Kinzler, K.W., and Vogelstein, B. 1993. Mutations of a *mutS* homolog in hereditary nonpolyposis colorectal cancer. *Cell* 75:1215–1225.

Miyoshi, Y., Ando, H., Nagase, H., Nishisho, I., Horii, A., Miki, Y., Mori, T., Utsunomiya, J., Baba, S., Petersen, G., Hamilton, S.R., Kinzler, K.W., Vogelstein, B., and Nakamura, Y. 1992. Germ-line mutations of the *APC* gene in 53 familial adenomatous polyposis patients. *Proc. Natl. Acad. Sci. USA* 89:4452–4456.

Nishisho, I., Nakamura, Y., Miyoshi, Y., Miki, Y., Ando, H., Horii, A., Koyama, K., Utsunomiya, J., Baba, S., Hedge, P., Markham, A., Krush, A.J., Petersen, G., Hamilton, S.R., Nilbert, M.C., Levy, D.B., Bryan, T.M., Preisinger, A.C., Smith, K.J., Su, L.-K., Kinzler, K.W., and Vogelstein, B. 1991. Mutations of chromosome 5q21 genes in FAP and colorectal cancer patients. *Science* 253:665–669.

Papadopoulos, N., Nicolaides, N.C., Wei, Y.-F., Ruben, S.M., Carter, K.C., Rosen, C.A., Haseltine, W.A., Fleischmann, R.D., Fraser, C.M., Adams, M.D., Venter, J.C., Hamilton, S.R., Petersen, G.M., Watson, P., Lynch, H.T., Peltomaki, P., Mecklin, J.-P., de la Chapelle, A., Kinzler, K.W., and Vogelstein, B. 1994. Mutation of a *mutL* homolog in hereditary colon cancer. *Science* 263:1625–1629.

Powell, S.M., Silz, N., Beazer-Barclay, Y., Bryan, T.M., Hamilton, S.R., Thibodeau, S.N., Vogelstein, B., and Kinzler, K.W. 1992. *APC* mutations occur early during colorectal tumorigenesis. *Nature* 359:235–237.

Su, L.-K., Kinzler, K.W., Vogelstein, B., Preisinger, A.C., Moser, A.R., Luongo, C., Gould, K.A., and Dove, W.F. 1992. Multiple intestinal neoplasia caused by a mutation in the murine homolog of the *APC* gene. *Science* 256:668–670.

Tumor Suppressor Genes in Other Inherited Cancers

Call, K.M., Glaser, T., Ito, C.Y., Buckler, A.J., Pelletier, J., Haber, D.A., Rose, E.A., Krai, A., Yeger, H., Lewis, W.H., Jones, C., and Housman, D.E. 1990. Isolation and characterization of a zinc finger polypeptide gene at the human chromosome 11 Wilms tumor locus. *Cell* 60:509–520.

Cawthon, R.M., Weiss, R., Xu, G., Viskochil, D., Culver, M., Stevens, J., Robertson, M., Dunn, D., Gesteland, R., O'Connell, P., and White, R. 1990. A major segment of the neurofibromatosis type 1 gene: cDNA sequence, genomic sequence, and point mutations. *Cell* 62:193–201.

Dowdy, S.F., Fasching, C.L., Araujo, D., Lai, K.-M., Livanos, E., Weissman, B.E., and Stanbridge, E.J. 1991. Suppression of tumorigenicity in Wilms tumor by the p15.5-p14 region of chromosome 11. *Science* 254:293–295.

Gessler, M., Poustka, A., Cavenee, W., Neve, R.L., Orkin, S.H., and Bruns, G.A.P. 1990. Homozygous deletion in Wilms tumours of a zinc-finger gene identified by chromosome jumping. *Nature* 343:774–778.

Haber, D.A., Buckler, A.J., Glaser, T., Call, K.M., Pelletier, J., Sohn, R.L., Douglass, E.C., and Housman, D.E. 1990. An internal deletion within an 11p13 zinc finger gene contributes to the development of Wilms' tumor. *Cell* 61:1257–1269.

Haber, D.A., Park, S., Maheswaran, S., Englert, C., Re, G.G., Hazen-Martin, D.J., Sens, D.A., and Garvin, A.J. 1993. *WT1*-mediated growth suppression of Wilms tumor cells expressing a *WT1* splicing variant. *Science* 262:2057–2059.

Haber, D.A., Timmers, H. Th. M., Pelletier, J., Sharp, P.A., and Housman, D.E. 1992. A dominant mutation in the Wilms tumor gene *WT1* cooperates with the viral oncogene *E1A* in transformation of primary kidney cells. *Proc. Natl. Acad. Sci. USA* 89:6010–6014.

Kamb, A., Gruis, N.A., Weaver-Feldhaus, J., Liu, Q., Harshman, K., Tavtigian, S.V., Stockert, E., Day, R.S., III, Johnson, B.E., and Skolnick, M.H. 1994. A cell cycle regulator potentially involved in genesis of many tumor types. *Science* 264:436–440.

Latif, F., Tory, K., Gnarra, J., Yao, M., Duh, F.-M., Orcutt, M.L., Stackhouse, T., Kuzmin, I., Modi, W., Geil, L., Schmidt, L., Zhou, F., Li, H., Wei, M.H., Chen, F., Glenn, G., Choyke, P., Walther, M.M., Weng, Y., Duan, D.-S.R., Dean, M., Glavac, D., Richards, F.M., Crossey, P.A., Ferguson-Smith, M.A., Le Paslier, D., Chumakov, I., Cohen, D., Chinault, A.C., Maher, E.R., Linehan, W.M., Zbar, B., and Lerman, M.I. 1993. Identification of the von Hippel-Lindau disease tumor suppressor gene. *Science* 260:1317–1320.

Li, Y., Bollag, G., Clark, R., Stevens, J., Conroy, L., Fults, D., Ward, K., Friedman, E., Samowitz, W., Robertson, M., Bradley, P., McCormick, F., White, R., and Cawthon, R. 1992. Somatic mutations in the neurofibromatosis gene in human tumors. *Cell* 69:275–281.

Little, M.H., Prosser, J., Condie, A., Smith, P.J., van Heyningen, V., and Hastie, N.D.

1992. Zinc finger point mutations within the *WT1* gene in Wilms tumor patients. *Proc. Natl. Acad. Sci. USA* 89:4791–4795.

Nobori, T., Miura, K., Wu, D.J., Lois, A., Takabayashi, K., and Carson, D.A. 1994. Deletions of the cyclin-dependent kinase-4 inhibitor gene in multiple human cancers. *Nature* 368:753–756.

Rose, E.A., Glaser, T., Jones, C., Smith, C.L., Lewis, W.H., Call, K.M., Minden, M., Champagne, E., Bonetta, L., Yeger, H., and Housman, D.E. 1990. Complete physical map of the WAGR region of 11p13 localizes a candidate Wilms tumor gene. *Cell* 60:495–508.

Rouleau, G.A., Merel, P., Lutchman, M., Sanson, M., Zucman, J., Marineau, C., Hoang-Xuan, K., Demczuk, S., Desmaze, C., Plougastel, B., Pulst, S.M., Lenoir, G., Bijisma, E., Fashold, R., Dumanski, J., de Jong, P., Parry, D., Eldrige, R., Aurias, A., Delattre, O., and Thomas, G. 1993. Alteration in a new gene encoding a putative membrane-organizing protein causes neurofibromatosis type 2. *Nature* 363:515–521.

Trofatter, J.A., MacCollin, M.M., Rutter, J.L., Murrell, J.R., Duyao, M.P., Parry, D.M., Eldridge, R., Kley, N., Menon, A.G., Pulaski, K., Haase, V.H., Ambrose, C.M., Munroe, D., Bove, C., Haines, J.L., Martuza, R.L., MacDonald, M.E., Seizinger, B.R., Short, M.P., Buckler, A.J., and Gusella, J.F. 1993. A novel moesin-, ezrin-, radixin-like gene is a candidate for the neurofibromatosis 2 tumor suppressor. *Cell* 72:791–800.

Viskochil, D., Buchberg, A.M., Xu, G., Cawthon, R.M., Stevens, J., Wolff, R.K., Culver, M., Carey, J.C., Copeland, N.G., Jenkins, N.A., White, R., and O'Connell, P. 1990. Deletions and a translocation interrupt a cloned gene at the neurofibromatosis type 1 locus. *Cell* 62:187–192.

Wallace, M.R., Marchuk, D.A., Andersen, L.B., Letcher, R., Odeh, H.M., Saulino, A.M., Fountain, J.W., Brereton, A., Nicholson, J., Mitchell, A.L., Brownstein, B.H., and Collins, F.S. 1990. Type 1 neurofibromatosis gene: identification of a large transcript disrupted in three NF1 patients. *Science* 249:181–186.

Candidates for Additional Tumor Suppressors

Brodeur, G.M., Sekhon, G., and Goldstein, M.N. 1977. Chromosomal aberrations in human neuroblastomas. *Cancer* 40:2256–2263.

Carlson, K.M., Dou, S., Chi, D., Scavarda, N., Toshima, K., Jackson, C.E., Wells, S.A., Jr., Goodfellow, P.J., and Donis-Keller, H. 1994. Single missense mutation in the tyrosine kinase catalytic domain of the *RET* protooncogene is associated with multiple endocrine neoplasia type 2B. *Proc. Natl. Acad. Sci. USA* 91:1579–1583.

Gailani, M.R., Bale, S.J., Leffell, D.J., DiGiovanna, J.J., Peck, G.L., Poliak, S., Drum, M.A., Pastakia, B., McBride, O.W., Kase, R., Greene, M., Mulvihill, J.J., and Bale, A.E. 1992. Developmental defects in Gorlin syndrome related to a putative tumor suppressor gene on chromosome 9. *Cell* 69:111–117.

Hall, J.M., Lee, M.K., Newman, B., Morrow, J.E., Anderson, L.A., Huey, B., and King, M.-C. 1990. Linkage of early-onset familial breast cancer to chromosome 17q21. *Science* 250:1684–1689.

Hofstra, R.M.W., Landsvater, R.M., Ceccherini, I., Stulp, R.P., Stelwagen, T., Luo, Y., Pasini, B., Hoppener, J.W.M., van Amstel, H.K.P., Romeo, G., Lips, C.J.M., and

Buys, C.H.C.M. 1994. A mutation in the *RET* proto-oncogene associated with multiple endocrine neoplasia type 2B and sporadic medullary thyroid carcinoma. *Nature* 367:375–376.

Kok, K., Osinga, J., Carritt, B., Davis, M.B., van der Hout, A.H., van der Veen, A.Y., Landsvater, R.M., de Leij, L.F.M.H., Berendsen, H.H., Postmus, P.E., Poppema, S., and Buys, C.H.C.M. 1987. Deletion of a DNA sequence at the chromosomal region 3p21 in all major types of lung cancer. *Nature* 330:578–581.

Larsson, C., Skogseid, B., Oberg, K., Nakamura, Y., and Nordenskjold, M. 1988. Multiple endocrine neoplasia type 1 gene maps to chromosome 11 and is lost in insulinoma. *Nature* 332:85–87.

Mulligan, L.M., Kwok, J.B.J., Healey, C.S., Elsdon, M.J., Eng, C., Gardner, E., Love, D.R., Mole, S.E., Moore, J.K., Pali, L., Ponder, M.A., Telenius, H., Tunnacliffe, A., and Ponder, B.A.J. 1993. Germ-line mutations of the *RET* proto-oncogene in multiple endocrine neoplasia type 2A. *Nature* 363:458–460.

Whang-Peng, J., Kao-Shan, C.S., Lee, E.C., Bunn, P.A., Carney, D.N., Gazdar, A.F., and Minna, J.D. 1982. Specific chromosome defect associated with human small-cell lung cancer; deletion 3p(14–23). *Science* 215:181–182.

❖ CHAPTER 11

Role of Oncogenes and Tumor Suppressor Genes in the Pathogenesis of Neoplasms

The preceding chapters have dealt with a number of cellular oncogenes that are activated in neoplasms by several distinct types of molecular alterations, as well as tumor suppressor genes whose inactivation contributes to tumorigenesis. The reproducibility and frequency of some of these genetic alterations strongly supports their role in the pathogenesis of malignant neoplasms. However, as indicated in chapter 1, the development of neoplasms *in vivo* is a multistep process in which normal cells gradually acquire the fully neoplastic phenotype through a series of progressive changes. The multistep nature of tumor pathogenesis is illustrated by the fact that the incidence of most human neoplasms increases dramatically with age, suggesting that the accumulation of multiple lesions is required to generate a tumor. The experimental induction of neoplasms by chemical carcinogens similarly requires a long latent period and can, in some experimental systems, be separated into distinct stages of tumor initiation and promotion. Additionally, in some human and animal neoplasms, it is possible to identify preneoplastic stages of the disease process, characterized by increased proliferation of nonmalignant cell populations.

The complete sequence of steps leading to the development of malignant neoplasms is not yet known. However, most tumors are clonal growths, consisting of the progeny of a single initially transformed cell. Development of a malignant neoplasm then results from a series of progressive alterations, culminating in unregulated autonomous cell proliferation, invasiveness, and the ability to metastasize to other sites.

How then does the activation of any single oncogene, or the inactivation of a tumor suppressor gene, fit into such a multistep pathway of tumorigenesis?

One clear possibility is that abnormalities resulting from mutation of a single oncogene or tumor suppressor gene are not sufficient for expression of the full neoplastic phenotype but are only one step in neoplasm pathogenesis. Consistent with this notion, a number of human and animal neoplasms have been found to contain different combinations of activated oncogenes plus inactivated tumor suppressor genes, suggesting the possibility that damage to different genes corresponds to distinct steps in tumor development.

This chapter deals with the potential roles of cellular oncogenes and tumor suppressor genes, both alone and in combination, in various stages of tumor pathogenesis. Animal model systems are described first because they are amenable to experimental manipulations that have provided direct data on the function of oncogenes during tumor development. We will then consider representative human neoplasms in which consistent patterns of oncogene activation and tumor suppressor gene inactivation suggest specific roles for some of these genes in tumor initiation or progression.

Biological Activities of Oncogenes *in Vitro*

The notion that activation of a single oncogene is not sufficient for tumorigenesis *in vivo* at first appears to contradict the ability of these genes to induce neoplastic transformation of cultured cells. However, this is presumably a reflection of more stringent requirements for the formation of a tumor in a whole animal, compared with transformation of cells in culture. For example, chapter 9 noted that hybrids between normal and malignant cells are generally nontumorigenic but still behave as transformed cells *in vitro*. Moreover, in most gene transfer experiments, the cells used as recipients are not truly normal cells but are established cell lines that have already acquired at least one phenotypic characteristic of tumor cells—immortality in culture. Such cells may already have undergone some of the events required for transformation of a normal cell and, therefore, may be particularly susceptible to transformation by single oncogenes. The notion that established cell lines may be partially transformed prompted investigators to utilize normal primary cells, rather than immortalized cell lines, as recipients in gene transfer experiments.

The initial studies of the transforming activities of cellular oncogenes in primary rat embryo fibroblasts were reported in 1983 from the laboratories of Robert Weinberg and Earl Ruley (Fig. 11.1). Under some conditions, single oncogenes (including activated *ras* and *myc* genes) failed to induce transformation of primary cells alone. However, under the same conditions, transformation was induced by combinations of the same oncogenes—for example, *ras* and *myc* together. These studies thus provided a more stringent cell culture model in which the combined actions of two different oncogenes were required

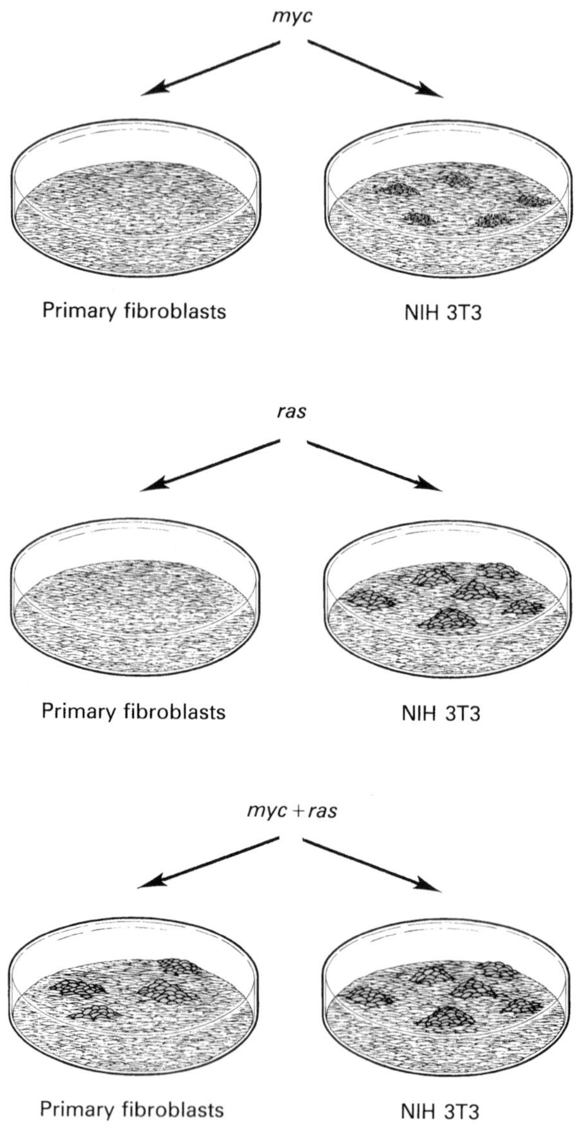

FIGURE 11.1
Cooperative transformation of primary embryo fibroblasts by two oncogenes. Either *ras* or *myc* oncogenes alone can transform established cell lines such as NIH 3T3 cells, although the morphological alterations induced by *myc* are less striking than those induced by *ras*. Transformation of primary fibroblasts, however, frequently requires the combined action of two different oncogenes.

for transformation, perhaps analogous to two different steps in the pathogenesis of neoplasms *in vivo*.

It was initially suggested that the *ras* and *myc* oncogenes functioned to alter different aspects of cell behavior and thus formed two distinct physiological complementation groups. In particular, *myc* was thought to function in maintaining proliferation of primary cells in culture (immortalization), whereas *ras* was thought to induce other aspects of the transformed phenotype, such as altered cell morphology and anchorage independence. Further experiments have indicated, however, that under other conditions either *ras* or *myc* alone can induce the full range of phenotypic alterations associated with transformation and can even induce full transformation of primary cells by themselves. For example, a *ras* oncogene alone can fully transform primary fibroblasts, provided that it is expressed at high levels. The requirement for two oncogenes to induce primary cell transformation under stringent conditions does not, therefore, unambiguously divide the cooperating oncogenes into groups with distinct physiological functions. Nonetheless, cooperative transforming activities of *ras* and *myc*, as well as other pairs of oncogenes, have been demonstrated in a number of *in vitro* culture systems and therefore appear to provide an important model for the possible combined effects of multiple oncogenes in neoplasia. In particular, these experiments clearly illustrate the potential contribution of multiple oncogene activation to multistep pathways of tumor development.

Induction of Neoplasms by Oncogenes *in Vivo*

Although cell culture experiments provide easily manipulated systems for analysis of oncogene activities, they are also obviously oversimplified approximations to the development of neoplasms in animals. In order to directly assess the role of oncogenes in tumor pathogenesis, it is clearly necessary to undertake experiments in whole animal systems. The most clear-cut analysis would be provided by the introduction of one or more activated oncogenes into normal cells *in vivo* so that the subsequent development of neoplasms directly induced by the oncogene(s) could be evaluated. This goal has been achieved by two approaches: the generation of transgenic mice and the infection of stem cells with retroviral vectors.

The first experiment to address cellular oncogene function in transgenic animals was reported in 1984 from the laboratory of Philip Leder (Fig. 11.2). As discussed earlier (chapters 6–8), the c-*myc* gene and other members of the *myc* family are activated in a variety of neoplasms as a consequence of DNA rearrangements or other alterations that lead to unregulated constitutive gene expression. Leder and his colleagues mimicked cellular *myc* activation by generating molecular constructs in which c-*myc* was expressed by the mouse mammary tumor virus (MMTV) LTR, a strong promoter inducible by glucocorticoid hormones. These constructs were microinjected into pronuclei of fertilized mouse eggs and several transgenic offspring, which contained the MMTV/*myc* constructs integrated into germ line DNA, were obtained. Two of the original MMTV/*myc* transgenic animals as well as their progeny developed

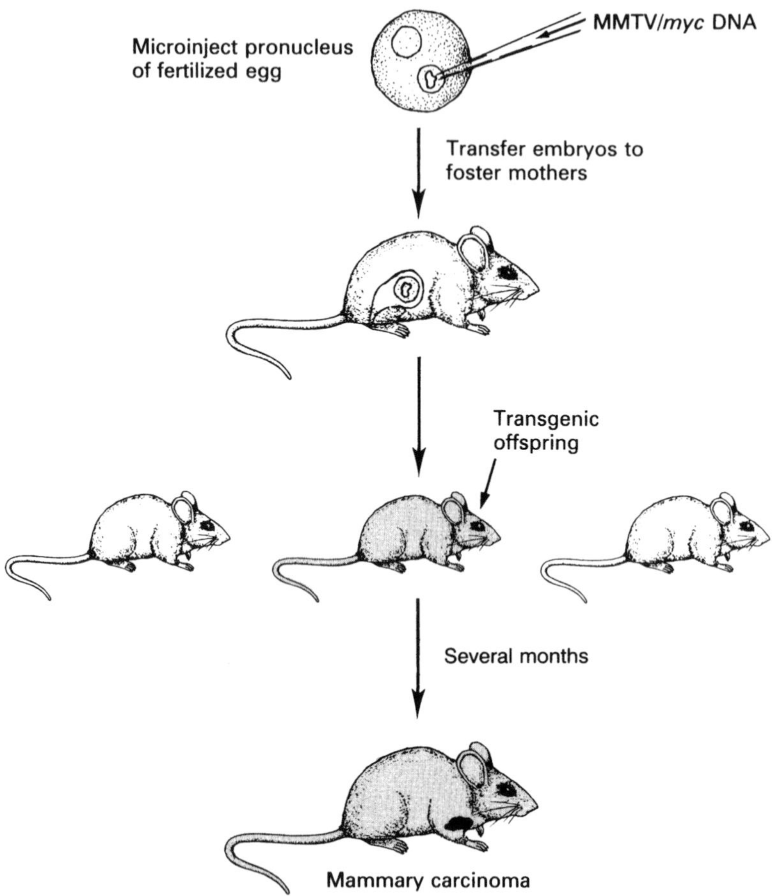

Figure 11.2
Tumor development in transgenic mice expressing the *myc* oncogene. A c-*myc* gene expressed from a mouse mammary tumor virus LTR (MMTV/*myc* DNA) is microinjected into pronuclei of fertilized mouse eggs. Injected eggs are transferred to foster mothers and some transgenic offspring that have incorporated the MMTV/*myc* DNA into their genomes are obtained. Mice that express the MMTV/*myc* transgene develop mammary carcinomas after a latent period of several months.

mammary carcinomas, providing the first direct experimental evidence for the ability of a cellular oncogene to induce neoplastic disease in an intact animal.

Importantly, however, the MMTV/*myc* transgene was clearly not the only factor required for tumor development in these experiments. Mammary carcinomas did not develop until several months of age, indicating that there was a substantial latent period for tumor formation, during which other pathogenic steps would be expected to occur. Furthermore, only a single tumor generally developed in one of the ten mammary glands of the transgenic animals, even though all of the cells in the animal contained the MMTV/*myc* transgene. Thus, the MMTV/*myc* transgene appeared to be necessary but not sufficient for development of mammary carcinomas in these mice. In this model system, expression of the *myc* oncogene thus appeared to be one step in a multistep pathway of carcinogenesis.

Translocation of the c-*myc* gene to the immunoglobulin loci is especially strongly associated with B-cell neoplasms, Burkitt's lymphoma in humans and plasmacytoma in mice (see chapter 7). These events have been mimicked in a transgenic mouse model, first developed in the laboratory of Jerry Adams and Suzanne Cory in collaboration with Richard Palmiter and Ralph Brinster in 1985. In these experiments, the c-*myc* gene was linked to the immunoglobulin heavy chain enhancer, equivalent to the *myc* rearrangements that occur in plasmacytomas as a result of chromosome translocation. Mice that were transgenic for this construct developed B-cell lymphomas, demonstrating a direct role for *myc* in the genesis of the same neoplasms in which activated *myc* genes have been reproducibly detected in several species. As was the case with *myc*-induced mammary carcinomas, these lymphomas developed as monoclonal neoplasms after latent periods of several months. In addition, further studies have indicated that the development of lymphomas follows a preneoplastic phase of enhanced B-cell proliferation. Thus, *myc* appears to induce the first preneoplastic stage of lymphomagenesis, which needs to be followed by additional events prior to development of the full neoplastic phenotype.

The transgenic mouse model has now been used to investigate the activities of a variety of oncogenes in addition to c-*myc*, and more than two dozen oncogenes have thus been shown to induce a variety of different types of tumors *in vivo*. As in the studies of c-*myc*, tumors generally develop as apparently monoclonal neoplasms after long latent periods during which preneoplastic cell proliferation can be observed, indicating that expression of the transgenic oncogene is only the first event in a multistep process leading to malignancy. In a number of cases, second activated oncogenes have been identified in the tumors arising in these transgenic mice. For example, B-cell lymphomas arising in *myc* transgenic mice have been found to contain activated *ras*, *pim*-1, or *bmi*-1 oncogenes, indicating that activation of these oncogenes is a second step in lymphomagenesis. Conversely, activation of c-*myc* is frequently a second step in the development of lymphomas in mice transgenic for the *bcl*-2 oncogene.

The transgenic mouse model has also been used to directly demonstrate cooperation between oncogenes *in vivo* (Fig. 11.3). An *in vivo* analogy of *myc* + *ras* cotransformation of primary fibroblasts in culture was provided by crossing two different strains of transgenic mice, one carrying MMTV/*myc* and the other carrying MMTV/*ras*H. The progeny of this cross, which inherited both

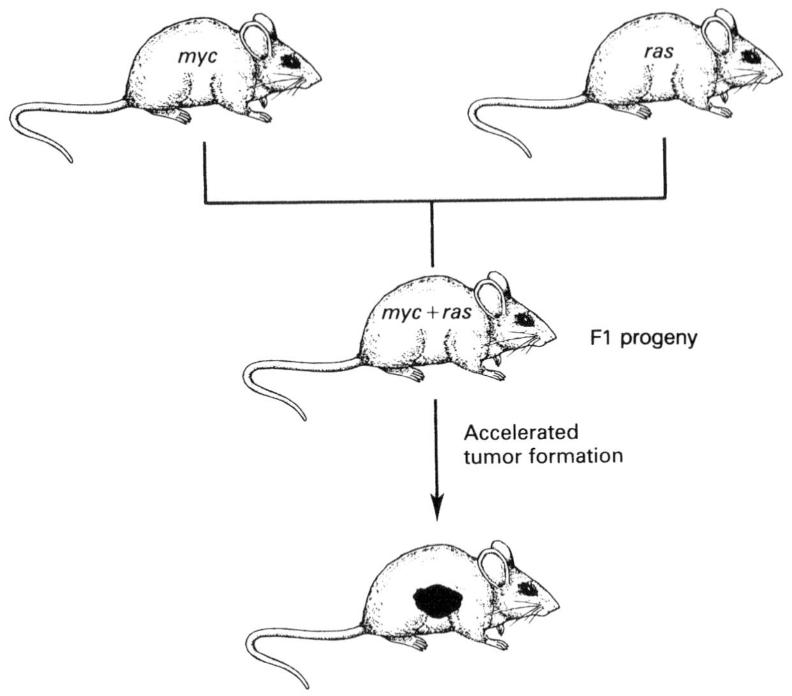

FIGURE 11.3
Cooperation between *myc* and *ras* oncogenes in transgenic mice. Strains of transgenic mice bearing either MMTV/*myc* or MMTV/*ras* transgenes develop tumors at several months of age. Crossing these two strains yields mice carrying both MMTV/*myc* and MMTV/*ras*. Cooperation between *myc* and *ras* oncogenes is indicated by accelerated tumor development in these mice compared with the parental strains.

myc and *ras* oncogenes, developed tumors with much shorter latent periods than either of the parental lines, indicating a cooperative interaction between these two oncogenes in the genesis of tumors in whole animals. However, the *myc* + *ras* induced tumors still develop over a period of about two months and are monoclonal, indicating the need for further steps in neoplasm pathogenesis beyond the activation of these two oncogenes. Such accelerated tumorigenesis is not restricted to the *myc* + *ras* combination but has been similarly documented for a number of other oncogene pairs.

The results that have emerged from studies of transgenic mice are fortified by experiments in which cellular oncogenes have been introduced into embryos or discrete stem cell populations with the use of retroviral vectors. For example, introduction of activated *myc* genes into stem cells used to reconstitute a chicken bursa, the system in which c-*myc* activation was first described, results in the outgrowth of preneoplastic lymphocyte nodules, only a small fraction of which progress to malignant lymphomas. Similarly, introduction of *ras* or *myc* oncogenes alone into cells used to reconstitute a mouse prostate gland results in the proliferation of preneoplastic cell populations, whereas the combination of *ras* + *myc* oncogenes induces carcinomas. Nonetheless, the tumors induced by the *myc* + *ras* combination are monoclonal, implying the need for still further events in tumorigenesis.

Taken together, these experiments clearly demonstrate that oncogenes can induce tumorigenesis in intact animals. Moreover, it appears that activation of a single oncogene corresponds to only one step in the multistep process of neoplasm development. The ability of two different oncogenes to contribute synergistically to tumorigenesis has been demonstrated in several experimental settings, suggesting that activation of multiple oncogenes in series could be involved in tumor evolution. For example, activation of one oncogene could lead to expansion of a preneoplastic cell population and activation of a second oncogene could result in the progression of preneoplastic cells to malignancy. However, it has not so far been possible to identify oncogene combinations that result in acute neoplasm formation in animals without the need for still further, currently undefined, events, likely including loss of tumor suppressor genes.

Oncogenes and Tumor Suppressor Genes in Tumors Induced by Chemical Carcinogens

In the transgenic mouse models discussed in the preceding section, the inherited oncogenes apparently initiate the process of tumorigenesis by driving the proliferation of a preneoplastic cell population from which neoplastic cells eventually arise. In these cases, the role of the oncogene in initiation of tumor development is clearly defined, because germ line transmission of an activated oncogene has been experimentally used to induce neoplasia. The roles of tumor suppressor genes in inherited cancers resulting from germ line mutations of these genes is similarly evident, as is the role of germ line mutations that activate the *ret* oncogene in cases of inherited multiple endocrine neoplasia type 2 (discussed in chapter 10). An independent question concerns the roles of oncogenes and tumor suppressor genes in noninherited neoplasms. Animal models have again provided important information because tumors can be induced reproducibly by known carcinogenic agents, allowing the analysis of oncogene activation at discrete stages of the neoplastic process.

Chemical carcinogenesis in particular has provided an experimental system in which activation of oncogenes, particularly members of the *ras* gene family, can be directly related to the initiation of neoplastic disease. One of the classic models for multistep carcinogenesis is the induction of tumors in mouse skin by the topical application of chemical carcinogens (Fig. 11.4). Tumorigenesis in this system can be divided experimentally into distinct stages of initiation and promotion. Initiation is an irreversible step resulting from a single carcinogen treatment. Tumors do not develop, however, unless the mice are subsequently treated with a tumor promoter, usually a phorbol ester. Treat-

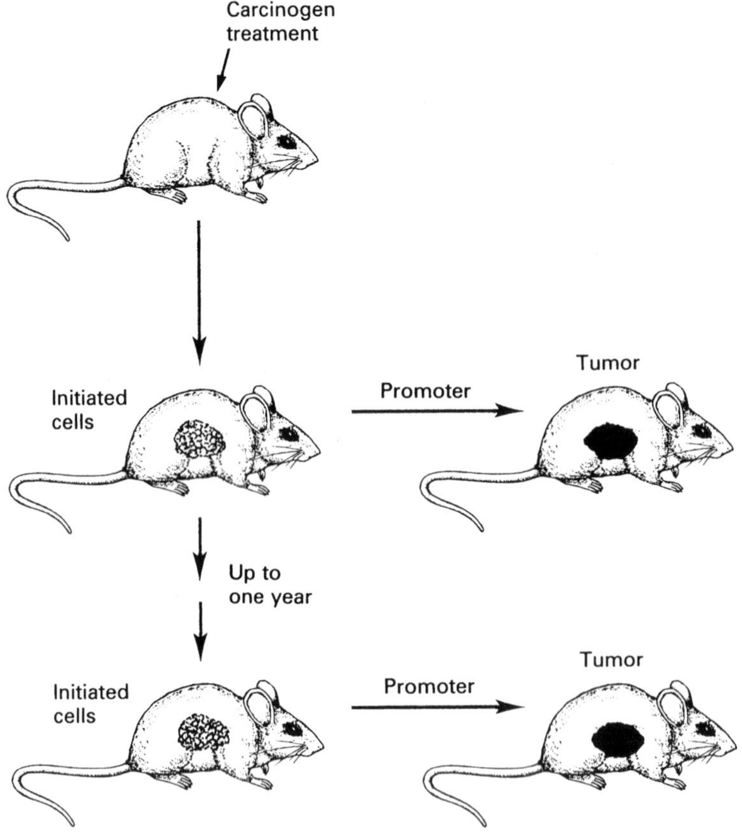

FIGURE **11.4**
Two-step carcinogenesis in mouse skin. Treatment with a carcinogen (for example, dimethylbenzanthracene) leads to the irreversible generation of initiated cells, presumably as a consequence of carcinogen-induced mutations. Neoplasm development, however, requires subsequent treatment with a tumor promoter (for example, 12-*O*-tetradecanoyl-phorbol-13-acetate, or TPA). Tumors will develop even if promoter treatment is delayed for as long as a year after carcinogen exposure.

ment with tumor promoters can take place up to a year after carcinogen treatment, indicating that cells with the initiated phenotype persist in the skin and are irreversibly committed to eventual neoplastic progression. Because the carcinogens active in initiation are mutagens, it is likely that initiation results from direct damage to DNA. In contrast, the tumor promoter phorbol esters are not mutagens but act to stimulate cell proliferation. They are therefore thought to lead to neoplasm development by stimulating the growth of initiated cells. The tumors that develop following initiation and promotion are primarily benign papillomas. Some of these evolve further to malignant carcinomas, illustrating yet another stage in tumor progression.

Given the utility of the mouse skin system in evaluating progressive preneoplastic and neoplastic stages, it was of considerable interest when Allan Balmain and his colleagues reported in 1983 that skin carcinomas induced in this system reproducibly contained activated *ras*H oncogenes. This was soon followed by the finding that *ras* activation was detected not only in malignant carcinomas, but also in benign skin papillomas. These results indicated that activation of *ras*H was an early event in carcinogenesis, occurring in premalignant lesions prior to their progression to malignancy.

The reproducible activation of *ras* genes in chemically induced tumors is not limited to skin carcinomas. Similar high frequencies of *ras* activation (50%–100% of tumors) have been found in other chemical carcinogen-induced neoplasms, including mammary and lung carcinomas, lymphomas, and hepatomas. The reproducibility of *ras* activation in tumors induced by these agents raised the possibility that *ras* proto-oncogenes could be the direct targets for carcinogen-induced mutation in these systems. This hypothesis has been supported by analysis of the specific activating mutations in *ras* genes isolated from tumors induced by different carcinogens. The first such study was reported by Mariano Barbacid and colleagues in 1985. These investigators found that multiple independent mammary carcinomas induced by the alkylating agent methylnitrosourea (MNU) reproducibly contained *ras*H genes activated by the same point mutation: a G→A substitution resulting in the change of codon 12 from GGA (glycine) to GAA (glutamic acid). Because MNU specifically induces G→A transitions, these results suggested that carcinogen induced mutation directly activated the *ras*H oncogene to initiate carcinogenesis. This conclusion was further supported by the finding that mammary carcinomas induced by a different carcinogen (DMBA, dimethylbenzanthracene) contained *ras* genes activated by distinct mutations, consistent with the known mutagenic activity of DMBA.

Similar observations have been made in other tumor systems, including skin carcinomas, hepatomas, and lung carcinomas. In each case, different chemical carcinogens activate *ras* genes by inducing specific base substitutions consistent with their known mutagenic activities (Table 11.1). In addition, the

TABLE 11.1
ras Oncogene Activation by Chemical Carcinogens

Neoplasm	Carcinogen	Activating Mutation
rat mammary carcinoma	MNU	*ras*H codon 12 G→A
	DMBA	*ras*H codon 61 A→N
mouse skin carcinoma	DMBA	*ras*H codon 61 A→T
mouse hepatoma	*N*-HO-AAF	*ras*H codon 61 C→A
	VC	*ras*H codon 61 A→T
	HO-DHE	*ras*H codon 61 A→T/G
mouse lung carcinoma	MNU	*ras*K codon 12 G→A
	BP	*ras*K codon 12 G→T/A
	EC	*ras*K codon 61 A→T/G

Abbreviations: MNU, methylnitrosourea; DMBA, 7,12- dimethylbenzanthracene; *N*-HO-AAF, *N*-hydroxy-2-acetylaminofluorene; VC, vinyl carbamate; HO-DHE, 1'-hydroxy-2',3'-dehydroestragole; BP, benzpyrene; EC, ethyl carbamate.

use of sensitive polymerase chain reaction (PCR) assays has allowed *ras* mutations to be detected as early as two weeks after carcinogen treatment, substantially before the appearance of neoplastic cells. These findings strongly indicate that *ras* genes are direct targets for mutations induced by chemical carcinogens at the initiation stage of tumor development in animal neoplasms.

The analysis of *ras* oncogenes in human tumors also suggests that *ras* mutations may be induced by the action of chemical carcinogens. For example, *ras*K oncogenes in lung cancers are often activated by different mutations from those in colon and pancreatic cancers, leading to the suggestion that different carcinogens induced distinct *ras* mutations in these tumors.

The clearest examples of carcinogen-induced mutations in human tumors, however, are provided by mutations of the *p53* tumor suppressor gene. Exposure to aflatoxin B_1, a potent chemical carcinogen produced by certain molds and found in contaminated food supplies, is a major risk factor for liver cancer in areas of China and southern Africa. A high proportion of liver cancers from patients in these areas have mutations of *p53* resulting from G→T transversions—the principal type of mutation produced by aflatoxin. Because *p53* genes in other types of tumors are inactivated by different mutations, it appears that *p53* mutations in these liver cancers are induced directly by aflatoxin. Moreover, it is notable that distinct types of *p53* mutations also occur in several other kinds of tumors, such as lung and colon carcinomas, consistent with the involvement of different carcinogens. In addition, *p53* mutations in skin cancers are characteristic of those induced by UV light—the principal causative agent of skin cancer. It thus appears that oncogenes and tumor suppressor genes in human tumors, as well as animal models, can be direct targets for the action of mutagenic carcinogens.

Oncogenes and Tumor Suppressor Genes in Development of Human Neoplasms

Understanding the role of oncogenes in the development of human neoplasms is of necessity based on less-direct data than are obtained from animal model systems because experimental manipulations are impossible. However, as discussed in earlier chapters, a number of oncogenes and tumor suppressor genes are reproducibly mutated in a variety of different human tumors. In some of these cases, sufficient data exist to correlate alterations in specific oncogenes and tumor suppressor genes with discrete stages of human tumor development.

The role of c-*myc* in the pathogenesis of Burkitt's lymphoma is particularly interesting in comparison with some of the animal models discussed earlier. Burkitt's lymphoma is a human B-cell neoplasm in which the c-*myc* gene is uniformly activated by translocation to immunoglobulin gene loci (see chapter 7). This disease occurs with high incidence in some areas of Africa, where it is strongly associated with infection by Epstein-Barr virus (EBV), a herpesvirus discussed in chapter 2. Not only is EBV implicated as a causative agent of Burkitt's lymphoma on this epidemiological basis, but EBV has also been shown to transform human lymphocytes in culture, apparently as a direct consequence of viral oncogene expression. Thus, both EBV infection and c-*myc* translocation figure in the pathogenesis of this neoplasm, most likely representing two distinct events in lymphomagenesis.

An interesting model for the possible roles of EBV and c-*myc* in Burkitt's lymphoma has been provided by cell culture experiments with human B lymphocytes. EBV infection of normal human B cells results in the establishment of immortalized B-cell lines that proliferate continuously in culture but, in contrast with Burkitt's lymphoma cell lines, are nontumorigenic in immunodeficient mice. Introduction of an activated c-*myc* oncogene into such EBV-immortalized cells converts them into the full tumorigenic phenotype, demonstrating a two-step transformation of human lymphocytes *in vitro* by EBV and c-*myc*. These results suggest the hypothesis that lymphomagenesis *in vivo* is likewise initiated by EBV, with c-*myc* translocation responsible for further progression to neoplasia. For example, EBV might induce proliferation of a preneoplastic cell population. The increased numbers of such proliferative lymphocytes might then result in an increased probability of chromosome translocations, leading to c-*myc* activation and progression to malignancy.

Amplification of a different member of the *myc* gene family, N-*myc* (see chapter 8), is involved in the pathogenesis of human neuroblastomas and is strongly associated with the progression of these neoplasms to a highly malignant state. Neuroblastomas are divided into four clinical stages, based on tumor size and metastasis, that are closely correlated with prognosis. For example, the majority of patients with stage I or II disease are likely to be free of further tumor progression over a two-year period, whereas patients with stage III or IV tumors have a high probability of developing still more advanced disease within this time frame. Amplified N-*myc* genes are detected frequently (~50%) in stage III and IV tumors but only rarely (<10%) in stage I and II tumors. Most significantly, N-*myc* amplification within each of these groups is closely correlated with rapid tumor progression. For example, those stage II tumors with N-*myc* amplification have a high likelihood of further progression compared with tumors of the same clinical stage without N-*myc* amplification. Amplification of N-*myc* therefore appears to be a biologically and clinically important event in the progression of human neuroblastoma to increasing malignancy.

Amplification of several oncogenes appears to be related to the progression of other tumors, including c-*myc*, N-*myc*, and L-*myc* in lung carcinomas and *erb*B-2 in breast and ovarian carcinomas. The amplification of *erb*B-2, in particular, is associated with more aggressive tumors and correlates with reduced patient survival.

The *ras* oncogenes appear to play a role in both early and late stages of the genesis of human tumors. As noted in chapter 5, activated *ras* oncogenes are detected in ~15% of human tumors. In most types of tumors, *ras* activation has been detected only sporadically in a small fraction of individual neoplasms. However, *ras* activation is a more frequent event in the development of some types of neoplasms. For example, activated *ras*N genes are detected in ~25% of acute myeloid leukemias; activated *ras*K genes have been found in ~50% of colon carcinomas, ~25% of lung carcinomas, and ~90% of pancreatic carcinomas; and all three members of the *ras* family are frequently activated in thyroid carcinomas. The activation of *ras* genes appears to occur as a relatively early event in the genesis of acute myeloid leukemias, colon carcinomas, pancreatic carcinomas, and thyroid carcinomas, because activated *ras* genes are readily detected at premalignant stages of these diseases. In other neoplasms, however, *ras* activation seems to be a late event in tumor progression. Thus, some neoplasms, including melanomas, hepatomas, leukemias, and colon carcinomas, have been found to contain activated *ras* genes in only a fraction of the malignant cells. The heterogeneity of activated *ras* genes in the cells of such tumors suggests that *ras* activation was a late event in tumor progression that occurred after the initiation of neoplasm development.

Colorectal carcinomas have provided the best characterized model for the role of multiple oncogenes and tumor suppressor genes in tumor development, emanating from the studies of Bert Vogelstein and his colleagues. Activation of *ras*K oncogenes and mutation or deletion of the *APC, DCC,* and *p53* tumor suppressor genes are all frequent events in the development of these tumors, occurring in a high fraction (>50%) of individual colorectal carcinomas. Importantly, colorectal cancers evolve over several years from small benign adenomas (polyps). Because lesions representing several stages of colorectal cancer development are regularly obtained as surgical specimens, it has been possible to correlate the genetic alterations characteristic of these tumors with a series of steps in tumor progression (Fig. 11.5). These studies indicate that mutations of the *APC* tumor suppressor gene occur at an early stage of tumor initiation, because these mutations are already present in the smallest detectable adenomas. Activation of *ras*K oncogenes then appears to occur, and mutated *ras*K genes are also frequently detected in small and intermediate size adenomas. In contrast, inactivation of the *DCC* and *p53* tumor suppressor genes appears to take place in later stages of tumor progression, because mutations of these genes are usually found only in advanced adenomas or, more frequently, in malignant carcinomas.

Although the order of genetic alterations just described is that most frequently observed in colorectal carcinoma development, it is important to note that this order of events is not invariant. For example, inactivation of *p53* has been detected in a few early adenomas. Perhaps of greater significance than the order in which mutations occur is the cumulative effect of damage to multiple growth regulatory genes. In particular, these studies of colorectal cancers clearly indicate that the progression of these tumors from early adenomas to metastatic carcinomas involves the accumulation of genetic damage to both oncogenes and tumor suppressor genes.

Like colorectal carcinomas, a number of other types of human tumors reproducibly contain mutations affecting multiple oncogenes and tumor suppressor genes. Coactivation of both *bcl*-2 and c-*myc* genes may contribute to the development of some human lymphomas, similar to the cooperation of these genes in transgenic mice. Among the most common types of human tumors, breast cancers are frequently associated with inactivation of the *p53* tumor suppressor gene, as well as with amplification of the c-*myc, PRAD*-1, and *erb*B-2 oncogenes. As discussed in chapter 10, a candidate tumor suppressor gene on chromosome 17q21 (*BRCA1*) may also play an important role in breast cancer pathogenesis. Lung cancers are associated with activation of *ras*K oncogenes, amplification of members of the *myc* oncogene family, inactivation of the *Rb* and *p53* tumor suppressor genes, as well as loss of an additional candidate tumor suppressor gene on chromosome 3p21. Although the roles of these genes in tumorigenesis are not yet clear, it would be surprising if the accumulation of such multiple genetic defects did not correlate with progression to malignancy, as in colorectal cancers.

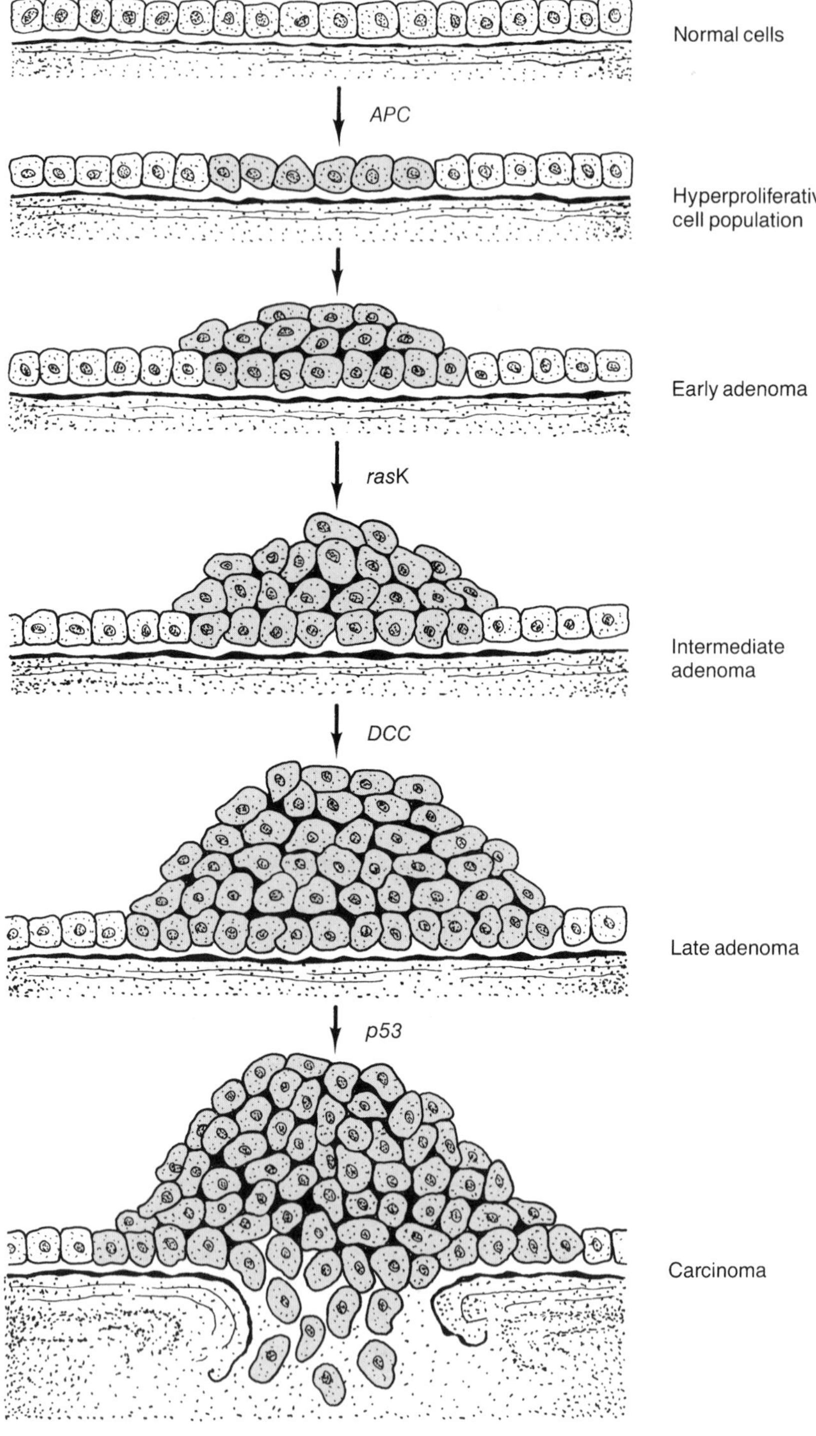
Normal cells
APC
Hyperproliferative cell population
Early adenoma
rasK
Intermediate adenoma
DCC
Late adenoma
p53
Carcinoma

◀ FIGURE 11.5
Genetic alterations in the development of colorectal carcinomas. Colorectal carcinomas develop over a period of several years from small benign adenomas. Mutations of the *APC*, *ras*K, *DCC*, and *p53* genes occur most frequently at the indicated stages of carcinogenesis. Progression to malignancy results from accumulated damage to both oncogenes and tumor suppressor genes.

Summary

Neoplasia is a multistep process in which normal cells become malignant through a series of progressive changes. The role of oncogene activation and tumor suppressor gene inactivation in this process is clearly illustrated in several animal models as well as in some human neoplasms. In these systems, activation of any single oncogene appears to correspond to one step in tumor pathogenesis. In different tumor systems, the same oncogene can participate in either early or late stages of tumor development, so that there is no apparent general association of particular oncogenes with specific stages of pathogenesis. In some neoplasms, however, the activation of specific oncogenes or the inactivation of tumor suppressor genes is a regular and reproducible event in tumor initiation or progression, indicating clear roles for these genes in tumor development. Cooperative effects of two distinct oncogenes have been demonstrated in a number of settings, and the multiple steps in development of malignant tumors appear to correspond to the accumulation of genetic damage affecting both oncogenes and tumor suppressor genes.

References

General Reference

Hunter, T. 1991. Cooperation between oncogenes. *Cell* 64:249–270.

Vogelstein, B., and Kinzler, K.W. 1993. The multistep nature of cancer. *Trends Genet.* 9:138–141.

Biological Activities of Oncogenes *in Vitro*

Hjelle, B., Liu, E., and Bishop, J.M. 1988. Oncogene v-*src* transforms and establishes embryonic rodent fibroblasts but not diploid human fibroblasts. *Proc. Natl. Acad. Sci. USA* 85:4355–4359.

Keath, E.J., Caimi, P.G., and Cole, M.D. 1984. Fibroblast lines expressing activated c-*myc* oncogenes are tumorigenic in nude mice and syngeneic animals. *Cell* 39:339–348.

Land, H., Parada, L.F., and Weinberg, R.A. 1983. Tumorigenic conversion of primary embryo fibroblasts requires at least two cooperating oncogenes. *Nature* 304:596–602.

Ruley, H.E. 1983. Adenovirus early region 1A enables viral and cellular transforming genes to transform primary cells in culture. *Nature* 304:602–606.

Spandidos, D.A., and Wilkie, N.M. 1984. Malignant transformation of early passage rodent cells by a single mutated human oncogene. *Nature* 310:469–475.

Induction of Neoplasms by Oncogenes *in Vivo*

Adams, J.M., and Cory, S. 1991. Transgenic models of tumor development. *Science* 254:1161–1167.

Adams, J.M., Harris, A.W., Pinkert, C.A., Corcoran, L.M., Alexander, W.S., Cory, S., Palmiter, R.D., and Brinster, R.L. 1985. The c-*myc* oncogene driven by immunoglobulin enhancers induces lymphoid malignancy in transgenic mice. *Nature* 318:533–538.

Alexander, W.S., Bernard, O., Cory, S., and Adams, J.M. 1989. Lymphomagenesis in Eμ-*myc* transgenic mice can involve *ras* mutations. *Oncogene* 4:575–581.

Clynes, R., Wax, J., Stanton, L.W., Smith-Gill, S., Potter, M., and Marcu, K.B. 1988. Rapid induction of IgM-secreting murine plasmacytomas by pristane and an immunoglobulin heavy-chain promoter/enhancer-driven c-*myc*/v-Ha-*ras* retrovirus. *Proc. Natl. Acad. Sci. USA* 85:6067–6071.

Compere, S.J., Baldacci, P., Sharpe, A.H., Thompson, T., Land, H., and Jaenisch, R. 1989. The *ras* and *myc* oncogenes cooperate in tumor induction in many tissues when introduced into midgestation mouse embryos by retroviral vectors. *Proc. Natl. Acad. Sci. USA* 86:2224–2228.

Haupt, Y., Alexander, W.S., Barri, G., Klinken, S.P., and Adams, J.M. 1991. Novel zinc finger gene implicated as *myc* collaborator by retrovirally accelerated lymphomagenesis in Eμ-*myc* transgenic mice. *Cell* 65:753–763.

Langdon, W.Y., Harris, A.W., Cory, S., and Adams, J.M. 1986. The c-*myc* oncogene perturbs B lymphocyte development in Eμ-*myc* transgenic mice. *Cell* 47:11–18.

McDonnell, T.J., and Korsmeyer, S.J. 1991. Progression from lymphoid hyperplasia to high-grade malignant lymphoma in mice transgenic for the t(14;18). *Nature* 349:254–256.

Sinn, E., Muller, W., Pattengale, P., Tepler, I., Wallace, R., and Leder, P. 1987. Coexpression of MMTV/v-Ha-*ras* and MMTV/c-*myc* genes in transgenic mice: synergistic action of oncogenes *in vivo*. *Cell* 49:465–475.

Stewart, T.A., Pattengale, P.K., and Leder, P. 1984. Spontaneous mammary adenocarcinomas in transgenic mice that carry and express MTV/*myc* fusion genes. *Cell* 38:627–637.

Strasser, A., Harris, A.W., Bath, M.L., and Cory, S. 1990. Novel primitive lymphoid tumors induced in transgenic mice by cooperation between *myc* and *bcl*-2. *Nature* 248:331–333.

Thompson, C.B., Humphries, E.H., Carlson, L.M., Chen, C.-L. H., and Neiman, P.E. 1987. The effect of alterations in *myc* gene expression on B cell development in the bursa of Fabricius. *Cell* 51:371–381.

Thompson, T.C., Southgate, J., Kitchener, G., and Land, H. 1989. Multistage carcinogenesis induced by *ras* and *myc* oncogenes in a reconstituted organ. *Cell* 56:917–930.

Van Lohuizen, M., Verbeek, S., Scheijen, B., Wientjens, E., van der Gulden, H., and Berns, A. 1991. Identification of cooperating oncogenes in Eμ-*myc* transgenic mice by provirus tagging. *Cell* 65:737–752.

Oncogenes and Tumor Suppressor Genes in Tumors Induced by Chemical Carcinogens

Balmain, A., and Pragnell, I.B. 1983. Mouse skin carcinomas induced *in vivo* by chemical carcinogens have a transforming Harvey-*ras* oncogene. *Nature* 303:72–74.

Balmain, A., Ramsden, M., Bowden, G.T., and Smith, J. 1984. Activation of the mouse cellular Harvey-*ras* gene in chemically induced benign skin papillomas. *Nature* 307:658–660.

Bos, J.L. 1989. *Ras* oncogenes in human cancer: a review. *Cancer Res.* 49:4682–4689.

Brash, D.E., Rudolph, J.A., Simon, J.A., Lin, A., McKenna, G.J., Baden, H.P., Halperin, A.J., and Pontén, J. 1991. A role for sunlight in skin cancer: UV-induced *p53* mutations in squamous cell carcinoma. *Proc. Natl. Acad. Sci. USA* 88:10124–10128.

Bressac, B., Kew, M., Wands, J., and Ozturk, M. 1991. Selective G to T mutations of *p53* gene in hepatocellular carcinoma from southern Africa. *Nature* 350:429–431.

Hollstein, M., Sidransky, D., Vogelstein, B., and Harris, C.C. 1991. *p53* mutations in human cancers. *Science* 253:49–53.

Hsu, I.C., Metcalf, R.A., Sun, T., Welsh, J.A., Wang, N.J., and Harris, C.C. 1991. Mutational hotspot in the *p53* gene in human hepatocellular carcinomas. *Nature* 350:427–428.

Kumar, R., Sukumar, S., and Barbacid, M. 1990. Activation of *ras* oncogenes preceding the onset of neoplasia. *Science* 248:1101–1104.

Nelson, M.A., Futscher, B.W., Kinsella, T., Wymer, J., and Bowden, G.T. 1992. Detection of mutant Ha-*ras* genes in chemically initiated mouse skin epidermis before the development of benign tumors. *Proc. Natl. Acad. Sci. USA* 89:6398–6402.

Quintanilla, M., Brown, K., Ramsden, M., and Balmain, A. 1986. Carcinogen-specific mutation and amplification of Ha-*ras* during mouse skin carcinogenesis. *Nature* 322:78–80.

Sukumar, S., Notario,V., Martin-Zanca, D., and Barbacid, M. 1983. Induction of mammary carcinomas in rats by nitroso-methylurea involves malignant activation of H-*ras*-1 locus by single point mutations. *Nature* 306:658–661.

Wiseman, R.W., Stowers, S.J., Miller, E.C., Anderson, M.W., and Miller, J.A. 1986. Activating mutations of the c-Ha-*ras* protooncogene in chemically induced hepatomas of the male B6C3 F1 mouse. *Proc. Natl. Acad. Sci. USA* 83:5825–5829.

You, M., Candrian, U., Maronpot, R.R., Stoner, G.D., and Anderson, M.W. 1989. Activation of the Ki-*ras* protooncogene in spontaneously occurring and chemically induced lung tumors of the strain A mouse. *Proc. Natl. Acad. Sci. USA* 86:3070-3074.

Zarbl, H., Sukumar, S., Arthur, A.V., Martin-Zanca, D., and Barbacid, M. 1985. Direct mutagenesis of Ha-*ras*-1 oncogenes by *N*-nitroso-*N*-methylurea during initiation of mammary carcinogenesis in rats. *Nature* 315:382–385.

Oncogenes and Tumor Suppressor Genes in Development of Human Neoplasms

Albino, A.P., Le Strange, R., Oliff, A.I., Furth, M.E., and Old, L.J. 1984. Transforming *ras* genes from human melanoma: a manifestation of tumour heterogeneity? *Nature* 308:69–72.

Almoguera, C., Shibata, D., Forrester, K., Martin, J., Arnheim, N., and Perucho, M. 1988. Most human carcinomas of the exocrine pancreas contain mutant c-K-*ras* genes. *Cell* 53:549–554.

Bos, J.L. 1989. *Ras* oncogenes in human cancer: a review. *Cancer Res.* 49:4682-4689.

Bos, J.L., Fearon, E.R., Hamilton, S.R., Verlaan-de Vries, M., van Boom, J.H., van der Eb, A.J., and Vogelstein, B. 1987. Prevalence of *ras* gene mutations in human colorectal cancers. *Nature* 327:293–297.

Fearon, E.R., and Vogelstein, B. 1990. A genetic model for colorectal tumorigenesis. *Cell* 61:759–767.

Forrester, K., Almoguera, C., Han, K., Grizzle, W.E., and Perucho, M. 1987. Detection of high incidence of K-*ras* oncogenes during human colon tumorigenesis. *Nature* 327:298–303.

Gauwerky, C.E., Haluska, F.G., Tsujimoto, Y., Nowell, P.C., and Croce, C.M. 1988. Evolution of B-cell malignancy: pre-B-cell leukemia resulting from *MYC* activation in a B-cell neoplasm with a rearranged *BCL2* gene. *Proc. Natl. Acad. Sci. USA* 85:8548–8552.

Hirai, H., Kobayashi, Y., Mano, H., Hagiwara, K, Maru, Y., Omine, M., Mizoguchi, H., Nishida, J., and Takaku, F. 1987. A point mutation at codon 13 of the N-*ras* oncogene in myelodysplastic syndrome. *Nature* 327:430–432.

Janssen, J.W.G., Steenvoorden, A.C.M., Lyons, J., Anger, B., Bohlke, J.U., Bos, J.L., Seliger, H., and Bartram, C.R. 1987. *RAS* gene mutations in acute and chronic myelocytic leukemias, chronic myeloproliferative disorders, and myelodysplastic syndromes. *Proc. Natl. Acad. Sci. USA* 84:9228–9232.

Johnson, B.E., Ihde, D.C., Makuch, R.W., Gazdar, A.F., Carney, D.N., Oie, H., Russell, E., Nau, M.M., and Minna, J.D. 1987. *myc* family oncogene amplification in tumor cell lines established from small cell lung cancer patients and its relationship to clinical status and course. *J. Clin. Invest.* 79:1629–1634.

Lombardi, L., Newcomb, E.W., and Dalla-Favera, R. 1987. Pathogenesis of Burkitt lymphoma: expression of an activated c-*myc* oncogene causes the tumorigenic conversion of EBV-infected human B lymphoblasts. *Cell* 49:161–170.

Powell, S.M., Zilz, N., Beazer-Barclay, Y., Bryan, T.M., Hamilton, S.R., Thibodeau, S.N., Vogelstein, B., and Kinzler, K.W. 1992. *APC* mutations occur early during colorectal tumorigenesis. *Nature* 359:235–237.

Seeger, R.C., Brodeur, G.M., Sather, H., Dalton, A., Siegel, S.E., Wong, K.Y., and Hammond, D. 1985. Association of multiple copies of the N-*myc* oncogene with rapid progression of neuroblastomas. *N. Engl. J. Med.* 313:1111–1116.

Slamon, D.J., Godolphin, W., Jones, L.A., Holt, J.A., Wong, S.G., Keith, D.E., Levin, W.J., Stuart, S.G., Udove, J., Ullrich, A., and Press, M.F. 1989. Studies of the *HER-2/neu* proto-oncogene in human breast and ovarian cancer. *Science* 244:707–712.

Yanagisawa, A., Ohtake, K., Ohashi, K., Hori, M., Kitagawa, T., Sugano, H., and Kato, Y. 1993. Frequent c-Ki-*ras* oncogene activation in mucous cell hyperplasias of pancreas suffering from chronic inflammation. *Cancer Res.* 53:953–956.

❖ PART IV

Functions of Oncogenes and Tumor Suppressor Genes

❖ CHAPTER 12

Oncogenes and Growth Factors

The preceding chapters of this book dealt with the identification of both cellular oncogenes, which encode proteins that actively induce transformation, and tumor suppressor genes, whose loss of function can lead to neoplasia. A large (and still growing) number of both oncogenes and tumor suppressor genes have been identified. How do the proteins encoded by these genes function to regulate cell growth, so that abnormalities in their activities ultimately lead to neoplastic disease? The next eight chapters will consider the biochemical and physiological activities of the proteins encoded by the oncogenes and tumor suppressor genes. Many of these proteins function as elements in signal transduction pathways that allow a cell to respond to extracellular stimuli. Consequently, studies of oncogene and tumor suppressor gene function have not only contributed to our understanding of neoplastic disease—they have also served to illuminate the mechanisms that control the growth and differentiation of normal cells.

Platelet-Derived Growth Factor and the *sis* Oncogene

The first link between oncogenes and cellular proteins with known physiological activities was the identification in 1983 of the *sis* oncogene product as platelet-derived growth factor (PDGF). The hypothesis that an oncogene could encode a growth factor is clearly attractive in principle. As discussed in chapter 1, normal cell growth and differentiation is regulated in part by extracellular factors, including a number of polypeptide growth factors that induce proliferation of appropriate target cells in culture. One could therefore easily envision that abnormal production of a growth factor by a responsive

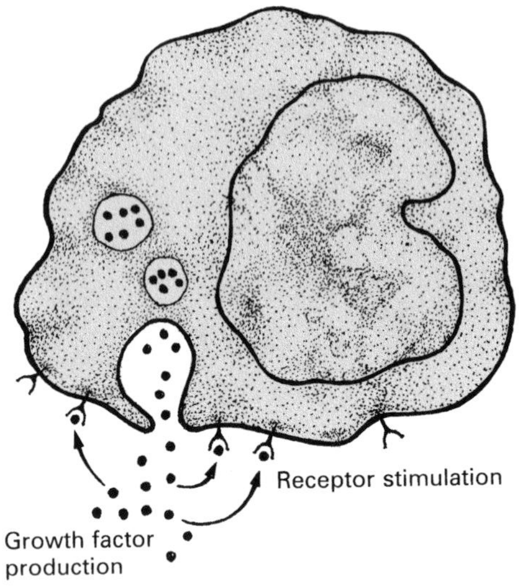

FIGURE 12.1
Autocrine stimulation of cell growth. A cell produces a growth factor to which it also responds, resulting in continuous cell proliferation.

cell would result in continual stimulation of cell growth, potentially culminating in transformation (Fig. 12.1). Indeed, many transformed cells have been found to produce growth factors, suggesting the hypothesis, proposed by George Todaro and colleagues in 1977, that autocrine stimulation of cell growth could lead to transformation. However, the link between oncogenes and growth factors did not come from a direct test of this hypothesis—instead it was an unexpected result of merging two independent sets of data by a computer search.

Sis was first identified as the oncogene of simian sarcoma virus, isolated from a fibrosarcoma of a woolly monkey in 1971. The *sis* oncogene was characterized at the molecular level by Stuart Aaronson and his colleagues, who reported its nucleotide sequence and the corresponding amino acid sequence of its predicted protein product in 1983. Antisera raised against peptides synthesized from this predicted sequence verified that *sis* encoded a polypeptide of 28 kd, but these data alone provided no insight into the protein's potential function.

Platelet-derived growth factor (PDGF) had been identified as a major growth factor utilized by fibroblasts in culture. It was purified from platelets (where it is stored and released during blood clotting) and was found to consist of polypeptides ranging from 28 to 35 kd. Partial amino acid sequence of PDGF was first reported in 1983 by Harry Antoniades and Michael Hunkapiller, who noted that the PDGF sequence was distinct from reported sequences of other growth factors. Shortly thereafter, Russell Doolittle analyzed the PDGF sequence for homology with other sequences in the protein sequence data base he had established in 1981. This search revealed extensive homology between the sequence of PDGF and that of the predicted *sis* protein—results published jointly by Doolittle, Hunkapiller, Aaronson, Antoniades, and their colleagues in 1983. At the same time, Michael Waterfield and his collaborators independently determined a partial amino acid sequence of PDGF and, by searching the Doolittle data base, discovered its relation to *sis*—results that also were reported in 1983. These papers provided for the first time a known physiological context in which to view the action of oncogenes—neoplastic transformation in this case was the apparent consequence of inappropriate expression of a protein whose function was to stimulate the proliferation of normal cells.

The relation between PDGF and the *sis* gene product has been substantiated and further elucidated by additional studies. Active platelet PDGF preparations consist of two related peptides, PDGF chains A and B, which share about 40% amino acid sequence identity. The biologically active PDGF molecule is a disulfide-linked dimer that may consist of only A chains, only B chains (A:A or B:B homodimer), or both A and B chains (A:B heterodimer). Cloning and sequencing the human *sis* proto-oncogene established that it encoded PDGF chain B. The gene encoding PDGF chain A has subsequently been isolated and localized to a different chromosome. Studies of the biological activity of the *sis* proto-oncogene indicated that it was activated to induce cell transformation by molecular constructs that led to abnormal constitutive expression of the normal *sis* gene product, PDGF-B. Importantly, only cells that are normally responsive to PDGF—and express PDGF receptor on their surface—are susceptible to transformation by *sis*. For example, fibroblasts and smooth muscle cells express PDGF receptors and are efficiently transformed by *sis*. In contrast, epithelial cells do not express the PDGF receptor and are refractory to *sis*-mediated transformation. Transformation by *sis* thus appears to require interaction of the *sis* gene product (PDGF) with its cognate receptor.

These findings provide strong support for the model that *sis*/PDGF-B induces cell transformation by continual autocrine stimulation of the normal growth factor response pathway, a series of events initiated by the binding of PDGF to its receptor. Interestingly, the *sis* gene product is able to bind to its receptor internally, during processing within the secretory pathway, although the activated receptor must reach the cell surface to initiate mitogenesis (cell division). Subsequent events in the mitogenic signal transduction pathway initiated by the PDGF receptor, and their relationship to other oncogenes, will be considered in subsequent chapters.

The transforming activity of *sis*/PDGF-B clearly raises the question as to whether the related gene encoding the A chain of PDGF can also induce cell transformation. The isolation of the PDGF-A gene allowed its activity as a potential oncogene to be tested by gene transfer assays. Not surprisingly, these experiments showed that overexpression of PDGF-A also was sufficient to induce cell transformation. The PDGF-A gene is thus an example of a gene that was shown to be an oncogene by *in vitro* experiments based on the known physiological activity of its polypeptide product. As will be discussed throughout this and subsequent chapters, similar *in vitro* experiments have demonstrated the transforming activity of a number of other genes first suspected to be potential oncogenes because of the biochemical or physiological activities of their protein products.

Several types of human tumors, particularly sarcomas, lung carcinomas, and astrocytomas, express both PDGF and its receptor, suggesting that autocrine stimulation by PDGF could play a role in tumor development. This has been supported by the use of dominant inhibitory mutants of PDGF, which have been found to revert the transformed phenotype of human astrocytoma cell lines in culture. It thus appears that *sis* functions as an oncogene in some human tumors, as well as in the acutely transforming retrovirus from which it was initially isolated.

A Family of Oncogenes Related to Fibroblast Growth Factors

Acidic and basic fibroblast growth factors (FGFs) are related polypeptides, sharing ~50% amino acid sequence identity, that stimulate growth and differentiation of a variety of cell types, including fibroblasts, endothelial cells, and neuronal cells. The relation between growth factors and oncogenes has been considerably extended by the identification of three oncogenes as additional members of the FGF family (Fig. 12.2).

The first oncogene recognized as a member of this family was *int*-2, an oncogene that had initially been identified as a target for activation by promoter insertion in mammary carcinomas induced by mouse mammary tumor virus (see chapter 6). Shortly thereafter, two oncogenes that had been isolated by gene transfer experiments (*hst* and *fgf*-5, see chapter 5) also were found to be relatives of FGF. All three of these oncogenes are related but distinct genes: they encode polypeptides that are 40% to 50% identical in sequence with each other as well as with acidic and basic FGFs. These genes thus define an FGF family with five members: acidic FGF (FGF-1), basic FGF (FGF-2), the *int*-2 and *hst* products (FGF-3 and FGF-4, respectively), and FGF-5. Additional members of the FGF family include FGF-6 and keratinocyte growth factor (KGF), which has been identified as a mitogen for epithelial cells.

Activation of the *int*-2, *hst,* and *fgf*-5 oncogenes is a result of transcriptional deregulation, resulting in constitutive overexpression of the normal polypeptide products. In addition, it has been shown that the proteins encoded by *hst* and *fgf*-5 are functional growth factors for fibroblasts. Thus, as was the case for *sis*/

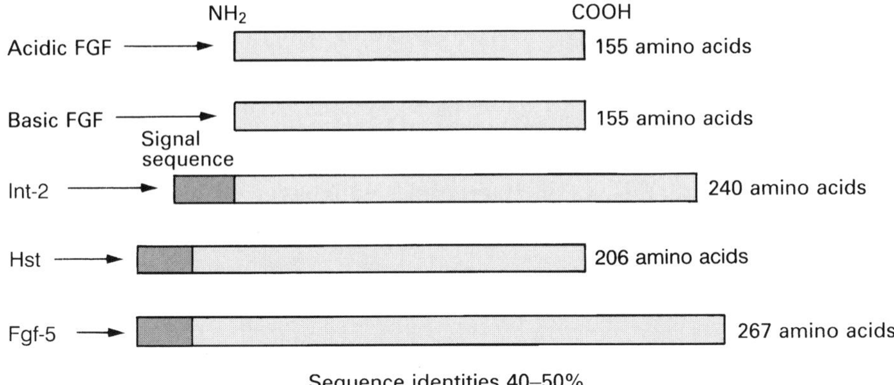

FIGURE 12.2
The fibroblast growth factor family. The Int-2, Hst, and Fgf-5 proteins have secretory signal sequences at their amino termini.

PDGF, transformation by the FGF family of oncogenes is a consequence of autocrine stimulation of a cell by a growth factor to which it normally responds.

An interesting difference between the polypeptides encoded by *int*-2, *hst*, and *fgf*-5 and acidic and basic FGF is the way in which these polypeptides are released by the cells in which they are synthesized. Like most growth factors (including PDGF), the Int-2, Hst, and Fgf-5 proteins (as well as KGF) have a hydrophobic signal sequence at the amino terminus that is responsible for insertion of the nascent polypeptide chain into the endoplasmic reticulum, the first step in the secretory pathway. In contrast, acidic and basic FGFs are unusual in that they lack a signal sequence and are not efficiently secreted by producing cells—release of these growth factors from the cells in which they are synthesized instead appears to be mediated by other mechanisms, which remain to be elucidated.

The significance of the secretory leader to the activity of this family of growth factors as potential oncogenes was revealed by experiments designed to investigate the transforming activity of the gene encoding basic FGF. Overexpression of the normal basic FGF polypeptide failed to induce cell transformation, in contrast with the consequences of similar overexpression of Hst or Fgf-5. However, transformation was induced by chimeric constructs in which the signal sequences of other secreted proteins (immunoglobulin or growth hormone) were fused to basic FGF, indicating that basic FGF could be converted into an oncogene by a structural alteration that allowed its product to enter the normal secretory pathway. These experiments further extend the relation between growth factors and oncogenes and support the need for such factors to interact with their receptors in order to elicit the transformed phenotype.

It is noteworthy that autocrine transformation by basic FGF may play a role in the development of human melanomas. The growth of normal melanocytes requires the exogenous addition of basic FGF to culture media, whereas melanomas are able to proliferate in the absence of exogenously added FGF. Instead, melanomas express basic FGF, suggesting that it acts by an autocrine mechanism. Consistent with this notion, interference with basic FGF expression by the use of antisense oligonucleotides is inhibitory to melanoma proliferation.

The *wnt* Gene Family

The prototype member of the *wnt* family was originally called *int*-1 because, like *int*-2, it was first identified as a target for retrovirus integration in mouse mammary carcinomas (chapter 6). The nucleotide sequence of *int*-1 indicated that it encoded a previously unknown protein of 370 amino acids with a signal sequence characteristic of secreted proteins. Moreover, further studies identified the *Drosophila* homolog of *int*-1 as the segment polarity gene known as *wingless,* which encodes a secreted protein that induces pattern formation during embryonic development. The *int*-1 gene was therefore renamed *wnt*-1, an amalgam of *wingless* and *int,* which clarifies its functional distinction from *int*-2 (a member of the FGF family).

Further studies have verifed that the Wnt-1 protein indeed acts as a secreted growth factor to stimulate proliferation of appropriate target cells in culture. Moreover, *wnt*-1 has proved to be the prototype of a large gene family, which now consists of at least 10 distinct *wnt* genes. One other member of the family, *wnt*-3, also is activated as an oncogene by retrovirus integration in mouse mammary carcinomas. Like *wnt*-1, other members of this interesting gene family play critical roles in development, which will be discussed in chapter 19.

Transforming Activity of Epidermal Growth Factor And TGF-α

Epidermal growth factor (EGF), first identified and isolated by Stanley Cohen in 1962, is a polypeptide of 53 amino acids that stimulates proliferation of a variety of different cell types. Several other growth factors related to EGF also have been identified, including transforming growth factor-α (TGF-α), amphiregulin, a growth factor encoded by vaccinia virus, the schwannoma-derived growth factor, the heparin-binding EGF-like factor, and betacellulin. All of these factors bind to the same cell surface receptor (the EGF receptor), which was identified in chapter 8 as the product of the *erb*B proto-oncogene. Additional members of the EGF family include the heregulins, which bind to related receptors encoded by other members of the *erb*B gene family (see chapter 13). It is noteworthy that members of the EGF family are examples of membrane-anchored growth factors: they are synthesized as higher molecular weight precursors that accumulate on the cell surface and can participate in cell–cell signaling. Soluble forms of these factors are then released by proteolytic cleavage of the precursor proteins.

TGF-α was initially of interest because it was found to be secreted by a variety of different types of transformed cells, suggesting its potential role as an autocrine growth factor important for cell transformation. The secreted TGF-α polypeptide of 50 amino acids is from 30% to 40% identical with EGF and has conserved cysteine residues that form disulfide bonds (Fig. 12.3). Because TGF-α and EGF bind to the same receptor, the possible autocrine growth factor activity of TGF-α stimulated experiments to test the potential transforming activities of the genes encoding both TGF-α and EGF. Gene transfer assays using appropriate molecular constructs indicated that constitutive overexpression of either TGF-α or EGF was sufficient to induce transformation of appropriate recipient cells. Moreover, overexpression of TGF-α results in the development of tumors in transgenic mice. These experiments thus extend the growth factor-oncogene relation to the EGF family, a conclusion further strengthened by studies of the *erb*B oncogene discussed in the next chapter. It is noteworthy that TGF-α is frequently expressed by human carcinomas that also express high levels of the EGF receptor, suggesting the functional importance of this autocrine pathway in human tumor development.

Hematopoietic Growth Factors as Oncogenes

The growth and differentiation of blood cells is regulated by a series of polypeptide factors that exert specific effects on different hematopoietic cell lineages. The first such factor to be identified was erythropoietin, which controls the growth and differentiation of red blood cells. At least twenty different hematopoietic growth factors have now been isolated, which regulate the development of cells belonging to each of the major blood cell lineages, including erythrocytes, platelets, monocytes, macrophages, granulocytes, and lymphocytes. Not surprisingly, four of these growth factors have been implicated as oncogenes that can contribute to hematopoietic cell transformation by autocrine stimulation of cell growth.

All four of these hematopoietic growth factors were discussed in chapter 6 as oncogenes activated by promoter insertion in retrovirus-induced leukemias. The targets of these retroviral integration events include the genes encoding interleukin-2 (T-cell growth factor), interleukin-3 (multipotential colony stimulating factor), granulocyte-macrophage colony stimulating factor, and colony stimulating factor-1 (macrophage colony stimulating factor). In each case, adjacent integration of a provirus results in constitutive expression of a growth factor gene, which then drives cell proliferation by an autocrine mechanism. In

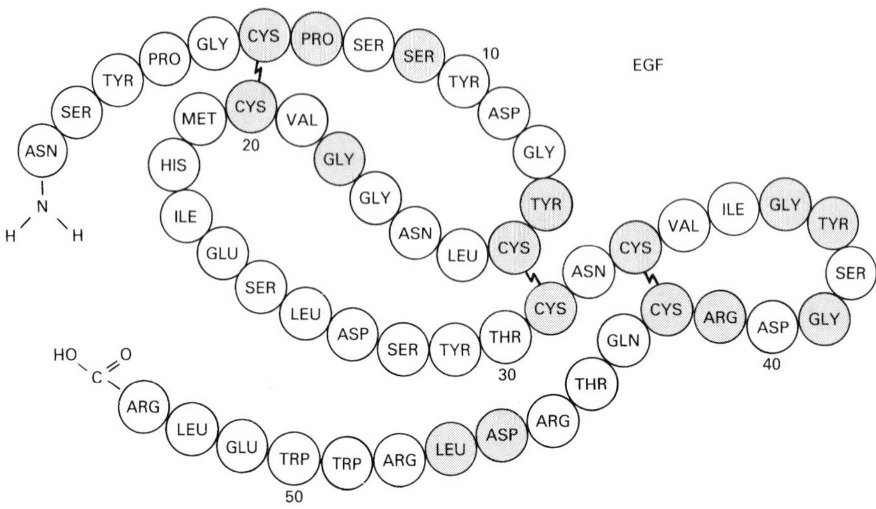

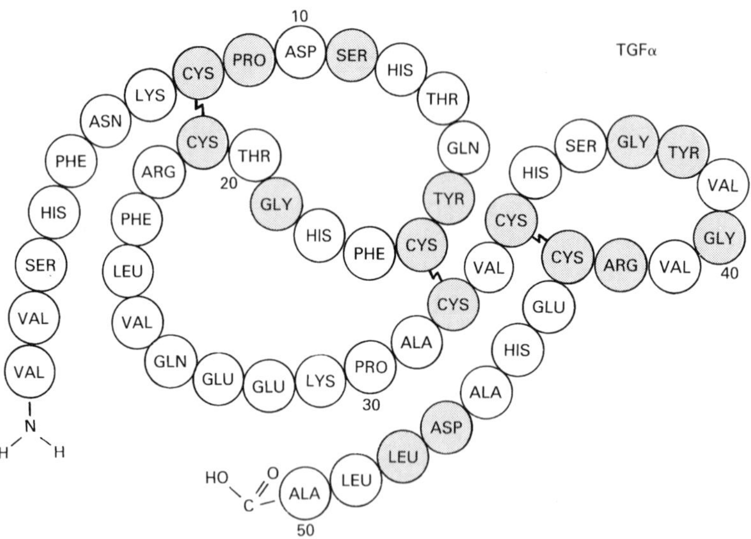

FIGURE 12.3
EGF and TGF-α. Identical amino acids are indicated by filled symbols. (From G. Carpenter and S. Cohen, *Ann. Rev. Biochem.* 48:193–216, 1979; and H. Marquardt, M.W. Hunkapiller, L.E. Hood, and G.J. Todaro, *Science* 223:1079–1082, 1984.)

addition, gene transfer experiments have directly established that constitutive expression of these hematopoietic growth factors can lead to transformation of appropriate target cells. The gene encoding interleukin-3 is also activated by translocation in some human acute lymphocytic leukemias, suggesting a role of autocrine growth factor production in this disease.

Summary

This chapter contains a number of examples in which (1) oncogenes have been identified as growth factors and (2) growth factors have been found experimentally to function as oncogenes (Table 12.1). It thus ap-

TABLE 12.1
Growth Factors with Oncogenic Potential

Proto-oncogene Product	Protein Molecular Weight
PDGF Family	
A chain	31,000
B chain (Sis)	28–35,000
FGF Family	
acidic FGF	16–18,000
basic FGF	16–18,000
Int-2	28–32,000
Hst	22–23,000
Fgf-5	32,000
EGF Family	
EGF	7,000
TGF-α	7,000
Wnt Family	
Wnt-1	36–44,000
Wnt-3	39,000
Hematopoietic Growth Factors	
Interleukin-2	15,000
Interleukin-3	14–28,000
M-CSF	35–45,000
GM-CSF	18–30,000

Note: The indicated molecular weights are those observed for the processed proto-oncogene proteins, with the exception of Fgf-5 and Wnt-3, for which the molecular weights are those predicted from the amino acid sequence. The variation in protein sizes is due to alternatively processed forms, which most frequently differ in extent of glycosylation.

pears that autocrine stimulation of cell growth provides a model of oncogene action that is applicable to a variety of different cell types and to a number of different polypeptide factors. Transformation in these instances apparently results from constitutive unregulated stimulation of a receptor that would normally signal proliferation of the target cell. These experiments provided the first link between oncogene action and normal cell physiology by establishing a clear-cut relation between molecular analysis of neoplastic transformation (oncogenes) and biochemical studies of the regulation of cell proliferation (growth factors).

References

General References

Aaronson, S.A. 1991. Growth factors and cancer. *Science* 254:1146–1153.

Cross, M., and Dexter, T.M. 1991. Growth factors in development, transformation, and tumorigenesis. *Cell* 64:271–280.

Platelet-Derived Growth Factor and the *sis* Oncogene

Beckmann, M.P., Betsholtz, C., Heldin, C.-H., Westermark, B., Di Marco, E., Di Fiore, P.P., Robbins, K.C., and Aaronson, S.A. 1988. Comparison of biological properties and transforming potential of human PDGF-A and PDGF-B chains. *Science* 241:1346–1349.

Bejcek, B.E., Li, D.Y., and Deuel, T.F. 1989. Transformation by v-*sis* occurs by an internal autoactivation mechanism. *Science* 245:1496–1499.

DeVare, S.G., Reddy, E.P., Law, J.D., Robbins, K.C., and Aaronson, S.A. 1983. Nucleotide sequence of the simian sarcoma virus genome: demonstration that its acquired cellular sequences encode the transforming gene product $p28^{sis}$. *Proc. Natl. Acad. Sci. USA* 80:731–735.

Doolittle, R.F., Hunkapiller, M.W., Hood, L.E., Devare, S.G., Robbins, K.C., Aaronson, S.A., and Antoniades, H.N. 1983. Simian sarcoma virus *onc* gene, v-*sis*, is derived from the gene (or genes) encoding a platelet-derived growth factor. *Science* 221:275–277.

Fleming, T.P., Matsui, T., Molloy, C.J., Robbins, K.C., and Aaronson, S.A. 1989. Autocrine mechanism for v-*sis* transformation requires cell surface localization of internally activated growth factor receptors. *Proc. Natl. Acad. Sci. USA* 86:8063–8067.

Shamah, S.M., Stiles, C.D., and Guha, A. 1993. Dominant-negative mutants of platelet-derived growth factor revert the transformed phenotype of human astrocytoma cells. *Mol. Cell. Biol.* 13:7203–7212.

Todaro, G.J., De Larco, J.E., Nissley, S.P., and Rechler, M.M. 1977. MSA and EGF receptors on sarcoma virus transformed cells and human fibrosarcoma cells in culture. *Nature* 267:526–528.

Waterfield, M.D., Scrace, G.T., Whittle, N., Stroobant, P., Johnsson, A., Wasteson, A., Westermark, B., Heldin, C.-H., Huang, J.S., and Deuel, T.F. 1983. Platelet-derived growth factor is structurally related to the putative transforming protein p28^{sis} of simian sarcoma virus. *Nature* 304:35–39.

Westermark, B., and Heldin, C.-H. 1991. Platelet-derived growth factor in autocrine transformation. *Cancer Res.* 51:5087–5092.

A Family of Oncogenes Related to Fibroblast Growth Factors

Becker, D., Meier, C.B., and Herlyn, M. 1989. Proliferation of human malignant melanomas is inhibited by antisense oligodeoxynucleotides targeted against basic fibroblast growth factor. *EMBO J.* 8:3685–3691.

Burgess, W.H., and Maciag, T. 1989. The heparin-binding (fibroblast) growth factor family of proteins. *Ann. Rev. Biochem.* 58:575–606.

Delli Bovi, P., Curatola, A.M., Kern, F.G., Greco, A., Ittmann, M., and Basilico, C. 1987. An oncogene isolated by transfection of Kaposi's sarcoma DNA encodes a growth factor that is a member of the FGF family. *Cell* 50:729–737.

Dickson, C., and Peters, G. 1987. Potential oncogene product related to growth factors. *Nature* 326:833.

Finch, P.W., Rubin, J.S., Miki, T., Ron, D., and Aaronson, S.A. 1989. Human KGF is FGF-related with properties of a paracrine effector of epithelial cell growth. *Science* 245:752–755.

Kiefer, P., Peters, G., and Dickson, C. 1991. The *Int*-2/*Fgf*-3 oncogene product is secreted and associates with the extracellular matrix: implications for cell transformation. *Mol. Cell. Biol.* 11:5929–5936.

Rogelj, S., Weinberg, R.A., Fanning, P., and Klagsbrun, M. 1988. Basic fibroblast growth factor fused to a signal peptide transforms cells. *Nature* 331:173–175.

Yoshida, T., Miyagawa, K., Odagira, H., Sakamoto, H., Little, P.F.R., Terada, M., and Sugimura, T. 1987. Genomic sequence of *hst,* a transforming gene encoding a protein homologous to fibroblast growth factors and the *int*-2-encoded protein. *Proc. Natl. Acad. Sci. USA* 84:7305–7309.

Zhan, X., Bates, B., Hu, X., and Goldfarb, M. 1988. The human *FGF*-5 oncogene encodes a novel protein related to fibroblast growth factors. *Mol. Cell. Biol.* 8:3487–3495.

The *wnt* Gene Family

Cabrera, C.V., Alonso, M.C., Johnston, P., Phillips, R.G., and Lawrence, P.A. 1987. Phenocopies induced with antisense RNA identify the *wingless* gene. *Cell* 50:659–663.

Jue, S.F., Bradley, R.S., Rudnicki, J.A., Varmus, H.E., and Brown, A.M.C. 1992. The mouse *Wnt*-1 gene can act via a paracrine mechanism in transformation of mammary epithelial cells. *Mol. Cell. Biol.* 12:321–328.

Nusse, R., and Varmus, H.E. 1992. *Wnt* genes. *Cell* 69:1073–1087.

Rijsewijk, F., Schuermann, M., Wagenaar, E., Parren, P., Weigel, D., and Nusse, R. 1987. The *Drosophila* homolog of the mouse mammary oncogene *int*-1 is identical to the segment polarity gene *wingless. Cell* 50:649–657.

Roelink, H., Wagenaar, E., da Silva, S.L., and Nusse, R. 1990. *Wnt*-3, a gene activated by proviral insertion in mouse mammary tumors, is homologous to *int*-1/*Wnt*-1 and is normally expressed in mouse embryos and adult brain. *Proc. Natl. Acad. Sci. USA* 87:4519–4523.

Van Leeuwen, F., Samos, C.H., and Nusse, R. 1994. Biological activity of soluble *wingless* protein in cultured *Drosophila* imaginal disc cells. *Nature* 368:342–344.

Transforming Activity of EGF and TGF-α

Derynck, R. 1992. The physiology of transforming growth factor-α. *Adv. Cancer Res.* 58:27–52.

Derynck, R., Goeddel, D.V., Ullrich, A., Gutterman, J.U., Williams, R.D., Bringman, T.S., and Berger, W.H. 1987. Synthesis of messenger RNAs for transforming growth factors α and ß and the epidermal growth factor receptor by human tumors. *Cancer Res.* 47:707–712.

Jhappan, C., Stahle, C., Harkins, R.N., Fausto, N., Smith, G.H., and Merlino, G.T. 1990. TGFα overexpression in transgenic mice induces liver neoplasia and abnormal development of the mammary gland and pancreas. *Cell* 61:1137–1146.

Marquardt, H., Hunkapiller, M.W., Hood, L.E., and Todaro, G.J. 1984. Rat transforming growth factor type 1: structure and relation to epidermal growth factor. *Science* 223:1079–1082.

Massagué, J., and Pandiella, A. 1993. Membrane-anchored growth factors. *Ann. Rev. Biochem.* 62:515–541.

Matsui, Y., Halter, S.A., Holt, J.T., Hogan, B.L.M., and Coffey, R.J. 1990. Development of mammary hyperplasia and neoplasia in MMTV-TGFα transgenic mice. *Cell* 61:1147–1155.

Plowman, G.D., Green, J.M., Culouscou, J.-M., Carlton, G.W., Rothwell, V.M., and Buckley, S. 1993. Heregulin induces tyrosine phosphorylation of HER4/p180^{erbB4}. *Nature* 366:473–475.

Rosenthal, A., Lindquist, P.B., Bringman, T.S., Goeddel, D.V., and Derynck, R. 1986. Expression in rat fibroblasts of a human transforming growth factor-α cDNA results in transformation. *Cell* 46:301–309.

Sandgren, E.P., Luetteke, N.C., Palmiter, R.D., Brinster, R.L., and Lee, D.C. 1990. Overexpression of TGFα in transgenic mice: induction of epithelial hyperplasia, pancreatic metaplasia, and carcinoma of the breast. *Cell* 61:1121–1135.

Shing, Y., Christofori, G., Hanahan, D., Ono, Y., Sasada, R., Igarashi, K., and Folkman, J. 1993. Betacellulin: a mitogen from pancreatic ß cell tumors. *Science* 259:1604–1607.

Stern, D.F., Hare, D.L., Cecchini, M.A., and Weinberg, R.A. 1987. Construction of a novel oncogene based on synthetic sequences encoding epidermal growth factor. *Science* 235:321–324.

Watanabe, S., Lazar, E., and Sporn, M.B. 1987. Transformation of normal rat kidney (NRK) cells by an infectious retrovirus carrying a synthetic rat type α transforming growth factor gene. *Proc. Natl. Acad. Sci. USA* 84:1258–262.

Hematopoietic Growth Factors as Oncogenes

Browder, T.M., Abrams, J.S., Wong, P.M.C., and Nienhuis, A.W. 1989. Mechanism of autocrine stimulation in hematopoietic cells producing interleukin-3 after retrovirus-mediated gene transfer. *Mol. Cell. Biol.* 9:204–213.

Lang, R.A., Metcalf, D., Gough, N.M., Dunn, A.R., and Gonda, T.J. 1985. Expression of a hemopoietic growth factor cDNA in a factor-dependent cell line results in autonomous growth and tumorigenicity. *Cell* 43:531–542.

Meeker, T.C., Hardy, D., Willman, C., Hogan, T., and Abrams, J. 1990. Activation of the interleukin-3 gene by chromosome translocation in acute lymphocytic leukemia with eosinophilia. *Blood* 76:285–289.

Metcalf, D. 1992. Hemopoietic regulators. *Trends Biochem. Sci.* 17:286–289.

Perkins, A., Kongsuwan, K., Visvader, J., Adams, J., and Cory, S. 1990. Homeobox gene expression plus autocrine growth factor production elicits myeloid leukemia. *Proc. Natl. Acad. Sci. USA* 87:8398–8402.

Sawyers, C.L., Denny, C.T., and Witte, O.N. 1991. Leukemia and the disruption of normal hematopoiesis. *Cell* 64:337–350.

❖ CHAPTER 13

Protein-Tyrosine Kinases and Growth Factor Receptors

The link between oncogenes and growth factors provides clear support for the notion that oncogenes act by perturbing the regulatory mechanisms that govern normal cell behavior. The response of a cell to an extracellular factor is mediated by the factor's binding to a cell surface receptor that then triggers a cascade of intracellular biochemical reactions that modulate both the function and the expression of target proteins. The signal imparted to the cell by a growth factor is thus transmitted by a regulatory network that ultimately leads to a programmed biological response. As will be seen in this and succeeding chapters, oncogenes can act to perturb such pathways of signal transduction at multiple levels.

One of the common modes of regulating protein function is phosphorylation, a classic example being provided by the function of cyclic AMP (cAMP) as a second messenger in the response of cells to a variety of hormones. The primary effect of many hormones is to increase the concentration of cAMP, which in turn activates cAMP-dependent protein kinase. This kinase then acts by phosphorylating appropriate target enzymes, which may themselves be kinases, to alter their catalytic activity and ultimately elicit a physiological response.

Given the importance of protein phosphorylation as an intracellular regulatory mechanism, it may not be surprising that many oncogenes encode protein kinases. In fact, the protein kinases constitute the largest group of oncogene products and include proteins that appear to function at several different levels of signal transduction pathways—from cell surface receptors to the targets of second messengers. Studies of these oncogenes have made a major contribution to our understanding of the role of protein phosphorylation in the control of normal cell growth and differentiation, as well as in the development of neoplastic disease.

This chapter deals with the function of those protein kinases that phosphorylate tyrosine residues on their substrate proteins, because this group of oncogene proteins includes kinases that are growth factor receptors and function in the initial steps of signal transduction after growth factor binding. In addition to the receptor protein-tyrosine kinases, other types of growth factor receptors that can act as oncogenes are discussed here. The second group of protein kinases, which phosphorylate serine and threonine on their substrates, generally function at later steps in signal transduction pathways and will be discussed in chapter 15.

Lessons from Src: Protein-Tyrosine Kinases

The *src* oncogene of Rous sarcoma virus (RSV) was discussed extensively in chapters 3 and 4 as the model that first defined both retroviral oncogenes and cellular proto-oncogenes. Although *src* has not figured as a cellular oncogene in the experiments discussed in chapters 5 through 8, it is again considered in detail here because the initial discovery of the protein-tyrosine kinase oncogenes emanated from studies of the Src protein.

The *src* gene product was first identified in the laboratory of Ray Erikson in 1977, before the application of recombinant DNA in tumor virology. Instead of identifying the Src protein by sequencing the *src* gene, Erikson exploited the genetics that initially defined *src* to detect its protein product by using an immunological approach. Specifically, sera from rabbits bearing RSV-induced tumors were found to react with a protein of 60 kd in either chicken or mammalian cells transformed by RSV. In contrast, the 60-kd protein was not detected in chicken cells infected with transformation-defective RSV mutants. The specific detection of this protein in RSV-transformed cells first suggested that it was the *src* gene product, a conclusion that was substantiated by *in vitro* translation of RSV RNA and, later, by cloning and sequencing of the *src* gene.

The demonstration that *src* encoded a protein of 60 kd against which antisera were available represented the first step in learning how an oncogene might act. Identifying the Src protein immunologically, however, was still far from determining its function. Nonetheless, the next step came surprisingly quickly: within a year after its identification, it was found that incubation of Src immunoprecipitates with radioactive ATP resulted in phosphorylation of immunoglobulin (Fig. 13.1). This protein kinase activity was specific for immunoprecipitates that contained functional Src protein and, importantly, was thermolabile in immunoprecipitates of a temperature-sensitive Src mutant. These results first indicated that the *src* gene product was a protein kinase, a conclusion thoroughly supported by further biochemical and molecular biological characterizations.

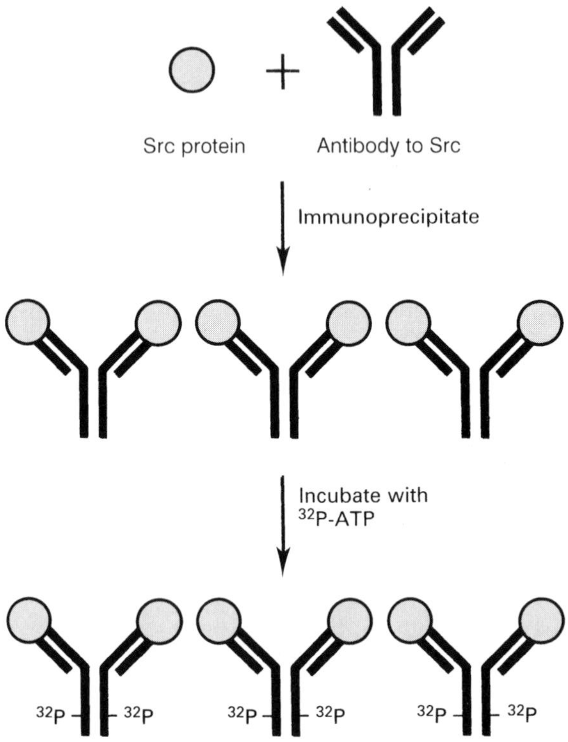

FIGURE 13.1
Demonstration of Src protein kinase activity. Incubation of Src protein immunoprecipitates with ^{32}P-ATP results in phosphorylation of the immunoglobulin heavy chain.

Src was not only the first oncogene protein found to be a protein kinase—it was also the first protein kinase found to phosphorylate tyrosine residues. The major amino acids that are phosphorylated in proteins from mammalian cells are serine and threonine, which had been found to be the substrates of previously studied protein kinases. Phosphorylation of proteins on tyrosine was first reported in 1979 by Walter Eckhart and Tony Hunter, who observed tyrosine phosphorylation in T antigen immunoprecipitates of polyomavirus transformed cells. Shortly thereafter, both the Src and Abl protein kinases were similarly found to specifically phosphorylate tyrosine residues of their substrate proteins. It is now known that the protein-tyrosine kinase activity of polyoma T antigen immunoprecipitates was due to the formation of complexes between polyoma middle T and the cellular *src* gene product—an interaction that is critical to the transforming activity of middle T (see chapter 2).

These findings established a unique protein kinase activity, leading to tyrosine phosphorylation, as being related to cell transformation. The importance of protein-tyrosine kinases was further emphasized by the results of other research groups that identified additional oncogene proteins—including Fes, Fgr, Fms, Ros, and Yes—with this enzymatic activity. Modulation of protein function by tyrosine-specific phosphorylation thus appeared to provide a biochemical mechanism by which a number of oncogene proteins functioned to induce neoplastic growth.

Growth Factor Receptors: Protein-Tyrosine Kinases and Oncogenes

The identification of oncogene products with protein-tyrosine kinase activity soon led to a further link between the function of oncogenes and the mechanisms that regulate normal cell growth—namely, the relation between oncogenes and growth factor receptors. The epidermal growth factor (EGF) receptor was first found to have protein kinase activity in 1978 in Stanley Cohen's laboratory. This enzymatic activity of the receptor was stimulated by EGF binding, suggesting that it played a role in signal transduction. In 1980, Cohen and his colleagues found that the amino acid phosphorylated by the EGF receptor kinase was tyrosine. Following these findings, several other growth factor receptors—including the receptors for PDGF, insulin, insulin-like growth factor-1 (IGF-1), and FGF—were similarly found to have protein-tyrosine kinase activity. Protein-tyrosine kinases were thus implicated as signal transducers in the response of cells to normal growth factors as well as in neoplastic transformation.

A more direct link between growth factor receptors and oncogenes was provided by Michael Waterfield and his collaborators in 1984. These investigators determined the amino acid sequence of several peptides derived from the human EGF receptor. When they entered these sequences in a computer homology search, several peptides of the EGF receptor were found to be nearly identical with the predicted amino sequence of the protein encoded by the *erb*B oncogene of avian erythroblastosis virus, suggesting that ErbB was derived from the EGF receptor. Further studies confirmed this relation—the ErbB oncogene product was a truncated version of the EGF receptor that contained its tyrosine kinase and transmembrane domains but lacked the extracellular ligand binding domain that comprises the amino-terminal half of the normal protein (Fig. 13.2). This relation indicated that aberrant function of a normal growth factor receptor with tyrosine kinase activity could result in cell transformation.

A similar relation between an oncogene and a known growth factor receptor was reported in 1985 by Charles Sherr, Richard Stanley, and their colleagues. These researchers found that the *fms* oncogene product was derived

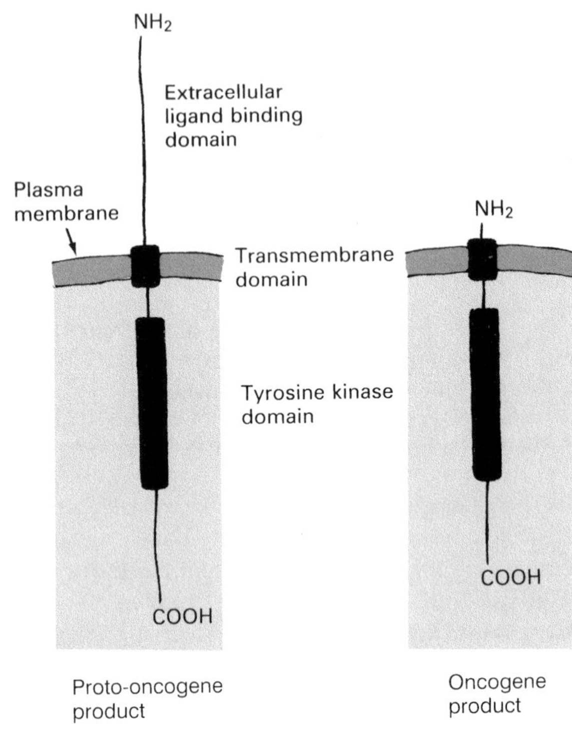

FIGURE 13.2
Comparison of the *erb*B proto-oncogene and oncogene proteins. The *erb*B proto-oncogene product (the EGF receptor) includes an extracellular ligand binding domain, a transmembrane domain, and an intracellular tyrosine kinase domain. The extracellular ligand binding domain has been deleted in the *erb*B oncogene product.

from the cell surface receptor for macrophage colony stimulating factor (CSF-1). Thus, two oncogenes corresponded to known growth factor receptors that displayed protein-tyrosine kinase activity.

Sequence analysis has been of major utility in further understanding the relation between oncogenes, protein kinases, and receptors (Fig. 13.3). Comparison of amino acid sequences has revealed that all protein kinases have similar catalytic domains of 250 to 300 amino acids. In addition, characteristic differences within this domain distinguish protein-tyrosine kinases from protein-serine/threonine kinases. Thus, the DNA sequence of a previously unknown gene can predict whether it encodes a protein kinase, the amino acid specificity of that kinase, and the location of the kinase catalytic domain in the protein sequence. In all cases so far analyzed, such predictions based on sequence have been confirmed by later biochemical analysis, indicating the validity of this structural approach to inferring protein function. In addition, the sequence similarities between different protein kinases have allowed the molecular cloning of novel protein kinases by using conserved sequences as probes.

```
Src   RESLRLEAKLGGGCFGEVWMGTWNDTTR-----VAIKTLKPGTMSPE--AFLQEAQVMKK
Abl   RTDITMKHKLGGGQYGEVYEGVWKKYSLT----VAVKTLKEDTMEVE--EFLKEAAVMKE
Fes   HEDLVLGEQIGRGNFGEVFSGRLRADNTL----VAVKSCRETLPPDIKAKFLQEAKILKQ
ErbB  ETEFKKVKVLGSGAFGTIYKGLWIPEGEKVKIPVAIKELREATSPKANKEILDEAYVMAS
Raf   ASEVMLSTRIGSGSFGIVYKGKWHGD-------VAVKILKVVDPTPEQLQAFRNEVAVLR
Mos   WEQVCLMHRLGSGGFGSVYKATYHGVP------VAIKQVNKCTEDLRASQRSFWAELNIA

Src   LR-HEKLVQLYAVVSEEP-IYIVIEYMSK-----GSLLDFLKGEMGKY--------LRLP
Abl   IK-HPNLVQLLGVCTREPPFYIITEFMTY-----GNLLDYLRECNRQE--------VSAV
Fes   YS-HPNIVRLIGVCTQKQPIYIVMELVQG-----GDFLTFLRTEGAR---------LRMK
ErbB  VD-NPHVCRLLGICLTST-VQLITQLMPY-----GCLLDYIREHKDN---------IGSQ
Raf   KTRHVNI--LLFMGYMTKDNLAIVTQWCEGSSLYKHLHVQETK-------------FQMF
Mos   GLRHDNIVRVVAASTRTPEDSNSLGTIIMEFGGNVTLHQVIYDATRSPEPLSCRKQLSLG

Src   QLVDMAAQIASGMAYVERMNYVHRDLRAANILVGENLVCKVADFGLARLIEDNEYTARQG
Abl   VLLYMATQISSAMEYLEKKNFIHRDLAARNCLVGENHLVKVADFGLSRLMTGDTYTAHAG
Fes   TLLQMVGDAAAGMEYLESKCCIHRDLAARNCLVTEKNVLKISDFGMSREEADGVYAASGG
ErbB  YLLNWCVQIAKGMNYLEERRLVHRDLAARNVLVKTPQHVKITDFGLAKLLGADEKEYHAE
Raf   QLIDIARQTAQGMDYLHAKNIIHRDMKSNNLFLHEGLTVKIGDFGLATVKSRWSGSQQVE
Mos   KCLKYSLDVVNGLLFLHSQSILHLDLKPANILISEQDVCKISDFGCSQKLQDLRGRQASP

Src   AKF--PIKWTAPEAALYGR---FTIKSDVWSFGILLTELTTKGRVPYPGMVNREVLDQVE
Abl   AKF--PIKWTAPESLAYNK---FSIKSDVWAFGVLLWEIATYGMSPYPGIDLSQVYELLE
Fes   LRLV-PVKWTAPEALNYGR---YSSESDVWSFGILLWETFSLGASPYPNLSNQQTREFVE
ErbB  GGKV-PIKWMALESILHRI---YTHQSDVWSYGVTVWELMTFGSKPYDGIPASEISSVLE
Raf   QPTG-SVLWMAPEVIRMQDDNPFSFQSDVYSYGIVLYELMA-GELPYAHINNRDQIIFMV
Mos   PHIGGTYTHQAPEILKGEI---ATPKADIYSFGITLWQMTT-REVPYSGEPQYVQYAVVA

Src   R--GYRMPCPPECPESLHD----LMCQCWRKDPEERPTFKYLGAQLLPA
Abl   K--DYRMERPEGCPEKVYE----LMRACWQWNPSDRPSFAEIHQAFETM
Fes   K--GGRLPCPELCPDAVFR----LMEQCWAYEPGQRPSFSAFYQELQSI
ErbB  K--GERLPQPPICTIDVYM----IMVKCWMIDADSRPKFRELIAEFSKM
Raf   GR-GYASPDLSRLYKNCPKAIKRLVADCVKKVKEERPLFPQILSSIELL
Mos   YNLRPSLAGAVFTASLTGKALQNIIQSCWEARGLQRPSAELLGRDLKAF
```

FIGURE 13.3
Protein kinase catalytic domains. Amino acid sequences of the catalytic domains of four protein-tyrosine kinases (Src, Abl, Fes, and ErbB) and two protein-serine/threonine kinases (Raf and Mos) are shown. Amino acids identical with those in Src are shaded. Single letter symbols for the amino acids are: A, alanine; C, cysteine; D, aspartic acid; E, glutamic acid; F, phenylalanine; G, glycine; H, histidine; I, isoleucine; K, lysine; L, leucine; M, methionine; N, asparagine; P, proline; Q, glutamine; R, arginine; S, serine; T, threonine; V, valine; W, tryptophan; Y, tyrosine.

The amino acid sequences of the EGF, PDGF, insulin, IGF-1, FGF, CSF-1, and other receptors indicate related structural organizations (Fig. 13.4). The *ligand binding domains* are encoded in the amino-terminal extracellular portions of the proteins. Then, a stretch of ~25 hydrophobic amino acids spans the plasma membrane (the *transmembrane domain*). Finally, the *tyrosine kinase catalytic domains* are on the cytosolic side of the membrane, encoded in the carboxy-terminal portions of the proteins beginning ~50 amino acids from the transmembrane region.

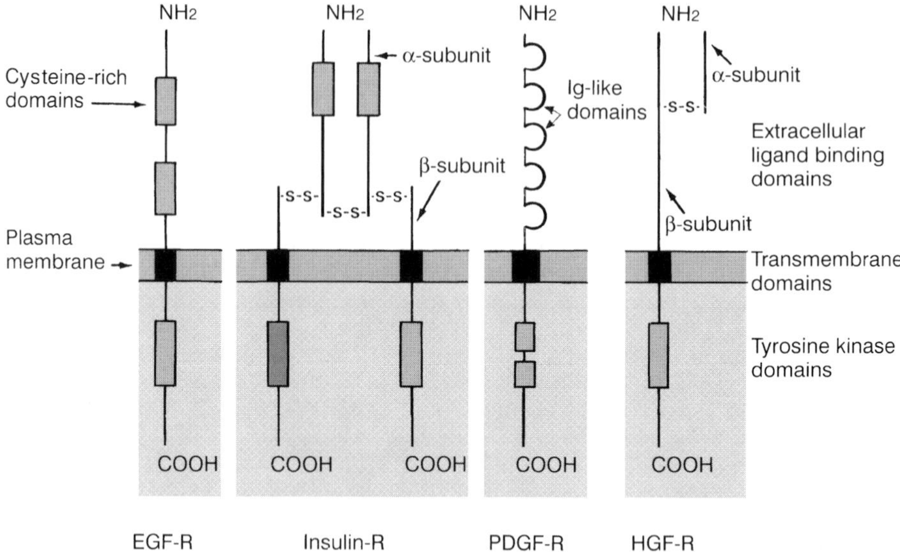

FIGURE 13.4
Organization of receptor protein-tyrosine kinases. These receptors consist of amino-terminal extracellular ligand binding domains, transmembrane domains, and carboxy-terminal intracellular tyrosine kinase domains. Different structural features then serve to characterize the various receptor subfamilies, only a few examples of which are shown here. The extracellular domains of the EGF, insulin receptor, and Eph/Elk families include cysteine-rich regions, whereas the extracellular domains of the PDGF, FGF, VEGF, and Axl families are characterized by the presence of immunoglobulin(Ig)-like domains. The insulin and HGF family receptors are composed of two types of subunits. The tyrosine kinase domains of the PDGF, FGF, and VEGF family receptors are interrupted by insert regions of amino acids that are not related to other protein kinase catalytic domain sequences.

Based on the structure of these known protein-tyrosine kinase receptors, the sequences of several other oncogene products have indicated that they also were derived from cell surface receptors (Table 13.1). For example, the *erb*B-2 oncogene encodes a transmembrane protein-tyrosine kinase that is closely related to ErbB (EGF receptor). A third member of the EGF receptor family (ErbB-3) also appears to have potential oncogenic activity, based on its constitutive activation in some breast cancers.

The sequences of a number of additional oncogene products—including Alk, Axl, K-Sam, Ros, Sea, Trk, Met, Kit, and Ret—also reveal tyrosine kinase domains that are preceded by transmembrane regions and extracellular domains, suggesting that they are likewise derived from growth factor receptors.

TABLE 13.1
Receptor Protein-Tyrosine Kinases

Receptor	Molecular Weight
EGF Receptor Family	
EGF receptor (ErbB)	175,000
ErbB-2	185,000
ErbB-3	180,000
ErbB-4	180,000
FGF Receptor Family	
FGF receptor-1	130,000
FGF receptor-2 (K-sam)	135,000
FGF receptor-3	125,000
FGF receptor-4	110,000
PDGF Receptor Family	
PDGF α-receptor	180,000
PDGF ß-receptor	180,000
CSF-1 receptor (Fms)	165,000
SL receptor (Kit)	145,000
Insulin Receptor Family	
Insulin receptor	α=135,000, ß=95,000
IGF-1 receptor	α=135,000, ß=90,000
Hepatocyte Growth Factor Receptor Family	
HGF receptor (Met)	α=50,000, ß=145,000
Sea (ligand unknown)	α=33,000, ß=133,000
Neurotrophin Receptor Family	
Trk (NGF receptor)	140,000
Trk-B (BDNF and NT4 receptor)	140,000
Trk-C (NT3 receptor)	145,000
Eph/Elk Family (B61; other ligands unknown)	130–135,000
VEGF Receptor	180,000
Ros (ligand unknown)	280,000
Ret (ligand unknown)	170,000
Axl (ligand unknown)	140,000
Alk (ligand unknown)	Not determined

Note: The indicated molecular weights are those observed for the processed (glycosylated) forms of the receptor proteins, with the exception of Sea, for which the molecular weight is that predicted by the amino acid sequence.

In some cases, the ligands for these receptors have been identified by subsequent studies. Thus, K-Sam is a member of the FGF receptor family, Trk is the receptor for nerve growth factor (NGF), Met is the receptor for hepatocyte growth factor, and Kit is the receptor for steel ligand (SLF). In other cases, the ligands for the proto-oncogene receptors remain to be identified.

The relations of a number of protein-tyrosine kinase oncogenes to growth factor receptors suggest that neoplastic transformation can result from the constitutive activity of receptor protein-tyrosine kinases, which would normally require growth factor binding for their activation. This is clearly consistent with the ability of growth factors themselves to function as oncogenes, as discussed in the preceding chapter, and supports the view that deregulation of the mechanisms that control normal cell growth can lead to abnormal cell proliferation and transformation. Comparisons of structure and function have elucidated the molecular alterations that convert the proto-oncogenes that encode normal growth factor receptors into oncogenes that potently induce cell transformation. These studies have further established the link between protein-tyrosine kinase activity and cell proliferation, as well as illustrating the ways in which the receptor protein-tyrosine kinases are normally regulated by growth factor binding.

The most common mode of activation of the receptor tyrosine kinase oncogenes, already noted for *erb*B, is deletion of the amino-terminal ligand binding domain of the proto-oncogene protein. Thus, the ErbB oncogene protein is a truncated version of the EGF receptor from which the EGF binding region has been deleted. Although the ErbB oncogene protein has also sustained other mutations relative to the normal EGF receptor, including carboxy-terminal deletion and a point mutation in the kinase domain, it is this amino-terminal deletion that appears to be primarily responsible for its transforming activity. The biochemical consequence of this deletion is to generate an oncogene protein that is a constitutively active protein-tyrosine kinase, whereas the kinase activity of the normal EGF receptor is expressed only following EGF binding (Fig. 13.5). Thus, the amino-terminal EGF binding domain appears to regulate the catalytic activity of the tyrosine kinase domain. In the absence of EGF, tyrosine kinase activity is suppressed. EGF binding relieves this negative regulation, resulting in activation of the receptor tyrosine kinase. In the ErbB oncogene protein, deletion of the EGF binding domain has abrogated this normal regulatory mechanism and the ErbB tyrosine kinase functions constitutively, independent of EGF binding. Thus it appears that constitutive activity of this protein-tyrosine kinase is responsible for driving abnormal cell proliferation leading to neoplasia.

Similar deletions of the amino-terminal ligand binding domains of the normal receptor proteins have been found to activate other oncogenes, including *erb*B-2, *kit, ros, met, ret,* and *trk*. In addition, *in vitro* molecular cloning manipula-

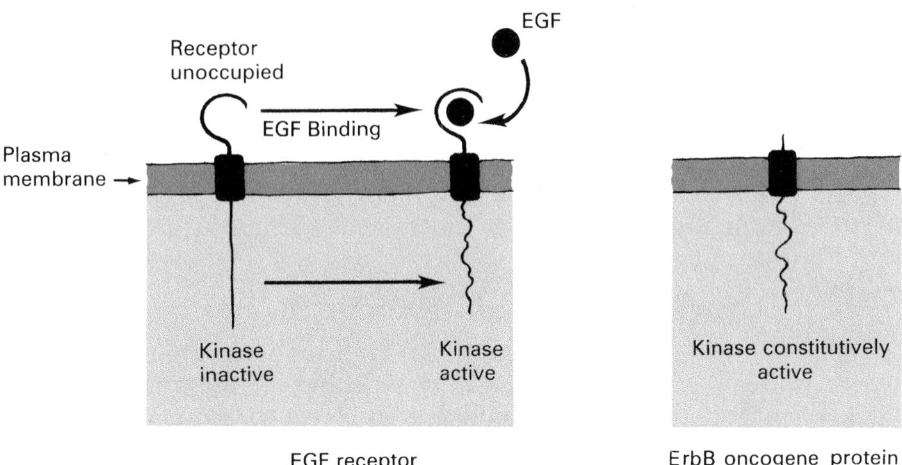

FIGURE 13.5
Mechanism of *erb*B oncogene activation. The tyrosine kinase activity of the EGF receptor (ErbB proto-oncogene protein) is controlled by EGF binding. In contrast, the ErbB oncogene kinase is constitutively active.

tions have demonstrated that deletion of the normal amino-terminal domain of the insulin receptor can convert it into a biologically active oncogene protein. As with *erb*B, these deletions appear to remove a regulatory ligand binding domain of the normal receptor proteins, generating instead a truncated oncogene protein with constitutive unregulated protein-tyrosine kinase activity.

The oncogenic potential of receptor tyrosine kinases can also be activated by other structural alterations, including point mutations in their extracellular or transmembrane domains. Like amino-terminal deletions, these mutations appear to result in ligand-independent activation of the intracellular tyrosine kinase domain. A germ line point mutation in the tyrosine kinase domain of *ret* is also responsible for inheritance of multiple endocrine neoplasia type 2B, although the effect of this mutation on activity of the Ret protein kinase is not yet clear.

Overexpression of structurally normal receptor tyrosine kinases is also frequently sufficient to induce transformation, and it is noteworthy that amplification of *erb*B and *erb*B-2 is a common event in some human malignancies, as discussed in preceding chapters. Finally, sequences encoding the extreme carboxy terminus (past the kinase domain) are deleted in several of the oncogenes of this group, including *erb*B, *fms, kit,* and *ret.* Molecular analysis indicates that these carboxy-terminal sequences comprise negative regulatory domains, whose deletion increases activity of the oncogene-encoded proteins.

Each of the molecular alterations that activate receptor protein-tyrosine kinase oncogenes thus has a common ultimate effect—namely, increased protein-tyrosine kinase activity. Elevated or deregulated expression of this activity, normally stimulated by the binding of growth factors to their receptors, therefore appears to be responsible for the transforming activity of a number of oncogenes that are related to surface receptors for either known or still-to-be-identified extracellular mediators of cell proliferation.

Nonreceptor Protein-Tyrosine Kinases

The cell surface receptors constitute only one group of protein-tyrosine kinases; a second large group of oncogenes encode protein-tyrosine kinases that are not membrane-spanning molecules but, instead, are intracellular proteins, some of which are associated with the cytoplasmic side of the plasma membrane (Table 13.2). The prototype of this group is *src*. Also included are a number of oncogenes initially derived from transforming retroviruses—*abl*, *yes*, *fgr*, and *fes*—as well as a number of genes that were isolated on the basis of sequence similarities with other protein-tyrosine kinases. These genes encode proteins that, in contrast with the receptor tyrosine kinases, lack extracellular or transmembrane domains. The nonreceptor tyrosine kinases, therefore, are not integral membrane proteins but, in some cases, are peripherally associated with the plasma membrane by lipid that is covalently added after their translation.

Nine of the nonreceptor tyrosine kinases—Src, Yes, Fgr, Lck, Fyn, Lyn, Hck, Blk, and Yrk—are closely related, constituting the Src subfamily of protein-tyrosine kinases. These ~60 kd proteins are homologous not only within the catalytic domain, but also throughout much of the rest of the molecule. Posttranslational processing and membrane association appears to proceed similarly for all of these proteins. The initiating methionine is removed and myristic acid (a 14 carbon fatty acid) is added to the amino-terminal glycine, which is encoded as the second amino acid. This modification is required for both membrane association and transforming activity, directly implicating the plasma membrane as the functional site of these protein kinases.

The kinase activities and transforming potentials of these proteins are regulated by phosphorylation at two important tyrosine residues that are conserved in all the members of the Src protein kinase subfamily. For example, the activity of Src is controlled by phosphorylation at tyrosine-416 and tyrosine-527 (Fig. 13.6). Tyrosine-416, which is within the catalytic domain, is autophosphorylated by the Src kinase *in vitro*. Phosphorylation of this residue apparently increases both kinase activity and transforming potential, indicating that Src is positively regulated by phosphorylation at this site. In contrast, phosphorylation of tyrosine-527 appears to down-regulate Src kinase activity. This resi-

TABLE 13.2
Nonreceptor Protein-Tyrosine Kinases

Protein Kinase	Molecular Weight
Src Subfamily	
Src	60,000
Yes	62,000
Fgr	55,000
Lck	56,000
Fyn	59,000
Lyn	56,000
Hck	59,000
Blk	55,000
Yrk	60,000
Abl Subfamily	
Abl	145,000
Arg	145,000
Fes Subfamily	
Fes	92,000
Fer	94,000
Jak Subfamily	
Jak1	132,000
Jak2	130,000
Tyk2	134,000
Itk Subfamily	
Itk	73,000
Bpk	77,000
Tec	68,000
Syk/Zap Subfamily	
Syk	72,000
Zap	70,000
Csk Subfamily	
Csk	50,000
Ctk	52,000
Fak	125,000

due lies within the 19 carboxy-terminal amino acids encoded by the *src* proto-oncogene but not by the RSV *src* oncogene, a deletion that activates transforming potential. Whereas the viral Src protein is phosphorylated on tyrosine-416 *in vivo,* the normal cellular Src protein is phosphorylated at tyrosine-527. This suggested the possible significance of tyrosine-527 in regulating Src activity, a

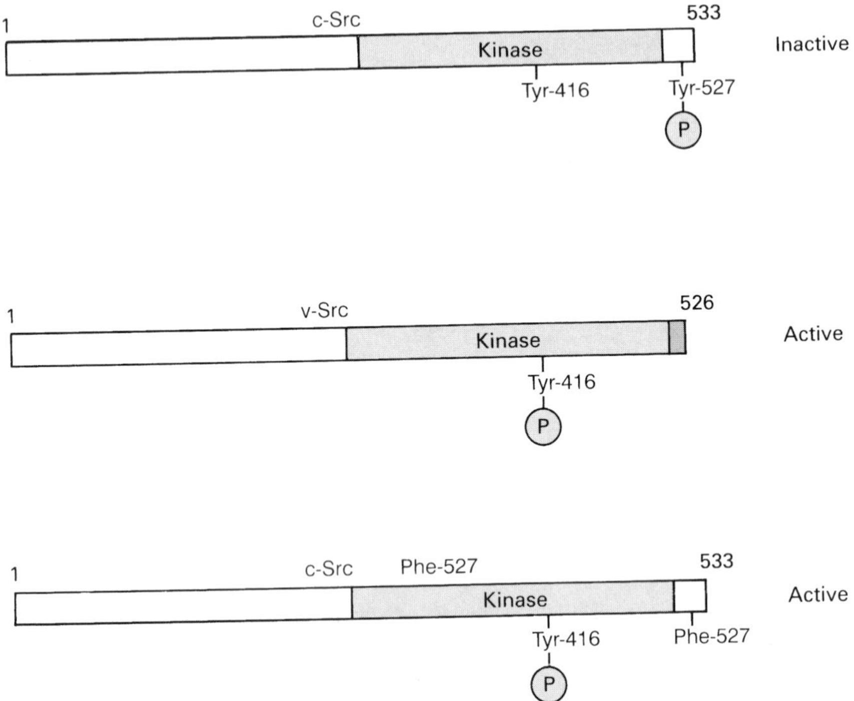

FIGURE 13.6
Regulation of Src kinase activity by carboxy-terminal phosphorylation. The normal c-Src proto-oncogene protein is phosphorylated *in vivo* at Tyr-527, which down-regulates kinase activity. This tyrosine has been deleted from the viral Src (v-Src) oncogene protein, which is active as a kinase and autophosphorylated at Tyr-416. The significance of Tyr-527 in regulation of Src kinase activity is demonstrated by mutation of c-*src* codon 527 to encode phenylalanine instead of tyrosine. A c- Src protein with a Phe-527 substitution is, like v-Src, activated as a kinase and autophosphorylated on Tyr-416.

hypothesis that was tested by *in vitro* mutagenesis experiments in which tyrosine-527 was changed to phenylalanine. This single amino acid substitution was found to increase both tyrosine kinase activity and transforming potential, directly demonstrating the significance of phosphorylation on tyrosine-527 as a negative regulator of Src activity.

The homologous carboxy-terminal tyrosines have similarly been implicated as determinants of the transforming potentials of other members of the Src subfamily of tyrosine kinases. It thus appears that the function of these kinases in normal cells is controlled, at least in part, by phosphorylation and dephosphorylation of this tyrosine moiety. Although the normal physiologic pathways that mediate this regulation remain unknown, it is clear that loss of this control leads to increased kinase activity resulting in cell transformation. Interestingly, phosphorylation of tyrosine-527 is not catalyzed by the Src protein itself, but by a distinct protein-tyrosine kinase called Csk (C-terminal Src kinase), the principal function of which may be regulation of the activities of Src family kinases.

In addition to the protein kinase catalytic domain, two other conserved domains of the Src family kinases are critical to their function. These domains, designated SH2 (Src homology 2) and SH3 (Src homology 3), are conserved in all members of the Src subfamily as well as being present in most other nonreceptor protein-tyrosine kinases and in a variety of other proteins participating in the transmission of intracellular signals from growth factor receptors (Fig. 13.7). Both SH2 and SH3 domains (which encompass about 100 and 60 amino acids, respectively) serve as binding sites that play critical roles in protein-protein interactions. SH2 domains bind phosphotyrosine-containing peptides; so proteins containing SH2 domains specifically interact with other

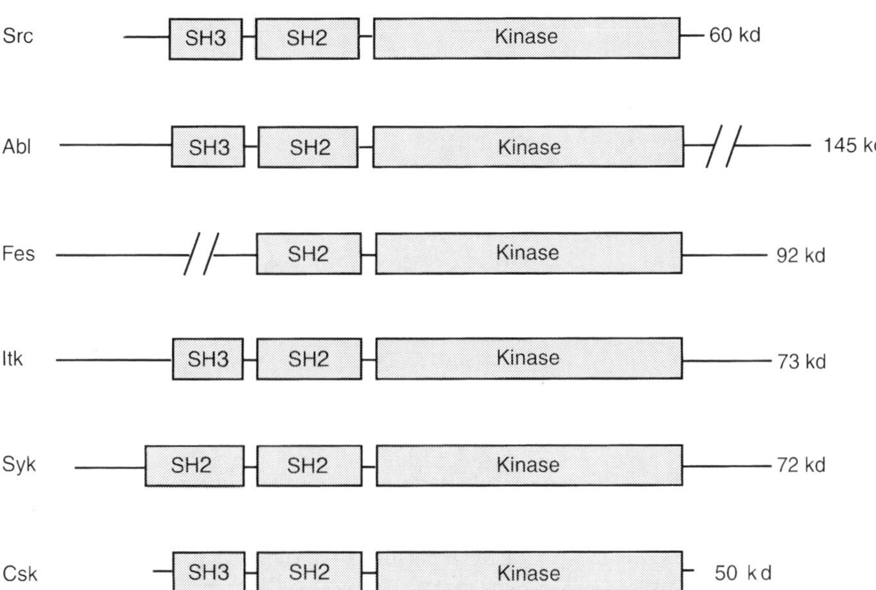

FIGURE 13.7
SH2 and SH3 domains of nonreceptor protein-tyrosine kinases. SH2 and SH3 are conserved noncatalytic domains that mediate protein–protein interactions and participate in regulation of kinase activity. Structures of the proto-oncogene proteins are shown.

proteins that have been modified by tyrosine phosphorylation. SH3 domains also serve as specific protein binding sites—in this case, recognizing peptides that are rich in proline residues. As discussed in detail in chapter 17, both SH2 and SH3 domains facilitate the formation of specific complexes between protein-tyrosine kinases and other proteins participating in intracellular signal transduction. As one example, the Src SH2 domain binds to the PDGF receptor, leading to activation of Src in response to mitogen stimulation. In addition, both the SH2 and the SH3 domains regulate Src kinase activity, and mutations in these regions can activate *src* transforming potential. The SH2 domain appears to bind to the negative regulatory phosphotyrosine-527, thereby folding the Src kinase into an inactive state. This inactive conformation of the Src protein may also be stabilized by SH3 interactions with C-terminal Src sequences.

Members of the Src subfamily normally function in the transmission of intracellular signals initiated by the binding of extracellular factors to cell surface receptors. For instance, as noted earlier, Src binds to activated protein-tyrosine kinase growth factor receptors (for example, the PDGF receptor) and serves to further transmit the signal initiated by growth factor binding. In addition, Src subfamily kinases are associated with a variety of cell surface receptors that do not have intrinsic protein kinase activity. The prototypical example is provided by Lck, which is normally expressed in T lymphocytes. In these cells, Lck is associated with the cell surface receptors CD4 and CD8, which are involved in cell–cell recognition during T-cell activation (Fig. 13.8). Stimulation of these receptors activates the Lck protein kinase, so the receptor/Lck complex functions analogously to receptors with intrinsic protein-tyrosine kinase activity. In both cases, the interaction of an extracellular ligand with its cell surface receptor stimulates a protein-tyrosine kinase, which then serves to transmit the signal initiated by ligand binding.

Like the oncogenes encoding the Src subfamily of protein-tyrosine kinases, the *abl* and *fes* oncogenes encode proteins whose transforming potential is activated as a consequence of increased protein-tyrosine kinase activity. As discussed earlier, activation of *abl* either by incorporation into a retrovirus (chapter 3) or by chromosome translocation (chapter 7) results in the formation of a recombinant fusion protein in which the normal amino-terminal sequences encoded by the proto-oncogene are deleted (Fig. 13.9). The deletion that occurs during generation of the viral *abl* oncogene results in loss of the SH3 domain, and *in vitro* mutagenesis experiments indicate that deletion of SH3 is sufficient both to increase intracellular kinase activity and to activate transforming potential of the *abl* proto-oncogene. In contrast with the viral *abl* oncogene, however, the normal Abl SH3 domain is retained in the Bcr/Abl fusion protein generated by the Philadelphia translocation. Moreover, it has been shown that the small deletion of *abl* sequences in the *bcr/abl* gene is not sufficient to activate *abl* transforming potential. Instead, the fused Bcr sequences directly activate both the

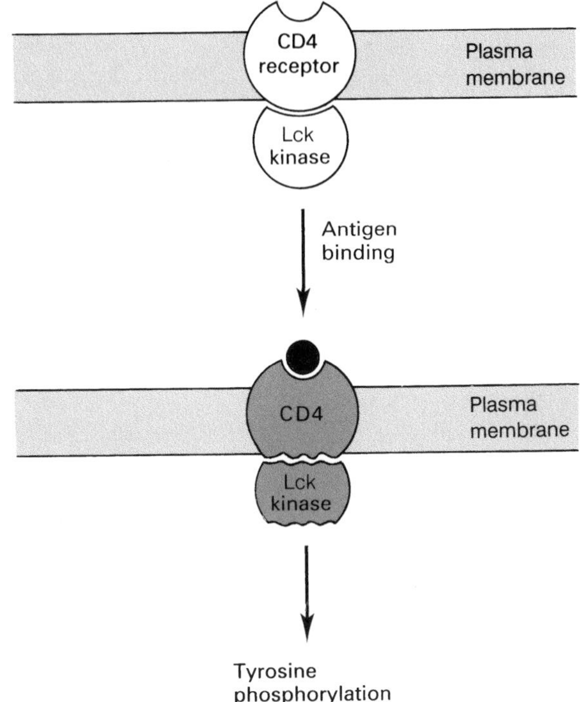

FIGURE 13.8
Association of the Lck protein kinase with the CD4 receptor. The Lck kinase is noncovalently associated with the CD4 receptor on the surface of T lymphocytes. Binding of ligand to CD4 stimulates Lck protein-tyrosine kinase activity.

tyrosine kinase activity and transforming potential of Bcr/Abl fusion proteins. Interestingly, the Bcr sequences required for Bcr/Abl transforming activity bind to SH2 domains, including that of Abl, and this interaction may disrupt negative regulation of the Abl tyrosine kinase.

In addition to increasing its tyrosine kinase activity, fusion with either Gag or Bcr sequences alters the subcellular localization of Abl. In contrast with members of the Src family, the normal Abl protein is localized primarily to the nucleus, where it may actually inhibit rather than stimulate cell proliferation. Fusion with Gag or Bcr sequences relocalizes Abl to the cytoplasm, presumably allowing it to phosphorylate cytoplasmic proteins that function in signal transduction pathways leading to cell proliferation. Such phosphorylation of alternative substrates is facilitated by the binding of Bcr sequences to SH2 domains, leading to the interaction of Bcr/Abl fusion proteins with other signal transducing molecules. Thus, the oncogenic Abl fusion proteins differ from normal Abl not only in tyrosine kinase activity, but also in subcellular localization and substrate specificity.

The viral *fes* oncogene, like *abl,* is expressed as a fusion protein with Gag sequences. In this case, it has been found that fusion of Gag sequences to the

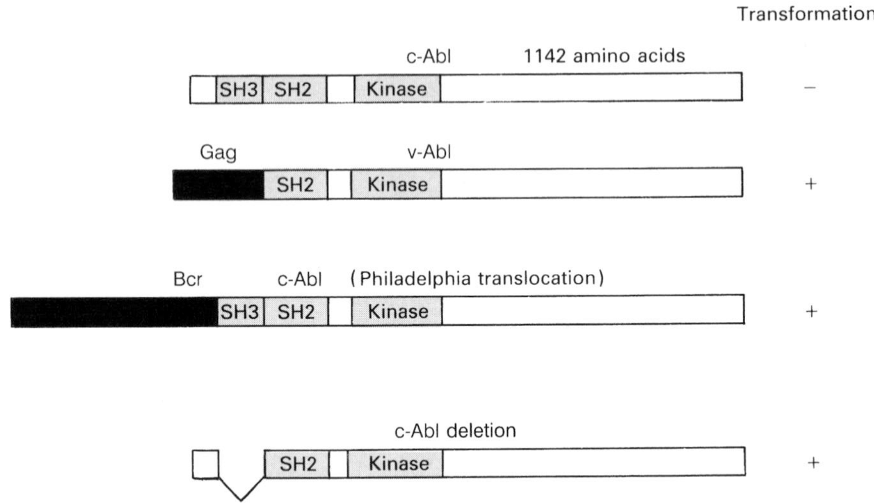

FIGURE 13.9
Activation of the *abl* oncogene. The *abl* oncogene is activated as a recombinant fusion protein both in Abelson leukemia virus (v-*abl*) and by the Philadelphia chromosome translocation. In both cases, oncogene activation involves deletion of the normal amino-terminal sequences of the proto-oncogene. *In vitro* mutagenesis experiments have demonstrated directly that deletion of SH3 is sufficient to activate transforming potential, but the fused Bcr sequences are required for oncogenic activity of the Bcr/Abl fusion protein.

normal amino terminus of the Fes protein is sufficient to activate both transforming potential and kinase activity. The normal function of Fes, however, is not understood; so details of the biochemical alterations that convert it into an oncogene remain to be elucidated.

Five other subfamilies of nonreceptor protein-tyrosine kinases (in addition to the Src, Abl, and Fes subfamilies) have been identified by biochemical isolation of proteins with tyrosine kinase activity or by molecular cloning of cDNAs by the use of conserved protein-tyrosine kinase sequences as probes (Table 13.2). One of these subfamilies is represented by Csk, which was discussed earlier as a negative regulator of Src family kinases. The others have not so far been shown to be directly involved in oncogenic transformation, although at least some normally function (like members of the Src subfamily) to transmit intracellular signals from cell surface receptors and might, therefore, be expected to have potential oncogenic activity in the appropriate cell types.

Oncogenes Derived from Nontyrosine Kinase Receptors

Although a number of growth factor receptors are protein-tyrosine kinases (Table 13.1), there are also a number of receptors that do not have intrinsic protein-tyrosine kinase activity. These receptors instead transmit intracellular signals by stimulating other proteins, which frequently include nonreceptor tyrosine kinases. Because stimulation of some of these nontyrosine kinase receptors can induce cell proliferation, it is not surprising that members of this group also act as oncogenes.

One example is the receptor for erythropoietin, a growth factor that induces proliferation and maturation of erythroid cells. The erythropoietin receptor is a member of the cytokine receptor family, which includes the receptors for most hematopoietic growth factors, including three of the potentially oncogenic hematopoietic growth factors discussed in chapter 12 (interleukin-2, interleukin-3, and GM-CSF). The cytokine receptors, like the tyrosine kinase receptors, have a single transmembrane domain and an extracellular ligand binding domain. However, the intracellular domain of the cytokine receptors lacks intrinsic protein kinase activity. Instead, the intracellular domains of these receptors are associated with nonreceptor protein-tyrosine kinases, particularly members of the Jak subfamily, which are activated in response to ligand binding.

The erythropoietin receptor was first implicated in oncogenesis by studies of Friend spleen focus-forming virus (SFFV), a murine retrovirus that rapidly induces erythroleukemia. The envelope glycoprotein encoded by SFFV was found to bind to the erythropoietin receptor and mimic the action of erythropoietin in stimulating cell proliferation, thereby contributing to the development of leukemia by acting as an autocrine growth factor. Subsequent studies identified a point mutation in the erythropoietin receptor that resulted in its constitutive ligand-independent activity and converted the normal receptor into an oncogene protein capable of inducing erythroleukemia in mice. This demonstration that a cytokine receptor could function as an oncogene is further fortified by the finding that the retroviral oncogene, *mpl*, also encodes a member of the cytokine receptor family, which has recently been identified as the thrombopoietin receptor.

The other class of receptors that have been found to act as oncogenes is coupled to intracellular targets not by protein-tyrosine kinases but by guanine nucleotide binding proteins (G proteins), which are discussed in the next chapter. The prototype of this group of receptor oncogenes is *mas*, which was identified as an oncogene in gene transfer assays (see chapter 5). The sequence of the *mas* oncogene predicted that it encoded a protein with seven transmembrane domains, which is characteristic of the structure of G protein-coupled receptors (Fig. 13.10). Further studies demonstrated that Mas is a receptor for the neurotransmitter angiotensin. Activation of Mas stimulates its associated G protein, which in turn stimulates hydrolysis of phosphatidylinositides. As discussed in chapter 14, this is an important second messenger system that triggers proliferation of a variety of cells, consistent with the overexpression of Mas resulting in oncogenic transformation. Like Mas, several other G protein-coupled receptors that stimulate phosphatidylinositide hydrolysis can also act as oncogenes, including the serotonin 1c receptor, certain muscarinic acetylcholine receptor subtypes, and the α_{1B}-adrenergic receptor.

Mutations of the gene encoding another G protein-coupled receptor, the thyrotropin receptor, appear to activate it as an oncogene in some human thyroid adenomas. Thyrotropin is a pituitary hormone that stimulates the growth of thyroid cells through a G protein-coupled receptor that stimulates adenylate cyclase, leading to increased synthesis of cAMP and activation of cAMP-dependent protein kinase. Analysis of a series of thyroid adenomas identified mutations in the thyrotropin receptor gene of three out of eleven tumors. These mutations resulted in constitutive activity of the receptor, leading to unregulated activation of adenylate cyclase and continual stimulation of thyroid cell proliferation. It thus appears that constitutive activity of G protein-coupled receptors, like that of tyrosine kinase-coupled receptors, can contribute to neoplastic transformation as a result of abnormal stimulation of cell division.

Summary

The protein-tyrosine kinases constitute a large functional group of oncogenes, with more than two dozen members divided into two broad families, receptor and nonreceptor kinases. Activation of the tyrosine kinases as oncogenes can result from a variety of different molecular alterations, with the common end result of increased unregulated catalytic activity. The fact that some of these oncogene products are derived from receptors for known growth factors clearly indicates that their normal function is to transduce information across the plasma membrane, coupling growth factor binding to an intracellular response. Deregulated expression of these tyrosine kinases results in neoplastic transformation, emphasizing that transformation can result from aberrant function of proteins that regulate normal cell proliferation. This is also illustrated by the ability of some cytokine and G protein-coupled receptors to act as oncogenes. Of still broader significance, however, is the fact that studies of the protein-tyrosine kinase oncogenes first demonstrated the role of protein phosphorylation in general, and protein-tyrosine phosphorylation in particular, in mitogenic signal transduction systems: a theme that remains dominant in our considerations of the control of cell proliferation.

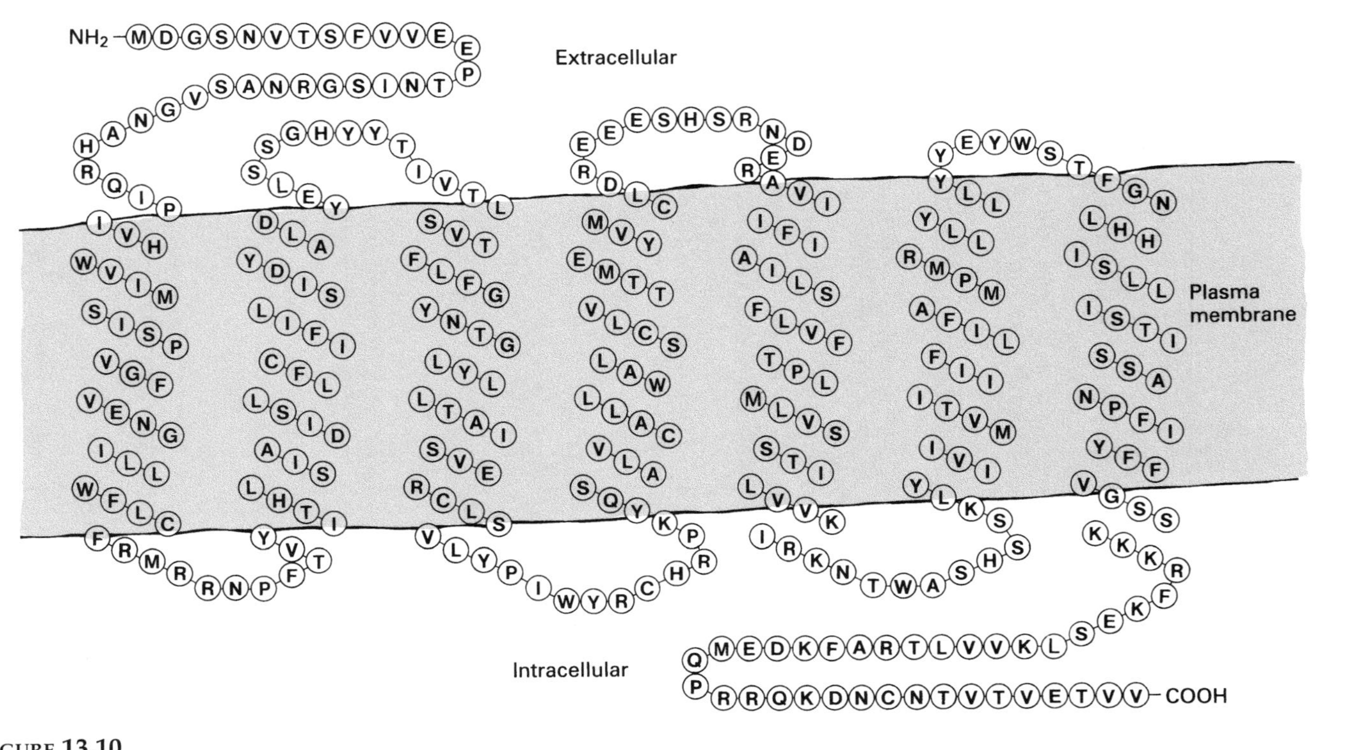

FIGURE 13.10
Structure of the Mas protein. The *mas* oncogene product contains seven helical transmembrane domains, characteristic of cell surface receptors that are coupled to second messenger metabolism by G proteins. Amino acids are designated by the single letter code, as in Figure 13.3. (Adapted from D. Young et al., *Cell* 45:711–719, 1986.)

References

General References

Aaronson, S.A. 1991. Growth factors and cancer. *Science* 254:1146–1153.

Hanks, S.K., Quinn, A.M., and Hunter, T. 1988. The protein kinase family: conserved features and deduced phylogeny of the catalytic domains. *Science* 241:42–52.

Hunter, T. 1991. Cooperation between oncogenes. *Cell* 64:249–270.

Taylor, S.S., Knighton, D.R., Zheng, J., Eyck, L.F.T., and Sowadski, J.M. 1992. Structural framework for the protein kinase family. *Ann. Rev. Cell. Biol.* 8:429–462.

Lessons from Src: Protein-Tyrosine Kinases

Brugge, J.S., and Erikson, R.L. 1977. Identification of a transformation-specific antigen induced by an avian sarcoma virus. *Nature* 269:346–348.

Collett, M.S., and Erikson, R.L. 1978. Protein kinase activity associated with the avian sarcoma virus *src* gene product. *Proc. Natl. Acad. Sci. USA* 75:2021–2024.

Collett, M.S., Purchio, A.F., and Erikson, R.L. 1980. Avian sarcoma virus-transforming protein, pp60^{src} shows protein kinase activity specific for tyrosine. *Nature* 285:167–169.

Eckhart, W., Hutchinson, M.A., and Hunter, T. 1979. An activity phosphorylating tyrosine in polyoma T antigen immunoprecipitates. *Cell* 18:925–933.

Hunter, T., and Sefton, B.M. 1980. Transforming gene product of Rous sarcoma virus phosphorylates tyrosine. *Proc. Natl. Acad. Sci. USA* 77:1311–1315.

Levinson, A.D., Oppermann, H., Levintow, L., Varmus, H.E., and Bishop, J.M. 1978. Evidence that the transforming gene of avian sarcoma virus encodes a protein kinase associated with a phosphoprotein. *Cell* 15:561–572.

Witte, O.N., Dasgupta, A., and Baltimore, D. 1980. Abelson murine leukaemia virus protein is phosphorylated *in vitro* to form phosphotyrosine. *Nature* 283:826–831.

Growth Factor Receptors: Protein-Tyrosine Kinases and Oncogenes

Barbacid, M. 1993. Nerve growth factor: a tale of two receptors. *Oncogene* 8:2033–2042.

Bargmann, C.I., Hung, M.-C., and Weinberg, R.A. 1986. Multiple independent activations of the *neu* oncogene by a point mutation altering the transmembrane domain of p185. *Cell* 45:649–657.

Bargmann, C.I., and Weinberg, R.A. 1988. Increased tyrosine kinase activity associated with the protein encoded by the activated *neu* oncogene. *Proc. Natl. Acad. Sci. USA* 85:5394–5398.

Bartley, T.D., Hunt, R.W., Welcher, A.A., Boyle, W.J., Parker, V.P., Lindberg, R.A., Lu, H.S., Colombero, A.M., Elliott, R.L., Guthrie, B.A., Holst, P.L., Skrine, J.D., Toso, R.J., Zhang, M., Fernandez, E., Trall, G., Varnum, B., Yarden, Y., Hunter, T., and Fox, G.M. 1994. B61 is a ligand for the ECK receptor protein-tyrosine kinase. *Nature* 368:558–560.

Birchmeier, C., O'Neill, K., Riggs, M., and Wigler, M. 1990. Characterization of *ROS1* cDNA from a human glioblastoma cell line. *Proc. Natl. Acad. Sci. USA* 87:4799–4803.

Bottaro, D.P., Rubin, J.S., Faletto, D.L., Chan, A. M.-L., Kmiecik, T.E., Vande Woude, G.F., and Aaronson, S.A. 1991. Identification of the hepatocyte growth factor receptor as the c-*met* proto-oncogene product. *Science* 251:802–804.

Browning, P.J., Bunn, H.F., Cline, A., Shuman, M., and Nienhuis, A.W. 1986. "Replacement" of COOH-terminal truncation of v-*fms* with c-*fms* sequences markedly reduces transformation potential. *Proc. Natl. Acad. Sci. USA* 83:7800–7804.

Carpenter, G., King, L., and Cohen, S. 1978. Epidermal growth factor stimulates phosphorylation in membrane preparations *in vitro*. *Nature* 276:409–410.

Chabot, B., Stephenson, D.A., Chapman, V.M., Besmer, P., and Bernstein, A. 1988. The proto-oncogene c-*kit* encoding a transmembrane tyrosine kinase receptor maps to the mouse *W* locus. *Nature* 335:88–89.

Dean, M., Park, M., LeBeau, M.M., Robins, T.S., Diaz, M.O., Rowley, J.D., Blair, D.G., and Vande Woude, G.F. 1985. The human *met* oncogene is related to the tyrosine kinase oncogenes. *Nature* 318:385–388.

Di Fiore, P.P., Pierce, J.H., Fleming, T.P., Hazan, R., Ullrich, A., King, C.R., Schlessinger, J., and Aaronson, S.A. 1987. Overexpression of the human EGF receptor confers an EGF-dependent transformed phenotype to NIH 3T3 cells. *Cell* 51:1063–1070.

Di Fiore, P.P., Pierce, J.H., Kraus, M.H., Segatto, O., King, C.R., and Aaronson, S.A. 1987. *erb*B-2 is a potent oncogene when overexpressed in NIH/3T3 cells. *Science* 237:178–182.

Downward, J., Yarden, Y., Mayes, E., Scrace, G., Totty, N., Stockwell, P., Ullrich, A., Schlessinger, J., and Waterfield, M.D. 1984. Close similarity of epidermal growth factor receptor and v-*erb*-B oncogene protein sequences. *Nature* 307:521–527.

Fantl, W.J., Johnson, D.E., and Williams, L.T. 1993. Signalling by receptor tyrosine kinases. *Ann. Rev. Biochem.* 62:453–482.

Geissler, E.N., Ryan, M.A., and Housman, D.E. 1988. The dominant-white spotting (*W*) locus of the mouse encodes the c-*kit* proto-oncogene. *Cell* 55:185–192.

Hattori, Y., Odagiri, H., Nakatani, H., Miyagawa, K., Naito, K., Sakamoto, H., Katoh, O., Yoshida, T., Sugimura, T., and Terada, M. 1990. K-*sam*, an amplified gene in stomach cancer, is a member of the heparin-binding growth factor receptor genes. *Proc. Natl. Acad. Sci. USA* 87:5983–5987.

Huff, J.L., Jelinek, M.A., Borgman, C.A., Lansing, T.J., and Parsons, J.T. 1993. The protooncogene c-*sea* encodes a transmembrane protein-tyrosine kinase related to the Met/hepatocyte growth factor/scatter factor receptor. *Proc. Natl. Acad. Sci. USA* 90:6140–6144.

Kaplan, D.R., Hempstead, B.L., Martin-Zanca, D., Chao, M.V., and Parada, L.F. 1991. The *trk* proto-oncogene product: a signal transducing receptor for nerve growth factor. *Science* 252:554–558.

Klein, R., Jing, S., Nanduri, V., O'Rourke, E., and Barbacid, M. 1991. The *trk* proto-oncogene encodes a receptor for nerve growth factor. *Cell* 65:189–197.

Kraus, M.H., Fedi, P., Starks, V., Muraro, R., and Aaronson, S.A. 1993. Demonstration of ligand-dependent signaling by the *erb*B-3 tyrosine kinase and its constitutive activation in human breast tumor cells. *Proc. Natl. Acad. Sci. USA* 90:2900–2904.

Martin-Zanca, D., Hughes, S.H., and Barbacid, M. 1986. A human oncogene formed by the fusion of truncated tropomyosin and protein tyrosine kinase sequences. *Nature* 319:743–748.

Matsushime, H., Wang, L-H., and Shibuya, M. 1986. Human c-*ros*-1 gene homologous to the v-*ros* sequence of UR2 sarcoma virus encodes for a transmembrane receptorlike molecule. *Mol. Cell. Biol.* 6:3000–3004.

Morris, S.W., Kirstein, M.N., Valentine, M.B., Dittmer, K.G., Shapiro, D.N., Saltman, D.L., and Look, A.T. 1994. Fusion of a kinase gene, *ALK,* to a nucleolar protein gene, *NPM,* in non-Hodgkin's lymphoma. *Science* 263:1281–1284.

O'Bryan, J.P., Frye, R.A., Cogswell, P.C., Neubauer, A., Kitch, B., Prokop, C., Espinosa, R., III, Le Beau, M.M., Eary, H.S., and Liu, E.T. 1991. *axl,* a transforming gene isolated from primary human myeloid leukemia cells, encodes a novel receptor tyrosine kinase. *Mol. Cell. Biol.* 11:5016–5031.

Poon, B., Dixon, D., Ellis, L., Roth, R.A., Rutter, W.J., and Wang, L.-H. 1991. Molecular basis of the activation of the tumorigenic potential of Gag-insulin receptor chimeras. *Proc. Natl. Acad. Sci. USA* 88:877–881.

Roussel, M.F., Downing, J.R., Rettenmier, C.W., and Sherr, C.J. 1988. A point mutation in the extracellular domain of the human CSF-1 receptor (c-*fms* proto-oncogene product) activates its transforming potential. *Cell* 55:979–988.

Segatto, O., King, C.R., Pierce, J.H., Di Fiore, P.P., and Aaronson, S.A. 1988. Different structural alterations upregulate *in vitro* tyrosine kinase activity and transforming potency of the *erb*B-2 gene. *Mol. Cell. Biol.* 8:5570–5574.

Sherr, C.J., Rettenmier, C.W., Sacca, R., Roussel, M.F., Look, A.T., and Stanley, E.R. 1985. The c-*fms* proto-oncogene product is related to the receptor for the mononuclear phagocyte growth factor, CSF-1. *Cell* 41:665–676.

Takahashi, M., Asai, N., Iwashita, T., Isomura, T., Miyazaki, K., and Matsuyama, M. 1993. Characterization of the *ret* proto-oncogene products expressed in mouse L cells. *Oncogene* 8:2925–2929.

Takahashi, M., Buma, Y., Iwamoto, T., Inaguma, Y., Ikeda, H., and Hiai, H. 1988. Cloning and expression of the *ret* proto-oncogene encoding a tyrosine kinase with two potential transmembrane domains. *Oncogene* 3:571–578.

Takahashi, M., and Cooper, G.M. 1987. *ret* transforming gene encodes a fusion protein homologous to tyrosine kinases. *Mol. Cell. Biol.* 7:1378–1385.

Ullrich, A., and Schlessinger, J. 1990. Signal transduction by receptors with tyrosine kinase activity. *Cell* 61:203–212.

Ushiro, H., and Cohen, S. 1980. Identification of phosphotyrosine as a product of epidermal growth factor-activated protein kinase in A431 cell membranes. *J. Biol. Chem.* 255:8363–8365.

Van Daalen Wetters, T., Hawkins, S.A., Roussel, M.F., and Sherr, C.J. 1992. Random mutagenesis of CSF-1 receptor (*FMS*) reveals multiple sites for activating mutations within the extracellular domain. *EMBO J.* 11:551–557.

Weiner, D.B., Liu, J., Cohen, J.A., Williams, W.V., and Greene, M.I. 1989. A point mutation in the *neu* oncogene mimics ligand induction of receptor aggregation. *Nature* 339:230–231.

Wells, A., and Bishop, J.M. 1988. Genetic determinants of neoplastic transformation by the retroviral oncogene v-*erb*B. *Proc. Natl. Acad. Sci. USA* 85:7597-7601.

Nonreceptor Protein-Tyrosine Kinases

Bolen, J.B. 1993. Nonreceptor tyrosine protein kinases. *Oncogene* 8:2025–2031.

Cartwright, C.A., Eckhart, W., Simon, S., and Kaplan, P.L. 1987. Cell transformation by $pp60^{c\text{-}src}$ mutated in the carboxy-terminal regulatory domain. *Cell* 49:83–91.

Cooper, J.A., and Howell, B. 1993. The when and how of Src regulation. *Cell* 73:1051–1054.

Foster, D.A., Shibuya, M., and Hanafusa, H. 1985. Activation of the transformation potential of the cellular *fps* gene. *Cell* 42:105–115.

Franz, W.M., Berger, P., and Wang, J.Y.J. 1989. Deletion of an N-terminal regulatory domain of the c-*abl* tyrosine kinase activates its oncogenic potential. *EMBO J.* 8:137–147.

Jackson, P., and Baltimore, D. 1989. N-terminal mutations activate the leukemogenic potential of the myristoylated form of c-*abl*. *EMBO J.* 8:449–456.

Klages, S., Adam, D., Class, K., Fargnoli, J.F., Bolen, J.B., and Penhallow, R.C. 1994. Ctk: a protein-tyrosine kinase related to Csk that defines an enzyme family. *Proc. Natl. Acad. Sci. USA* 91:2597–2601.

Kmiecik, T.E., and Shalloway, D. 1987. Activation and suppression of $pp60^{c\text{-}src}$ transforming ability by mutation of its primary sites of tyrosine phosphorylation. *Cell* 49:65–73.

Koch, C.A., Anderson, D., Moran, M.F., Ellis, C., and Pawson, T. 1991. SH2 and SH3 domains: elements that control interactions of cytoplasmic signaling proteins. *Science* 252:668–674.

McWhirter, J.R., and Wang, J.Y. 1991. Activation of tyrosine kinase and microfilament-binding functions of c-*abl* by *bcr* sequences. *Mol. Cell. Biol.* 11:1553–1565.

Muller, A.J., Young, J.C., Pendergast, A.-M., Pondel, M., Landau, N.R., Littman, D.R., and Witte, O.N. 1991. *BCR* first exon sequences specifically activate the *BCR/ABL* tyrosine kinase oncogene of Philadelphia chromosome-positive human leukemias. *Mol. Cell. Biol.* 11:1785–1792.

Nada, S., Okada, M., MacAuley, A., Cooper, J.A., and Nakagawa, H. 1991. Cloning of a complementary DNA for a protein-tyrosine kinase that specifically phosphorylates a negative reglatory site of $p60^{c\text{-}src}$. *Nature* 351:69–72.

O'Brien, M., Fukui, Y., and Hanafusa, H. 1990. Activation of the proto-oncogene $p60^{c\text{-}src}$ by point mutations in the SH2 domain. *Mol. Cell. Biol.* 10:2855–2862.

Pawson, T., and Gish, G.D. 1993. SH2 and SH3 domains: from structure to function. *Cell* 71:359–362.

Pendergast, A.M., Muller, A.J., Havlik, M.H., Maru, Y., and Witte, O.N. 1991. BCR sequences essential for transformation by the *BCR-ABL* oncogene bind to the ABL SH2 regulatory domain in a non-phosphotyrosine-dependent manner. *Cell* 66:161–171.

Pendergast, A.M., Quilliam, L.A., Cripe, L.D., Bassing, C.H., Dai, Z., Li, N., Batzer, A., Rabun, K.M., Der, C.J., Schlessinger, J., and Gishizky, M.L. 1993. BCR-ABL-induced oncogenesis is mediated by direct interaction with the SH2 domain of the GRB-2 adaptor protein. *Cell* 75:175–185.

Piwnica-Worms, H., Saunders, K.B., Roberts, T.M., Smith, A.E., and Cheng, S.H. 1987. Tyrosine phosphorylation regulates the biochemical and biological properties of $pp60^{c\text{-}src}$. *Cell* 49:75–82.

Ren, R., Mayer, B.J., Cicchetti, P., and Baltimore, D. 1993. Identification of a ten-amino acid proline-rich SH3 binding site. *Science* 259:1157–1161.

Resh, M.D. 1994. Myristylation and palmitylation of Src family members: the fats of the matter. *Cell* 76:411–413.

Roussel, R.R., Brodeur, S.R., Shalloway, D., and Laudano, A.P. 1991. Selective binding of activated pp60$^{c\text{-}src}$ by an immobilized synthetic phosphopeptide modeled on the carboxyl terminus of pp60$^{c\text{-}src}$. *Proc. Natl. Acad. Sci. USA* 88:10696–10700.

Sawyers, C.L., McLaughlin, J., Goga, A., Havlik, M., and Witte, O. 1994. The nuclear tyrosine kinase c-Abl negatively regulates cell growth. *Cell* 77:121–131.

Seidel-Dugan, C., Meyer, B.E., Thomas, S.M., and Brugge, J.S. 1992. Effects of SH2 and SH3 deletions on the functional activities of wild-type and transforming variants of c-Src. *Mol. Cell. Biol.* 12:1835–1845.

Songyang, Z., Shoelson, S.E., Chaudhuri, M., Gish, G., Pawson, T., Haser, W.G., King, F., Roberts, T., Ratnofsky, S., Lechleider, R.J., Neel, B.G., Birge, R.B., Fajardo, J.E., Chou, M.M., Hanafusa, H., Schaffhausen, B., and Cantley, L.C. 1993. SH2 domains recognize specific phosphopeptide sequences. *Cell* 72:767–778.

Superti-Funga, G., Fumagilli, S., Koegl, M., Courtneidge, S.A., and Draetta, G. 1993. Csk inhibition of c-Src activity requires both the SH2 and SH3 domains of Src. *EMBO J.* 12:2625–2634.

Van Etten, R.A., Jackson, P., and Baltimore, D. 1989. The mouse type IV *c-abl* gene product is a nuclear protein, and activation of transforming ability is associated with cytoplasmic localization. *Cell* 58:669–678.

Veillette, A., Bookman, M.A., Horak, E.M., Samelson, L.E., and Bolen, J.B. 1989. Signal transduction through the CD4 receptor involves the activation of the internal membrane tyrosine-protein kinase p56^{lck}. *Nature* 338:257–259.

Oncogenes Derived from Nontyrosine Kinase Receptors

Allen, L.F., Lefkowitz, R.J., Caron, M.G., and Cotecchia, S. 1991. G-protein-coupled receptor genes as protooncogenes: constitutively activating mutation of the α_{1B}-adrenergic receptor enhances mitogenesis and tumorigenicity. *Proc. Natl. Acad. Sci. USA* 88:11354–11358.

Gutkind, J.S., Novotny, E.A., Brann, M.R., and Robbins, K.C. 1991. Muscarinic acetylcholine receptor subtypes as agonist-dependent oncogenes. *Proc. Natl. Acad. Sci. USA* 88:4703–4707.

Jackson, T.R., Blair, L.A.C., Marshall, J., Goedert, M., and Hanley, M.R. 1988. The *mas* oncogene encodes an angiotensin receptor. *Nature* 335:437–440.

Julius, D., Livelli, T.J., Jessell, T.M., and Axel, R. 1989. Ectopic expression of the serotonin 1c receptor and the triggering of malignant transformation. *Science* 244:1057–1062.

Li, J.-P., D'Andrea, A.D., Lodish, H.F., and Baltimore, D. 1990. Activation of cell growth by binding of Friend spleen focus-forming virus gp55 glycoprotein to the erythropoietin receptor. *Nature* 343:762–764.

Longmore, G.D., and Lodish, H.F. 1991. An activating mutation in the murine erythropoietin receptor induces erythroleukemia in mice: a cytokine receptor superfamily oncogene. *Cell* 67:1089–1102.

Parma, J., Duprez, L., Van Sande, J., Cochaux, P., Gervy, C., Mockel, J., Dumont, J., and

Vassart, G. 1993. Somatic mutations in the thyrotropin receptor gene cause hyperfunctioning thyroid adenomas. *Nature* 365:649–651.

Skoda, R.C., Seldin, D.C., Chiang, M.-K., Peichel, C.L., Vogt, T.F., and Leder, P. 1993. Murine c-*mpl:* a member of the hematopoietic growth factor receptor subfamily that transduces a proliferative signal. *EMBO J.* 12:2645–2653.

Souyri, M., Vigon, I., Penciolelli, J.-F., Heard, J.-M., Tambourin, P., and Wendling, F. 1990. A putative truncated cytokine receptor gene transduced by the myeloproliferative leukemia virus immortalizes hematopoietic progenitors. *Cell* 63:1137–1147.

Witthuhn, B.A., Quelle, F.W., Silvennoinen, O., Yi, T., Tang, B., Miura, O., and Ihle, J.N. 1993. JAK2 associates with the erythropoietin receptor and is tyrosine phosphorylated and activated following stimulation with erythropoietin. *Cell* 74:227–236.

Yoshimura, A., Longmore, G., and Lodish, H.F. 1990. Point mutation in the exoplasmic domain of the erythropoietin receptor resulting in hormone-independent activation and tumorigenicity. *Nature* 348:647–649.

Young, D., Waitches, G., Birchmeier, C., Fasano, O., and Wigler, M. 1986. Isolation and characterization of a new cellular oncogene encoding a protein with multiple potential transmembrane domains. *Cell* 45:711–719.

❖ CHAPTER 14

Guanine Nucleotide Binding Proteins

In addition to protein-tyrosine kinases, the last chapter discussed cell surface receptors that are coupled to intracellular targets by a family of guanine nucleotide binding proteins called G proteins. Because some G protein-coupled receptors can act as oncogenes, it may be expected that the G proteins also have potential oncogenic activity. Moreover, the G proteins are part of a larger superfamily of proteins whose function is regulated by GTP binding and hydrolysis. This group of proteins (called the *GTPase superfamily*) is of particular interest because it includes the *ras* oncogene products, which are not only frequently involved in human cancers but also play a central role in the propagation of signals initiated by growth factor binding to receptor protein-tyrosine kinases.

G Proteins

The G proteins function as physiological switches that regulate the activity of target enzymes in response to a variety of signals. A prototypical example is provided by hormonal stimulation of adenylate cyclase (Fig. 14.1). The binding of an appropriate hormone to its receptor activates the protein G_s, which binds to and stimulates its effector, adenylate cyclase. This results in increased synthesis of the second messenger, cAMP, which in turn activates cAMP-dependent protein kinase. Other receptors are coupled to intracellular effectors by different G proteins that can act to inhibit adenylate cyclase or to regulate the activities of enzymes affecting different second

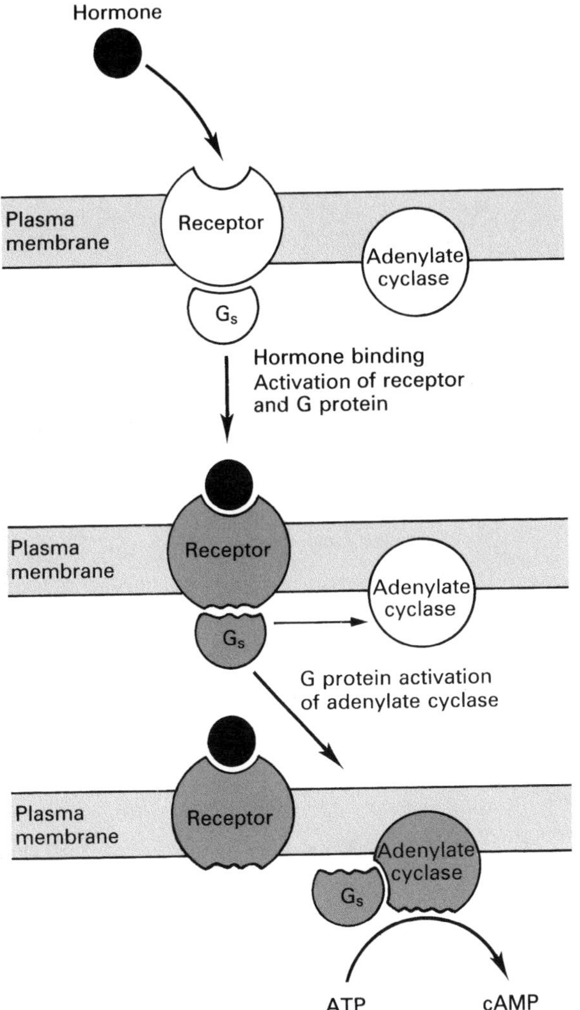

FIGURE 14.1
Hormonal activation of adenylate cyclase. Hormone binding to its receptor activates the G protein, G_s. Activated G_s stimulates adenylate cyclase, which catalyzes the conversion of ATP into cAMP.

messengers. These additional effectors of G proteins include cGMP phosphodiesterase, ion channels, and phospholipase C-ß.

The regulation of phospholipase C-ß is of particular interest because this enzyme catalyzes the hydrolysis of phosphatidylinositides to generate second messengers that signal proliferation of a variety of cell types (Fig. 14.2). In particular, phospholipase C (PLC) catalyzes the hydrolysis of phosphatidylinositol 4, 5-bisphosphate (PIP_2) to yield two second messengers, diacylglycerol (DAG)

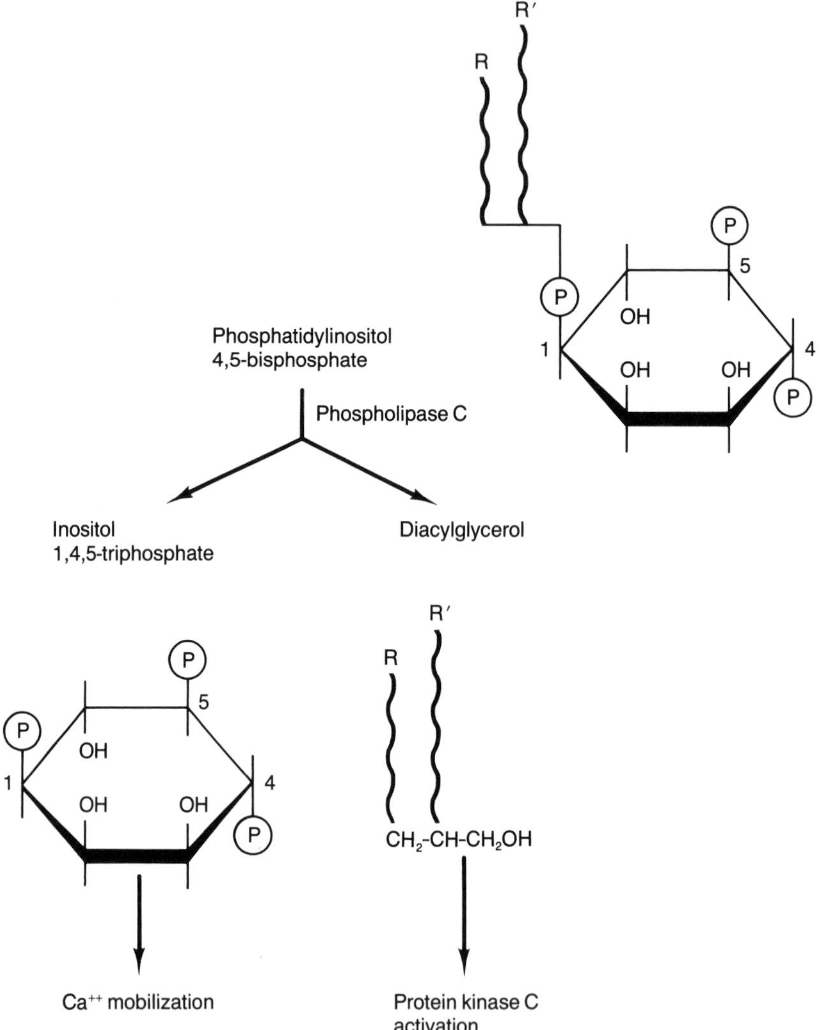

FIGURE 14.2
Phosphatidylinositol 4,5-bisphosphate hydrolysis generates two second messengers. The side chains designated R and R′ are usually stearate and arachidonate, respectively. Inositol 1,4,5-triphosphate stimulates release of calcium from intracellular stores. Diacylglycerol activates protein kinase C.

and inositol triphosphate (IP_3). DAG activates protein kinase C, a family of protein-serine/threonine kinases that will be discussed in detail in chapter 15. IP_3 functions to stimulate release of calcium from the endoplasmic reticulum. The mobilized calcium serves to activate a number of protein kinases, including a group of calcium/calmodulin-dependent protein kinases as well as some members of the protein kinase C family.

The importance of phosphatidylinositide hydrolysis in regulation of cell proliferation is indicated by the fact that a number of G protein-coupled receptors that stimulate PLC-ß can function as oncogenes, as discussed in the preceding chapter. In addition, a different isotype of phospholipase C, PLC-γ, is stimulated by receptor protein-tyrosine kinases; so both tyrosine kinase and G protein-coupled receptors activate this second messenger system. Another indication of the importance of this pathway is provided by the tumor promoter TPA (12-*O*-tetradecanoyl-phorbol-13-acetate), which is an analog of DAG that activates protein kinase C and stimulates cell division. The relation of the phosphatidylinositide second messenger system to cell proliferation makes it a prominent figure in considerations of oncogene function in this and later chapters.

The G proteins consist of three subunits, designated α, ß, and γ. There are at least sixteen different α subunits, which are proteins of ~40 kd that bind guanine nucleotides and function as signal transducing moieties by interacting with their target enzymes (for example, adenylate cyclase or PLC-ß). The ß and γ subunits (37 and 8 kd, respectively) are tightly bound to each other and function as a ßγ complex, which also interacts with some effector proteins. The activity of the α subunits is regulated by guanine nucleotide binding: GTP activates and GDP inhibits α function (Fig. 14.3). In the resting state, the α subunit is tightly bound to GDP in a complex with ßγ and a cell surface receptor. Ligand binding to the receptor results in an interaction with the G protein that increases the rate of dissociation of bound GDP from the α subunit. Because intracellular GTP concentrations are in excess of GDP, this usually results in exchange of the bound GDP for GTP, leading to α subunit activation. The activated α subunit (bound to GTP) dissociates from ßγ and the receptor and interacts with its target molecule to affect the programmed second messenger response, such as cAMP synthesis or PIP_2 hydrolysis. Activity of the α subunit is subsequently terminated by hydrolysis of the bound GTP to GDP through an intrinsic GTPase activity. The inactive GDP-bound form of α then reassociates with ßγ and receptor, ready for the cycle to start anew with ligand binding.

The α subunits of G proteins that stimulate PLC-ß can act as oncogenes, as would be predicted by the oncogenic activity of several receptors coupled to these G proteins (see chapter 13). In particular, the α subunit of the PLC-ß-stimulatory G_q protein can be converted into an oncogene product by muta-

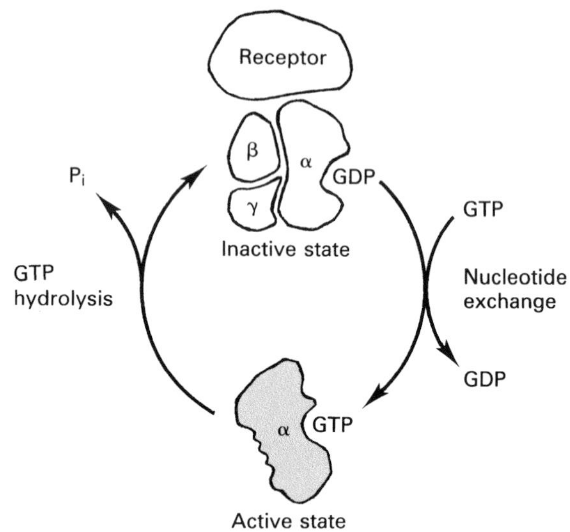

FIGURE 14.3
Regulation of G proteins by guanine nucleotide binding. In the inactive state, the G protein α subunit is bound to GDP in a complex with the ß and γ subunits. Ligand binding to a cell surface receptor induces the exchange of GDP for GTP. The activated GTP-bound α subunit dissociates from the ß and γ subunits and the receptor to interact with its target molecule (for example, adenylate cyclase). Activity of the α subunit is terminated by hydrolysis of the bound GTP to GDP.

tions that inhibit its GTPase activity, so that it is constitutively maintained in the active GTP-bound form. The effect of such activating mutations in $G_{q\alpha}$ is continual stimulation of PLC-ß, resulting in unregulated cell proliferation and transformation.

The proliferation of some cell types is controlled by cAMP and, for these cells, the G proteins that regulate adenylate cyclase can act as oncogenes. For example, hormones that stimulate cAMP synthesis promote the proliferation of cells of the pituitary and thyroid glands. As noted in the preceding chapter, a receptor coupled to G_s can act as an oncogene in some thyroid tumors. Not surprisingly then, the α subunit of G_s can also be activated as an oncogene in pituitary and thyroid tumors by mutations that maintain it in the constitutively active GTP-bound state, resulting in continual stimulation of adenylate cyclase. Conversely, activating mutations of a G protein α subunit that inhibits adenylate cyclase (αi2) convert it into an oncogene product in some adrenal cortical and ovarian tumors, as well as for some types of cells in culture. It is not clear, however, whether the transforming activity of αi2 results directly from inhibition of cAMP synthesis or from interaction of αi2 with another second messenger pathway. Another G protein α subunit, α12, also has been identified as an oncogene product that can induce transformation of cultured cells, although its target has not been established.

Similarities between Ras and the G Proteins

The *ras* oncogenes were first identified in the murine retroviruses, Harvey and Kirsten sarcoma viruses (see chapter 3). Analysis of these genes and their encoded proteins was pioneered in the laboratory of Edward Scolnick. By 1980, Scolnick and his colleagues had found that the *ras*H and *ras*K viral oncogenes encoded related proteins of 21 kd that were localized to the inner face of the plasma membrane. Importantly, these investigators also determined that the Ras proteins specifically bound guanine nucleotides—GDP and GTP—with high affinity and specificity. The guanine nucleotide binding activity of Ras proteins led Scolnick to initially suggest in 1979 that they might function analogously to the G proteins in coupling receptors to intracellular second messengers.

Studies of Ras intensified in a number of laboratories following the findings in 1982 of cellular *ras* oncogene activation in human tumors (see chapter 5). Three members of the cellular *ras* gene family have been identified—*ras*H, *ras*K, and *ras*N (Fig. 14.4). These genes encode closely related proteins of 188 or 189 amino acids, which differ primarily in a region of 20 amino acids near the carboxy terminus. Downstream of this variable region, a conserved cysteine (position 186) is the site of posttranslational addition of lipid (a farnesyl isoprenoid), a processing step that is required for both plasma membrane localization and biological activity of the Ras proteins. Following prenylation, the three amino acids C-terminal to Cys-186 are removed by proteolysis and Cys-186 is further modified by carboxymethylation. In addition, some Ras proteins are further modified by palmitoylation of upstream cysteines. In contrast with prenylation, however, these later steps in Ras processing may not be obligatory for either correct subcellular localization or function of the Ras proteins.

Further studies have identified more than forty additional genes that encode small GTP-binding proteins related to Ras. Most of these Ras-related proteins, which are from 30% to 50% identical with Ras, have not been found to induce oncogenic transformation and instead function to regulate other cellular processes. For example, members of the largest group of these Ras-related proteins (the Rab proteins) regulate vesicular transport in both the secretory and endocytic pathways. Other Ras-related proteins (the Rho and Rac proteins) regulate organization of the cytoskeleton. The functions of the closest relatives of Ras (Ral, Rap, and R-Ras), however, are unclear. At least one member of the R-Ras family is capable of inducing transformation and may function similarly to Ras in mitogenic signal transduction. It is also interesting that overexpression of one member of the Rap family (Rap1A) suppresses *ras*-induced cell transformation, possibly by competing with Ras for interaction with its target proteins.

```
                                               Splice
                       Exon 1                    ↓
RasH      MTEYKLVVVGAGGVGKSALTIQLIQNHFVDEYDPTIEDSYRKQVVIDGET  50
RasN      ..................................................
RasK      ..................................................
                                                       Splice
                       Exon 2                            ↓
RasH      CLLDILDTAGQEEYSAMRDQYMRTGEGFLCVFAINNTKSFEDIHQYREQI  100
RasN      ....................................S...A..NL.....
RasK      ............................................H.....
                                                          Splice
                                Exon 3                      ↓
RasH      KRVKDSDDVPMVLVGNKCDLAARTVESRQAQDLARSYGIPYIETSAKTRQ  150
RasN      ....................PT...DTK..HE..K.....F.........
RasK      ......E.............PS...DTK............F.........
                   Exon 4
RasH      GVEDAFYTLVREIRQHKLRKLNPPDESGPGCMSCK-CVLS
RasN      ...............YRMK...SS.DGTQ...GLP-..VM
RasK-4a   R..............YR.K.I-SKE.KT...VKI.K.IIM
RasK-4b   ..D...........K..-E.M-SK.GKKKKKK.KTK..IM
```

FIGURE 14.4
Amino sequences of Ras proteins. The *ras* genes contain four coding exons. Alternative splicing generates two RasK proteins, designated RasK-4a and RasK-4b, with different fourth exons. Identical amino acids are indicated by dots and alignment gaps are indicated by dashes. Each of the RasH, RasN, and RasK-4a proteins consist of 189 amino acids, whereas the RasK-4b protein is 188 amino acids. Amino acids are designated by single letters: A, alanine; C, cysteine; D, aspartic acid; E, glutamic acid; F, phenylalanine; G, glycine; H, histidine; I, isoleucine; K, lysine; L, leucine; M, methionine; N, asparagine; P, proline; Q, glutamine; R, arginine; S, serine; T, threonine; V, valine; W, tryptophan; Y, tyrosine.

The analogy between Ras and the G proteins was further strengthened in 1984, when several research groups reported that the Ras proteins not only specifically bound GDP and GTP, but also possessed an intrinsic GTPase activity that hydrolyzed bound GTP to GDP. The interactions of Ras with guanine nucleotides thus paralleled those of the G protein α subunits, suggesting that the functional activity of the Ras proteins might be regulated by a similar cycle of guanine nucleotide binding and hydrolysis. In contrast to the G protein α subunits, however, it should be noted that Ras proteins are not associated with ßγ subunits.

Studies of the mechanism of *ras* oncogene activation have further clarified the regulation of Ras function by GTP/GDP binding and strengthened the analogy between Ras and G protein α subunits. As discussed in chapter 5, the transforming activity of *ras* oncogenes in human tumors is a consequence of point mutations resulting in single amino acid substitutions. The first such mutation described was the substitution of valine for glycine at position 12. Subsequent studies of *ras* oncogenes isolated from a number of different human and rodent tumors, as well as *ras* oncogenes isolated from retroviruses and activated *in vitro* by mutagenesis of cloned *ras* proto-oncogenes, have revealed that a number of single amino acid substitutions in critical positions will activate Ras transforming potential. In addition to amino acid 12, activating mutations have been identified that alter amino acids 13, 59, 61, 63, 116, 117, 119, and 146. At several of these positions, particularly amino acids 12 and 61, substitutions of multiple different amino acids are capable of activating Ras transforming potential.

Biochemical studies have revealed that all of these activating mutations alter the interaction of Ras proteins with guanine nucleotides, and they do so as would be predicted by the G protein analogy (Fig. 14.5). Activating mutations of *ras* either decrease the GTPase activity of the Ras protein or they increase the rate of exchange of bound GDP for free GTP. Both of these alterations would be predicted to increase the amount of GTP-bound, and therefore presumably active, Ras protein. An increased rate of nucleotide exchange would increase the rate of formation of the Ras-GTP complex, whereas a decreased rate of GTP hydrolysis would decrease the rate of its physiological inactivation. The point

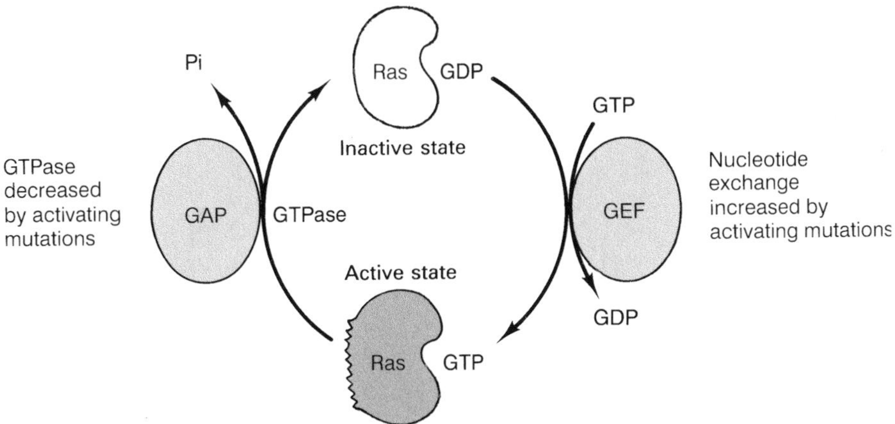

FIGURE 14.5
Interaction of Ras protein with guanine nucleotides. In the inactive state, Ras is bound to GDP. Exchange of bound GDP for GTP is associated with conversion of Ras into the active state, which is terminated by GTP hydrolysis. Guanine nucleotide exchange factors (GEFs) stimulate GDP/GTP exchange and GTPase activating proteins (GAPs) stimulate GTP hydrolysis. Mutations that activate *ras* oncogenes increase the rate of GDP/GTP exchange or decrease GTPase activity, thereby increasing the level of active GTP-bound Ras.

mutations that convert Ras into an oncogene product thus alter its interactions with guanine nucleotides so as to maintain the Ras protein in the constitutively active GTP-bound state.

Structural analysis of Ras proteins by X-ray crystallography has further demonstrated that the amino acids affected by activating mutations are part of a guanine nucleotide binding pocket of the Ras protein (Fig. 14.6). This pocket includes three regions of the polypeptide chain (Ras amino acids 10–17, 57–60, and 116–119) that are highly conserved throughout the GTPase superfamily, in addition to Ras amino acids 144–146. Alterations in GTPase result from mutations affecting amino acids that interact with the phosphates of bound guanine nucleotides (for example, amino acids 12 and 61), whereas increases in nucleotide exchange result from mutations affecting amino acids that interact with the guanine base (amino acids 116–119 and 146).

Comparisons of the structures of GDP- and GTP-bound forms of Ras also have elucidated the effect of GTP binding on the three-dimensional conforma-

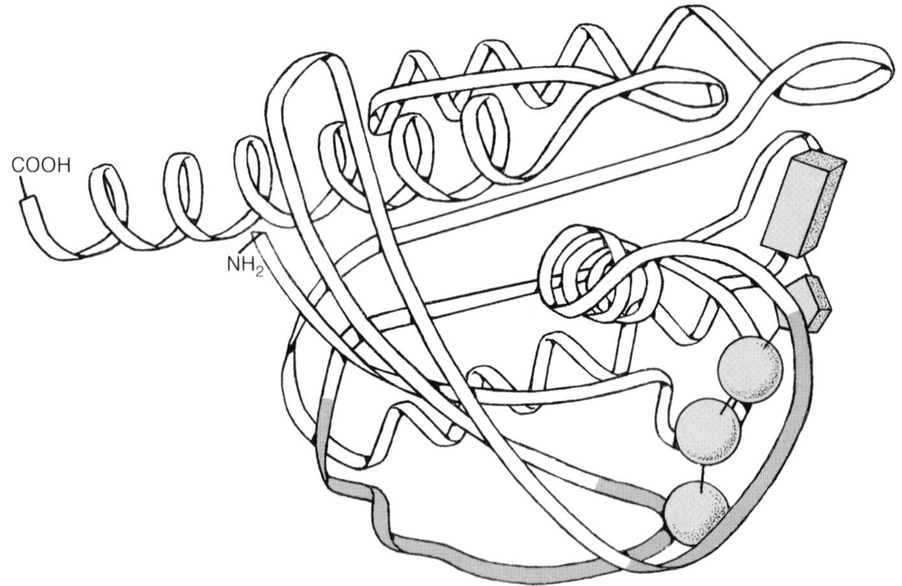

Figure 14.6
Structure of Ras bound to GTP. The backbone of the polypeptide chain is represented as a ribbon. The guanine base, ribose, and phosphates are depicted as rectangular box, pentagonal box, and circles, respectively. Two regions of Ras (indicated by shading) undergo a conformational change as a result of GDP/GTP exchange. (Courtesy of Dr. S.-H. Kim.)

tion of the Ras protein. The structures of Ras-GTP and Ras-GDP complexes differ in two principal regions: amino acids 30–38 and 60–76 (Fig. 14.6). Both of these regions are exposed on the surface of the folded protein, as would be expected if conformational changes induced by GTP binding regulate the interaction of Ras with its target molecule. Further consistent with this hypothesis, earlier studies had shown that mutations altering amino acids 30–40 and surrounding residues inactivated *ras* transforming potential, as predicted if this region of the protein is an effector domain. The combined results of mutagenesis and structural analysis thus provide strong support for the conclusion that GTP binding induces a conformational change that controls the interaction of Ras with its target proteins.

Proteins That Regulate Ras

The ability of Ras to alternate between inactive (GDP-bound) and active (GTP-bound) forms is clearly a major determinant of cell growth and proliferation; so it is to be expected that a number of cellular proteins take part in regulating the guanine nucleotide exchange and GTPase activities of the Ras proteins. In particular, the activation of Ras proteins by exchange of GTP for bound GDP is stimulated by guanine nucleotide exchange factors (GEFs), whereas the inactivation of the Ras-GTP complex is stimulated by GTPase activating proteins (GAPs) (see Fig. 14.5). Given the oncogenic activity of mutant Ras proteins resulting from aberrant guanine nucleotide exchange and GTPase activities, it is not surprising that mutations affecting Ras-GEFs and Ras-GAPs also can contribute to cell transformation.

The first of these Ras-regulatory proteins to be characterized was GAP, which was identified in Frank McCormick's laboratory as an activity in cell extracts that substantially increased Ras GTPase activity. Although GAP increases the GTP hydrolysis activity of normal Ras proteins by several orders of magnitude, it does not affect the GTPase-deficient Ras proteins encoded by mutationally activated *ras* oncogenes. It thus appears that GAP specifically acts as a negative regulator of normal Ras and that the GAP resistance of Ras oncogene proteins is a critical determinant of their transforming activity.

Purification and cloning of GAP from mammalian cells have shown it to be a protein of ~120 kd, and related proteins also have been identified in *Drosophila* and yeast. Interestingly, mammalian GAP contains SH2 and SH3 domains that, as discussed in the preceding chapter, serve as sites of protein–protein interactions. SH2 domains in particular serve as binding sites for phosphotyrosine-containing peptides and, as discussed further in chapter 17, the SH2 domains of GAP lead to its association with protein-tyrosine kinases.

As expected from its ability to stimulate Ras GTPase, GAP acts as a negative regulator of Ras activity. Such negative regulatory activity of GAP has been directly demonstrated by mutational inactivation of GAP in yeast and *Drosophila,* which results in constitutive Ras signaling. In mammalian cells, overexpression of GAP also down-regulates Ras activity—it increases the fraction of Ras in the inactive GDP-bound state and suppresses transformation resulting from overexpression of normal Ras or from the action of oncogenes (for example, *src*) that act upstream of Ras in signal transduction pathways. It has been further suggested, however, that GAP may play a dual role, serving as an effector as well as a negative regulator of Ras function. Consistent with this hypothesis, GAP interacts with the Ras effector domain (amino acids 30–40). In addition, the targets of other G proteins, such as PLC-ß, have been found to stimulate GTPase activity of the G_α subunits with which they interact. Whether GAP similarly serves as an effector for Ras signaling in mammalian cells remains an open question. It is clear, however, that GAP is not the only target of Ras and that other effectors (discussed in the next section) are required for Ras activity.

A second protein with GAP activity is encoded by the *NF1* tumor suppressor gene, which is inactivated in neurofibromatosis type 1 (see chapter 10). The *NF1* gene product, called neurofibromin, is a protein of ~280 kd containing an internal domain of 350 amino acids that is related to the catalytic (GTPase activating) domain of GAP. In tumor cell lines derived from patients with neurofibromatosis type 1, the lack of neurofibromin activity has been shown to result in aberrant regulation of Ras, with high levels of Ras being present in the active GTP-bound form. At least one function of the *NF1* tumor suppressor gene product thus appears to be negative regulation of Ras proto-oncogene proteins. It is noteworthy, however, that neurofibromin also functions as a tumor suppressor in some types of cells without affecting the proportion of Ras in the GTP-bound form and, therefore, can apparently act independently of its GAP activity. Such GAP-independent action of neurofibromin as a tumor suppressor might result from competition for the binding of effector molecules to Ras or it might indicate that neurofibromin plays an additional role in the control of cell growth and differentiation.

In contrast to GAP and neurofibromin, the guanine nucleotide exchange factors (GEFs) function as positive regulators of Ras by stimulating the exchange of bound GDP for free GTP. The first of these proteins (CDC25) was identified in yeast and subsequently used to isolate a cDNA encoding a related protein from rat brain (Ras-GRF or $CDC25^{Mm}$). Genetic studies in *Drosophila* identified a distinct guanine nucleotide exchange factor (Sos), which was also used to isolate mammalian homologs. Consistent with the ability of these factors to stimulate the conversion of Ras-GDP into the active Ras-GTP complex, both a yeast exchange factor related to CDC25 (SDC25) and *Drosophila* Sos are able to induce transformation of mammalian cells in culture.

As discussed in detail in chapter 17, Sos plays a critical role in activating Ras proteins in response to growth factor stimulation of protein-tyrosine kinases. In this role, Sos associates with phosphotyrosine-containing proteins through an adaptor molecule, Grb2, which contains SH2 and SH3 domains. The prototype of such adaptor molecules is the protein encoded by the *crk* oncogene, initially isolated from an avian sarcoma virus, which contains only SH2 and SH3 domains with no catalytic sequences. Other members of this group of adaptor proteins, including Shc, Nck, and Grb2, also appear to have potential oncogenic activity. Rather than acting as enzymes, these molecules function by facilitating protein–protein interactions, including interactions that result in Ras activation by guanine nucleotide exchange factors. In growth factor-stimulated cells, for example, Grb2 binds to Sos through its SH3 domains and to phosphotyrosine-containing peptides of growth factor receptors through its SH2 domain (Fig. 14.7). Grb2 thus serves to recruit Sos to the plasma membrane in association with an activated protein-tyrosine kinase. Sos then acts to stimulate Ras guanine nucleotide exchange, resulting in the conversion of Ras into its active GTP-bound state. The oncogenic activity of both gua-

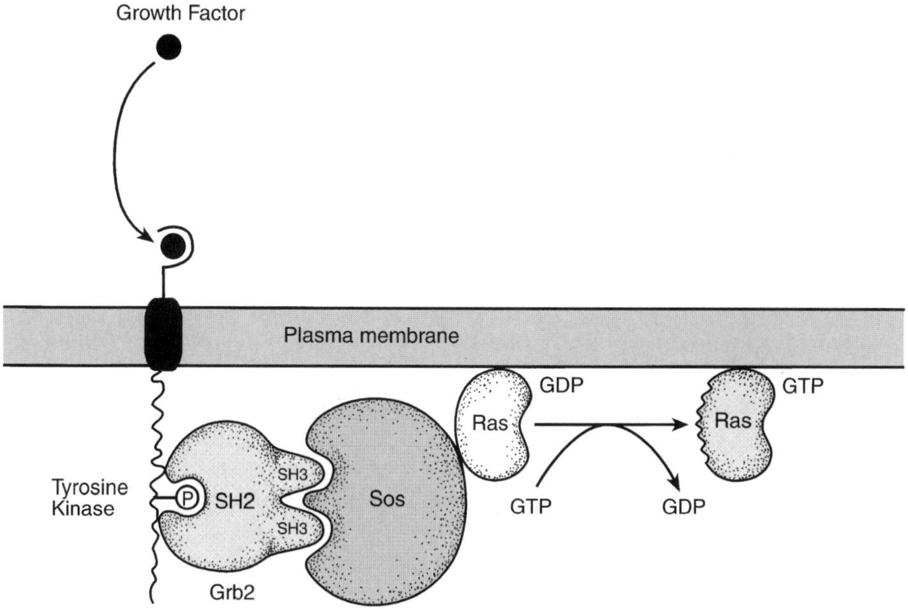

FIGURE 14.7
Role of adaptor proteins in Ras regulation. Grb2 is an example of an adaptor protein containing one SH2 and two SH3 domains. It functions in Ras activation by recruiting the guanine nucleotide exchange factor Sos to an activated receptor protein-tyrosine kinase. The Grb2 SH2 domain binds to a phosphotyrosine- containing peptide of the receptor and the SH3 domains bind to Sos, which then stimulates Ras GDP/GTP exchange.

nine nucleotide exchange factors and adaptor molecules thus results from uncontrolled activation of Ras, further emphasizing the critical importance of Ras in control of cell proliferation.

It is also of interest that several other oncogene products are related to guanine nucleotide exchange factors that may act on other Ras-related proteins, such as Rac and Rho. The Dbl oncogene protein, in particular, is a guanine nucleotide exchange factor for CDC42—a member of the Rho family of GTP-binding proteins that act to regulate the cytoskeleton. The Ect2 and Vav oncogene proteins also have GEF domains similar to that of Dbl, leading to the prediction that these proteins may function as guanine nucleotide exchange factors for Rac and Rho proteins. A similar GEF domain is present in Bcr, which fuses with Abl to form the Bcr/Abl oncogene protein as a result of the Philadelphia chromosome translocation in chronic myelogenous leukemia (see chapter 7). It is at present unclear, however, how the GEF domains of these oncogene proteins contribute to cell transformation. Somewhat surprisingly on the basis of its structure, Vav has been reported to stimulate Ras guanine nucleotide exchange, suggesting that it may function directly to activate Ras, rather than members of the Rac and Rho families.

Targets of Ras

If Ras proteins act like the G proteins to couple cell surface receptors to second messengers, the next apparent question is the identity of the receptor-second messenger systems in which they function. This question was first productively studied in yeast, where the powerful genetic approaches available in this unicellular organism could be applied to the problem of understanding Ras function.

The *ras* genes of yeast (in particular *Saccharomyces cerevisiae*) have been extensively studied in the laboratories of Edward Scolnick and Michael Wigler. Yeast contains two *ras*-related genes, designated *RAS*-1 and *RAS*-2, which were identified by hybridization with mammalian *ras* probes. Both of these yeast genes encode proteins that are substantially larger than the mammalian Ras proteins—~40 kd rather than 21 kd. However, the mammalian Ras proteins are quite similar to the amino-terminal half of the yeast proteins. The difference in size is due to a large carboxy-terminal region of ~120 amino acids in the yeast proteins that is absent from the mammalian *ras* gene products. Moreover, the yeast RAS proteins are functionally similar to the mammalian Ras proteins in their interactions with guanine nucleotides, and mutations in homologous residues of yeast and mammalian genes have similar biochemical consequences. The yeast *RAS* genes thus appear to represent a reasonable model for their mammalian relatives.

The biological function of genes in yeast can be readily determined by gene disruption methods. Under appropriate conditions, there is a high frequency of homologous recombination between a transfected cloned DNA segment and its resident chromosomal homolog. Consequently, transfection with a modified or partial gene copy will lead to efficient inactivation of its chromosomal allele. Such experiments with yeast *RAS* genes indicated that either *RAS*-1 or *RAS*-2 could be inactivated without loss of viability. However, inactivation of both of these *RAS* genes was incompatible with proliferation. Spores containing disruptions of both *RAS*-1 and *RAS*-2 were incapable of growth, indicating that RAS function was required for normal yeast proliferation.

The first insight into the target for RAS came from searching for mutations in other genes that would allow growth of yeast bearing mutations in both *RAS* genes. Such experiments indicated that yeast lacking any RAS function could grow if a mutation in the gene *bcy*-1 was also present. Further studies established that the *bcy*-1 mutation resulted in constitutive activity of cAMP-dependent protein kinase even in the absence of cAMP. The ability of the *bcy*-1 mutation to allow yeast to grow in the absence of RAS function therefore suggested that yeast RAS proteins might regulate adenylate cyclase and be required for cAMP synthesis.

The interaction of RAS proteins with adenylate cyclase predicted by these genetic analyses was directly demonstrated by biochemical experiments. Either yeast RAS or mammalian Ras proteins functioned to activate adenylate cyclase in yeast membranes. As predicted by the G protein analogy, the RAS proteins were active when bound to GTP but not to GDP. It thus appears that, in *Saccharomyces cerevisiae,* RAS proteins function analogously to the mammalian protein G_s to regulate adenylate cyclase activity.

Surprisingly, however, adenylate cyclase does not appear to be the effector of Ras in other organisms, particularly in mammalian cells. Consequently, these studies in *S. cerevisiae* failed to reveal the identity of the mammalian Ras targets. The initial insights into this question came instead from microinjection experiments in which an anti-Ras monoclonal antibody was used to block Ras function. Microinjection of cells with this antibody was found to inhibit transformation by *src* or *ras* oncogenes, but not by *raf,* which encodes a protein-serine/threonine kinase discussed in the next chapter. Similar results were obtained using a dominant inhibitory Ras mutant (Ras N-17), which preferentially binds GDP rather than GTP and appears to interfere with normal Ras activity by blocking the action of guanine nucleotide exchange factors. These experiments thus suggested that these oncogene proteins functioned in a signal transduction cascade (Fig. 14.8). Protein-tyrosine kinases, such as Src, required Ras to induce transformation and therefore appeared to act upstream of Ras. On the other hand, transformation by protein-serine/threonine kinases (for example, Raf) did not require

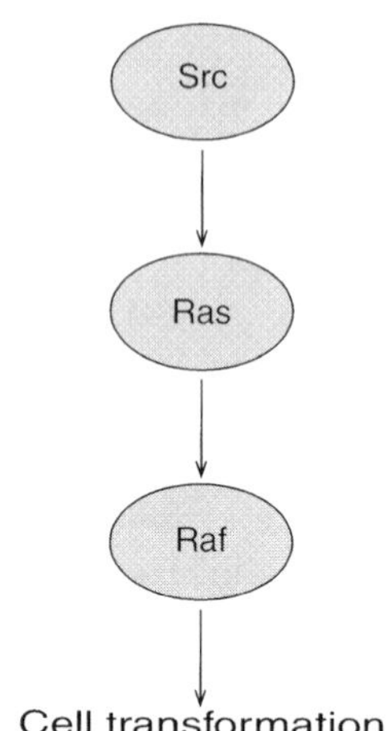

FIGURE 14.8
The Ras signaling pathway in mammalian cells. Ras is activated downstream of protein-tyrosine kinases, such as Src, by the action of guanine nucleotide exchange factors (GEFs). Ras then activates the Raf protein-serine/threonine kinase, leading to cell transformation.

Ras activity, suggesting that Raf functioned downstream of Ras. As predicted by this model, expression of a dominant inhibitory Raf mutant (Raf-301) interferes with transformation by both *src* and *ras.*

The hypothesis that Raf functions downstream of Ras was then supported by biochemical analysis of Raf activation in response to growth factor treatment. As discussed further in chapter 17, growth factor stimulation of protein-tyrosine kinase receptors leads to activation of the Raf protein kinase. Importantly, such growth factor-stimulated activation of Raf was blocked by expression of the dominant inhibitory Ras N-17 mutant, indicating that Ras function was required for Raf activation. Taken together with the ability of a *raf* oncogene to induce transformation even when normal Ras activity was blocked, these results indicated that Raf was activated downstream of Ras in a signaling pathway leading to cell proliferation.

Finally, a number of research groups have demonstrated that the Ras and Raf proteins form a complex through direct protein–protein interactions. This association between Ras and Raf is specific for the active GTP-bound form of Ras and is dependent on the presence of a functional Ras effector domain. It thus appears that the conformational change induced by GTP binding allows Ras to bind directly to Raf. Although the effect of the Ras–Raf interaction on Raf activity is not yet clear, it is apparent that the Raf protein-serine/threonine kinase is a critical target of Ras in mammalian cells.

Summary

The G proteins and Ras are members of a large superfamily of proteins whose activities are regulated by GTP binding and hydrolysis. The G proteins function to couple a variety of cell surface receptors to intracellular second messengers and can act as oncogenes in several cell types. The *ras* oncogenes, which are prevalent in human tumors, similarly arise from mutations that maintain the Ras proteins in the constitutively active GTP-bound state. A number of other proteins also act as oncogenes or tumor suppressor genes by regulating Ras function. These include guanine nucleotide exchange factors and adaptor proteins, which normally activate Ras in response to growth factor stimulation, as well as the *NF1* tumor suppressor gene product, which down-regulates Ras by stimulating its GTPase activity. Further studies have established that Ras plays a central role in signaling cell proliferation in response to stimulation of growth factor receptors. It is activated downstream of protein-tyrosine kinases and serves in turn to activate the Raf protein-serine/threonine kinase, which is discussed in the next chapter.

References

General References

Bourne, H.R., Sanders, D.A., and McCormick, F. 1990. The GTPase superfamily: a conserved switch for diverse cell functions. *Nature* 348:125–132.

Bourne, H.R., Sanders, D.A., and McCormick, F. 1991. The GTPase superfamily: conserved structure and molecular mechanism. *Nature* 349:117–127.

Lowy, D.R., and Willumsen, B.M. 1993. Function and regulation of Ras. *Ann. Rev. Biochem.* 62:851–891.

G Proteins

Birnbaumer, L. 1992. Receptor-to-effector signaling through G proteins: roles for $\beta\gamma$ dimers as well as α subunits. *Cell* 71:1069–1072.

Chan, A.M.-L., Fleming, T.P., McGovern, E.S., Chedid, M., Miki, T., and Aaronson, S.A. 1993. Expression cDNA cloning of a transforming gene encoding the wild-type $G_{\alpha 12}$ gene product. *Mol. Cell. Biol.* 13:762–768.

Conklin, B.R., and Bourne, H.R. 1993. Structural elements of G_α subunits that interact with $G_{\beta\gamma}$, receptors, and effectors. *Cell* 73:631–641.

Hepler, J.R., and Gilman, A.G. 1992. G proteins. *Trends Biochem. Sci.* 17:383–387.

Kalinec, G., Nazarali, A.J., Hermouet, S., Xu, N., and Gutkind, J.S. 1992. Mutated α subunit of the G_q protein induces malignant transformation in NIH 3T3 cells. *Mol. Cell. Biol.* 12:4687–4693.

Landis, C.A., Masters, S.B., Spada, A., Pace, A.M., Bourne, H.R., and Vallar, L. 1989. GTPase inhibiting mutations activate the α chain of G_s and stimulate adenylyl cyclase in human pituitary tumors. *Nature* 340:692–696.

Lyons, J., Landis, C.A., Harsh, G., Vallar, L., Grunewald, K., Feichtinger, H., Duh, Q.-Y., Clark, O.H., Kawasaki, E., Bourne, H.R., and McCormick, F. 1990. Two G protein oncogenes in human endocrine tumors. *Science* 249:655–659.

Pace, A.M., Wong, Y.H., and Bourne, H.R. 1991. A mutant α subunit of G_{i2} induces neoplastic transformation of Rat-1 cells. *Proc. Natl. Acad. Sci. USA* 88:7031–7035.

Simon, M.I., Strathmann, M.P., and Gautman, N. 1991. Diversity of G proteins in signal transduction. *Science* 252:802–808.

Weinstein, L.S., Shenker, A., Gejman, P.V., Merino, M.J., Friedman, E., and Spiegel, A.M. 1991. Activating mutations of the stimulatory G protein in the McCune-Albright syndrome. *N. Engl. J. Med.* 325:1688–1695.

Xu, N., Bradley, L., Ambdukar, I., and Gutkind, J.S. 1993. A mutant α subunit of G_{12} potentiates the eicosanoid pathway and is highly oncogenic in NIH 3T3 cells. *Proc. Natl. Acad. Sci. USA* 90:6741–6745.

Similarities between Ras and the G Proteins

Casey, P.J., Solski, P.A., Der, C.J., and Buss, J.E. 1989. p21^{ras} is modified by a farnesyl isoprenoid. *Proc. Natl. Acad. Sci. USA* 86:8323–8327.

Der, C.J., Finkel, T., and Cooper, G.M. 1986. Biological and biochemical properties of human *ras*H genes mutated at codon 61. *Cell* 44:167–176.

DeVos, A.M., Tong, L., Milburn, M.V., Matias, P.M., Jancarik, J., Noguchi, S., Nishimura, S., Miura, K., Ohtsuka, E., and Kim, S.-H. 1988. Three-dimensional structure of an oncogene protein: catalytic domain of human c-H-*ras* p21. *Science* 239:888–893.

Feig, L.A., and Cooper, G.M. 1988. Relationship among guanine nucleotide exchange, GTP hydrolysis, and transforming potential of mutated *ras* proteins. *Mol. Cell. Biol.* 8:2472–2478.

Gibbs, J.B., Sigal, I.S., Poe, M., and Scolnick, E.M. 1984. Intrinsic GTPase activity distinguishes normal and oncogenic *ras* p21 molecules. *Proc. Natl. Acad. Sci. USA* 81:5704–5708.

Graham, S.M., Cox, A.D., Drivas, G., Rush, M.G., D'Eustachio, P., and Der, C.J. 1994. Aberrant function of the Ras-related protein TC21/R-Ras2 triggers malignant transformation. *Mol. Cell. Biol.* 14:4108–4115.

Hall, A. 1990. The cellular functions of small GTP-binding proteins. *Science* 249:635–640.

Hancock, J.F., Magee, A.I., Childs, J.E., and Marshall, C.J. 1989. All *ras* proteins are polyisoprenylated but only some are palmitoylated. *Cell* 57:1167–1177.

Kringel, U., Schlichting, I., Scherer, A., Schumann, R., Frech, M., John, J., Kabsch, W., Pai, E.F., and Wittinghofer, A. 1990. Three-dimensional structures of H-*ras* p21 mutants: molecular basis for their inability to function as signal switch molecules. *Cell* 62:539–548.

Manne, V., Bekesi, E., and Kung, H. 1985. Ha-*ras* proteins exhibit GTPase activity: point mutations that activate Ha-*ras* gene products result in decreased GTPase activity. *Proc. Natl. Acad. Sci. USA* 82:376–380.

McGrath, J.P., Capon, D.J., Goeddel, D.V., Levinson, A.D. 1984. Comparative biochemical properties of normal and activated human *ras* p21 protein. *Nature* 310:644–649.

Milburn, M.V., Tong, L., DeVos, A.M., Brunger, A., Yamaizumi, Z., Nishimura, S., and Kim, S.-H. 1990. Molecular switch for signal transduction: structural differences between active and inactive forms of protooncogenic *ras* proteins. *Science* 247:939–945.

Pai, E.F., Kabsch, W., Krengel, U., Holmes, K.C., John, J., and Wittinghofer, A. 1989. Structure of the guanine-nucleotide-binding domain of the Ha-*ras* oncogene product p21 in the triphosphate conformation. *Nature* 341:209–214.

Ridley, A.J., and Hall, A. 1992. The small GTP-binding protein rho regulates the assembly of focal adhesions and actin stress fibers in response to growth factors. *Cell* 70:389–399.

Ridley, A.J., Paterson, H.F., Johnston, C.L., Diekmann, D., and Hall, A. 1992. The small GTP-binding protein rac regulates growth factor-induced membrane ruffling. *Cell* 70:401–410.

Schafer, W.R., Kim, R., Sterne, R., Thorner, J., Kim, S.-H., and Rine, J. 1989. Genetic and pharmacological suppression of oncogenic mutations in *RAS* genes of yeast and humans. *Science* 245:379–385.

Shih, T.Y., Papageorge, A.G., Stokes, P.E., Weeks, M.O., and Scolnick, E.M. 1980. Guanine nucleotide-binding and autophosphorylating activities associated with the $p21^{src}$ protein of Harvey murine sarcoma virus. *Nature* 287:686–691.

Sigal, I.S., Gibbs, J.B., D'Alonzo, J.S., and Scolnick, E.M. 1986. Identification of effector residues and a neutralizing epitope of Ha-*ras*-encoded p21. *Proc. Natl. Acad. Sci. USA* 83:4725–4729.

Sigal, I.S., Gibbs, J.B., D'Alonzo, J.S., Temeles, G.L., Wolanski, B.S., Socher, S.H., and Scolnick, E.M. 1986. Mutant *ras* encoded proteins with altered nucleotide binding exert dominant biological effects. *Proc. Natl. Acad. Sci. USA* 83:952–956.

Sweet, R.W., Yokoyama, S., Kamata, T., Feramisco, J.R., Rosenberg, M., and Gross, M. 1984. The product of *ras* is a GTPase and the T24 oncogenic mutant is deficient in this activity. *Nature* 311:273–275.

Walter, M., Clark, S.G., and Levinson, A.D. 1986. The oncogenic activation of human $p21^{ras}$ by a novel mechanism. *Science* 233:649–652.

Willumsen, B.M., Papageorge, A.G., Kung, H., Bekesi, E., Robins, T., Johnsen, M., Vass, W.C., and Lowy, D.R. 1986. Mutational analysis of a Ras catalytic domain. *Mol. Cell. Biol.* 6:2646–2654.

Proteins That Regulate Ras

Ballester, R., Marchuk, D., Boguski, M., Saulino, A., Letcher, R., Wigler, M., and Collins, F. 1990. The *NF1* locus encodes a protein functionally related to mammalian GAP and yeast *IRA* proteins. *Cell* 63:851–859.

Barlat, I., Schweighoffer, F., Chevallier-Multon, M.C., Duchesne, M., Fath, I., Landais, D., Jacquet, M., and Tocque, B. 1993. The *Saccharomyces cerevisiae* gene product SDC25 C-domain functions as an oncoprotein in NIH 3T3 cells. *Oncogene* 8:215–218.

Basu, T.N., Gutmann, D.H., Fletcher, J.A., Glover, T.W., Collins, F.S., and Downward, J. 1992. Aberrant regulation of *ras* proteins in malignant tumour cells from type 1 neurofibromatosis patients. *Nature* 356:713–715.

Boguski, M.S., and McCormick, F. 1993. Proteins that regulate Ras and its relatives. *Nature* 366:643–654.

Buday, L., and Downward, J. 1993. Epidermal growth factor regulates $p21^{ras}$ through the formation of a complex of receptor, Grb2 adapter protein, and Sos nucleotide exchange factor. *Cell* 73:611–620.

Chardin, P., Camonis, J.H., Gale, N.W., Van Aelst, L., Schlessinger, J., Wigler, M.H., and Bar-Sagi, D. 1993. Human Sos1: a guanine nucleotide exchange factor for Ras that binds to GRB2. *Science* 260:1338–1343.

Chou, M.M., Fajardo, J.E., and Hanafusa, H. 1992. The SH2- and SH3-containing Nck protein transforms mammalian fibroblasts in the absence of elevated phosphotyrosine levels. *Mol. Cell. Biol.* 12:5834–5842.

DeClue, J.E., Papageorge, A.G., Fletcher, J.A., Diehl, S.R., Ratner, N., Vass, W.C., and Lowy, D.R. 1992. Abnormal regulation of mammalian p21^{ras} contributes to malignant tumor growth in von Recklinghausen (type 1) neurofibromatosis. *Cell* 69:265–273.

Egan, S.E., Giddings, B.W., Brooks, M.W., Buday, L., Sizeland, A.M., and Weinberg, R.A. 1993. Association of Sos Ras exchange protein with Grb2 is implicated in tyrosine kinase signal transduction and transformation. *Nature* 363:45–51.

Gale, N.W., Kaplan, S., Lowenstein, E.J., Schlessinger, J., and Bar-Sagi, D. 1993. Grb2 mediates the EGF-dependent activation of guanine nucleotide exchange on Ras. *Nature* 363:88–92.

Gulbins, E., Coggeshall, K.M., Baier, G., Katzav, S., Burn, P., and Altman, A. 1993. Tyrosine kinase-stimulated guanine nucleotide exchange activity of Vav in T cell activation. *Science* 260:822–825.

Hart, M.J., Eva, A., Evans, T., Aaronson, S.A., and Cerione, R.A. 1991. Catalysis of guanine nucleotide exchange on the CDC42Hs protein by the *dbl* oncogene product. *Nature* 354:311–314.

Johnson, M.R., DeClue, J.E., Felzmann, S., Vass, W.C., Xu, G., White, R., and Lowy, D.R. 1994. Neurofibromin can inhibit Ras-dependent growth by a mechanism independent of its GTPase-accelerating function. *Mol. Cell. Biol.* 14:641–645.

Li, N., Batzer, A., Daly, R., Yajnik, V., Skolnik, E., Chardin, P., Bar-Sagi, D., Margolis, B., and Schlessinger, J. 1993. Guanine-nucleotide-releasing factor hSos1 binds to Grb2 and links receptor tyrosine kinases to Ras signalling. *Nature* 363:85–88.

Li, W., Hu, P., Skolnik, E.Y., Ullrich, A., and Schlessinger, J. 1992. The SH2 and SH3 domain-containing Nck protein is oncogenic and a common target for phosphorylation by different surface receptors. *Mol. Cell. Biol.* 12:5824–5833.

Lowenstein, E.J., Daly, R.J., Batzer, A.G., Li, W., Margolis, B., Lammers, R., Ullrich, A., Skolnik, E.Y., Bar-Sagi, D., and Schlessinger, J. 1992. The SH2 and SH3 domain-containing protein GRB2 links receptor tyrosine kinases to ras signaling. *Cell* 70:431–442.

Martin, G.A., Viskochil, D., Bollag, G., McCabe, P.C., Crosier, W.J., Haubruck, H., Conroy, L., Clark, R., O'Connell, P., Cawthon, R.M., Innis, M.A., and McCormick, F. 1990. The GAP-related domain of the neurofibromatosis type 1 gene product interacts with *ras* p21. *Cell* 63:843–849.

Mayer, B., Hamaguchi, M., and Hanafusa, H. 1988. A novel viral oncogene with structural similarity to phospholipase C. *Nature* 332:272–275.

Miki, T., Smith, C.L., Long, J.E., Eva, A., and Fleming, T.P. 1993. Oncogene *ect2* is related to regulators of small GTP-binding proteins. *Nature* 362:462–465.

Pelicci, G., Lanfrancone, L., Grignani, F., McGlade, J., Cavallo, F., Forni, G., Nicoletti, I., Grignani, F., Pawson, T., and Pelicci, P.G. 1992. A novel transforming protein (SHC) with an SH2 domain is implicated in mitogenic signal transduction. *Cell* 70:93–104.

Rozakis-Adcock, M., Fernley, R., Wade, J., Pawson, T., and Bowtell, D. 1993. The SH2 and SH3 domains of mammalian Grb2 couple the EGF receptor to the Ras activator mSos1. *Nature* 363:83–85.

Rozakis-Adcock, M., McGlade, J., Mbamalu, G., Pelicci, G., Daly, R., Li, W., Batzer, A., Thomas, S., Brugge, J., Pelicci, P.G., Schlessinger, J., and Pawson, T. 1992. Association of the Shc and Grb2/Sem5 SH2-containing proteins is implicated in activation of the Ras pathway by tyrosine kinases. *Nature* 360:689–692.

Shou, C., Farnsworth, C.L., Neel, B.G., and Feig, L.A. 1992. Molecular cloning of cDNAs encoding a guanine-nucleotide-releasing factor for Ras p21. *Nature* 358:351–354.

Trahey, M., and McCormick, F. 1987. A cytoplasmic protein stimulates normal N-*ras* p21 GTPase, but does not affect oncogenic mutants. *Science* 238:542–545.

Trahey, M., Wong, G., Halenbeck, R., Rubinfeld, B., Matrin, G.A., Ladner, M., Long, C.M., Crosier, W.J., Watt, K., Koths, K., and McCormick, F. 1988. Molecular cloning of two types of GAP complementary DNA from human placenta. *Science* 242:1697–1700.

Vogel, U.S, Dixon, R.A.F., Schaber, M.D., Diehl, R.E., Marshall, M.S., Scolnick, E.M., Sigal, I.S., and Gibbs, J.B. 1988. Cloning of bovine GAP and its interaction with oncogenic *ras* p21. *Nature* 335:90–93.

Xu, G., Lin, B., Tanaka, K., Dunn, D., Wood, D., Gesteland, R., White, R., Weiss, R., and Tamanoi, F. 1990. The catalytic domain of the neurofibromatosis type 1 gene product stimulates *ras* GTPase and complements *ira* mutants of *S. cerevisiae*. *Cell* 63:835–841.

Zhang, K., DeClue, J.E., Vass, W.C., Papageorge, A.G., McCormick, F., and Lowy, D.R. 1990. Suppression of c-*ras* transformation by GTPase-activating protein. *Nature* 346:754–756.

Targets of Ras

Feig, L.A., and Cooper, G.M. 1988. Inhibition of NIH 3T3 cell proliferation by a mutant *ras* protein with preferential affinity for GDP. *Mol. Cell. Biol.* 8:3235–3243.

Field, J., Broek, D., Kataoka, T., and Wigler, M. 1987. Guanine nucleotide activation of, and competition between, RAS proteins from *Saccharomyces cerevisiae*. *Mol. Cell. Biol.* 7:2128–2133.

Kataoka, T., Powers, S., McGill, C., Fasano, O., Strathern, J., Broach, J., and Wigler, M. 1984. Genetic analysis of yeast *RAS*1 and *RAS*2 genes. *Cell* 37:437–445.

Kolch, W., Heidecker, G., Lloyd, P., and Rapp, U.R. 1991. Raf-1 protein kinase is required for growth of induced NIH 3T3 cells. *Nature* 349:249–252.

Moodie, S.A., Willumsen, B.M., Weber, M.J., and Wolfman, A. 1993. Complexes of Ras-GTP with Raf-1 and mitogen-activated protein kinase. *Science* 260:1658–1661.

Mulcahy, L.S., Smith, M.R., and Stacey, D.W. 1985. Requirement for *ras* proto-oncogene function during serum-stimulated growth of NIH 3T3 cells. *Nature* 313:241–243.

Schweighoffer, F., Cai, H., Chevallier-Multon, M.C., Fath, I., Cooper, G., and Tocque, B. 1993. The *Saccharomyces cerevisiae* SDC25 C-domain gene product overcomes the dominant inhibitory activity of Ha-Ras Asn-17. *Mol. Cell. Biol.* 13:39–43.

Smith, M.R., DeGudicibus, S.J., and Stacey, D.W. 1986. Requirement for c-*ras* proteins during viral oncogene transformation. *Nature* 320:540–543.

Tatchell, K., Chaleff, D.T., DeFeo, J.D., and Scolnick, E.M. 1984. Requirement of either of a pair of *ras*-related genes of *Saccharomyces cerevisiae* for spore viability. *Nature* 309:523–527.

Toda, T., Uno, I., Ishikawa, T., Powers, S., Kataoka, T., Broek, D., Cameron, S., Broach, J., Matsumoto, K., and Wigler, M. 1985. In yeast, RAS proteins are controlling elements of adenylate cyclase. *Cell* 40:27–36.

Troppmair, J., Bruder, J.T., App, H., Cai, H., Liptak, L., Szeberenyi, J., Cooper, G.M., and Rapp, U.R. 1992. Ras controls coupling of growth factor receptors and protein kinase C in the membrane to Raf-1 and B-Raf protein serine kinases in the cytosol. *Oncogene* 7:1867–1873.

Van Aelst, L., Barr, M., Marcus, S., Polverino, A., and Wigler, M. 1993. Complex formation between RAS and RAF and other protein kinases. *Proc. Natl. Acad. Sci. USA* 90:6213–6217.

Vojtek, A.B., Hollenberg, S.M., and Cooper, J.A. 1993. Mammalian Ras interacts directly with the serine/threonine kinase Raf. *Cell* 74:205–214.

Warne, P.H., Viciana, P.R., and Downward, J. 1993. Direct interaction of Ras and the amino-terminal region of Raf-1 *in vitro*. *Nature* 364:352–355.

Wood, K.W., Sarnecki, C., Roberts, T.M., and Blenis, J. 1992. Ras mediates nerve growth factor receptor modulation of three signal-transducing protein kinases: MAP kinase, Raf-1, and RSK. *Cell* 68:1041–1050.

Zhang, X., Settleman, J., Kyriakis, J.M., Takeuchi-Suzuki, E., Elledge, S.J., Marshall, M.S., Bruder, J.T., Rapp, U.R., and Avruch, J. 1993. Normal and oncogenic $p21^{ras}$ proteins bind to the amino-terminal regulatory domain of c-Raf-1. *Nature* 364:308–313.

❖ CHAPTER 15

Protein-Serine/ Threonine Kinases

Chapter 13 dealt with the identification of protein-tyrosine kinases as oncogene products. Most of the protein kinase activity in mammalian cells, however, results in phosphorylation of serine and threonine, rather than tyrosine, residues. More than one hundred protein-serine/threonine kinases have been identified and these enzymes appear to regulate a wide variety of metabolic processes. For example, two of the major second messenger systems utilized in eucaryotic cells are the cAMP and inositol phospholipid pathways. In both of these intracellular signal transduction systems, protein-serine/threonine kinases play a central role, as discussed in chapter 14.

The classic example of signal transduction by the cAMP system is provided by the control of glycogen metabolism in muscle (Fig. 15.1). In this case, two protein-serine/threonine kinases (cAMP-dependent protein kinase and phosphorylase kinase) act in series to couple a metabolic response (production of glucose from glycogen) to epinephrine-stimulated production of a second messenger (cAMP). The binding of epinephrine to its cell surface receptor results in G protein-mediated activation of adenylate cyclase, resulting in increased cAMP levels and activation of cAMP-dependent protein kinase. The target for cAMP-dependent kinase is a second protein-serine/threonine kinase, phosphorylase kinase. Once activated by phosphorylation, phosphorylase kinase in turn phosphorylates its target enzyme, glycogen phosphorylase, which then catalyzes the conversion of glycogen into glucose-1-phosphate.

The second messengers generated by hydrolysis of inositol phospholipids likewise lead to activation of protein-serine/threonine kinases. Protein kinase C is activated by diacylglycerol, and release of calcium from intracellular stores

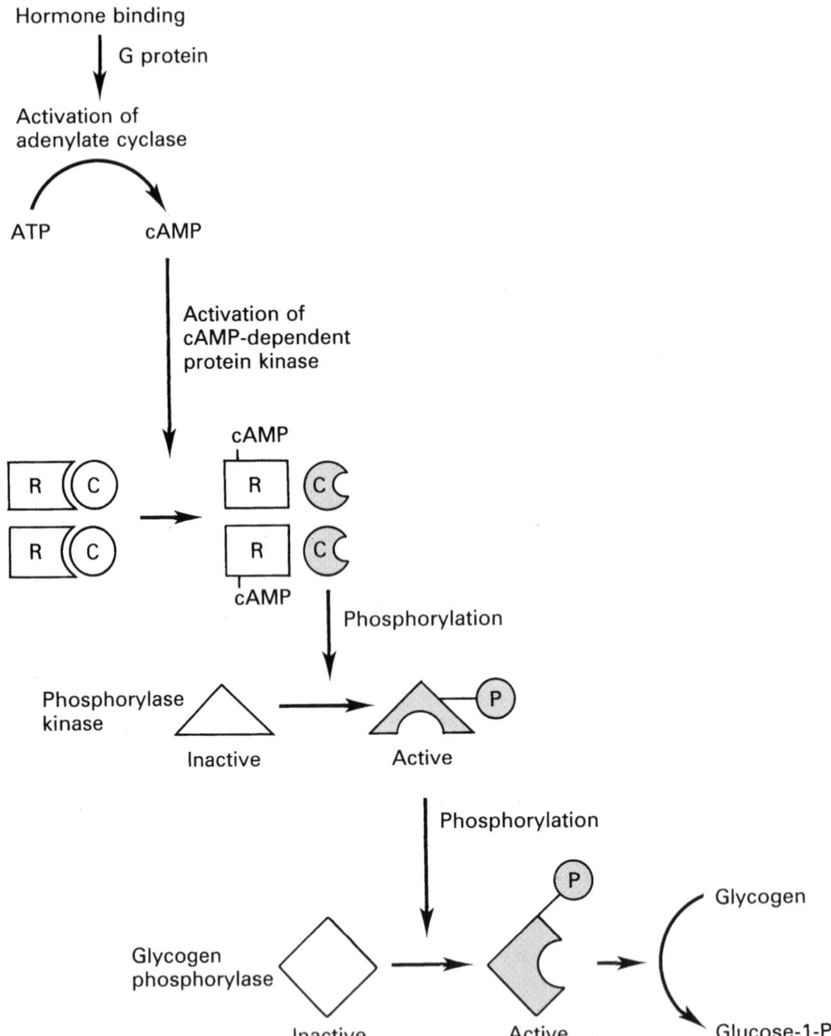

FIGURE 15.1
cAMP regulation of glycogen metabolism by a cascade of protein phosphorylation. Hormone (epinephrine) binding results in G protein-mediated activation of adenylate cyclase, leading to increased production of cAMP. cAMP then binds to the regulatory subunit (R) of cAMP-dependent protein kinase, resulting in dissociation and activation of the catalytic subunit (C). cAMP-dependent protein kinase then phosphorylates and activates phosphorylase kinase which, in turn, phosphorylates and activates glycogen phosphorylase.

also results in activation of an additional family of calcium/calmodulin-dependent protein kinases. The activity of protein-serine/threonine kinases, as well as protein-tyrosine kinases, is thus clearly implicated in the regulation of mammalian cell physiology, including control of cell proliferation. It will therefore be no surprise that several oncogenes encode proteins with protein-serine/threonine kinase activity. Among these are members of the Raf family, discussed in the preceding chapter as targets of Ras, as well as several isotypes of protein kinase C.

The *raf* Oncogene Family

The *raf* oncogenes were first identified in chicken and mouse retroviruses (chapter 3) and have since been frequently found as oncogenes activated by DNA rearrangements in gene transfer experiments with human and rodent DNAs (chapter 5). The identification of Raf proteins as protein-serine/threonine kinases is apparent from the amino acid sequence of their kinase domain, which is characteristic of serine/threonine rather than tyrosine kinases (see chapter 13). This inference has been supported by direct demonstration of the kinase activity of Raf proteins, coupled with identification of serine and threonine as the phosphorylated amino acids.

There are three members of the *raf* oncogene family, designated c-*raf*-1, A-*raf*, and B-*raf*, which encode proteins of 648 to 766 amino acids. The three Raf proteins are ~75% identical in their kinase domains, which are located in the carboxy-terminal half of each protein. In addition, there are two conserved domains in the amino-terminal halves of the Raf proteins that, as will be discussed, serve to regulate Raf kinase activity.

The molecular alterations that convert normal *raf* proto-oncogenes into active oncogenes generally consist of the deletion of amino-terminal regulatory sequences, resulting in unregulated activity of the carboxy-terminal protein kinase catalytic domains (Fig. 15.2). All of the activated *raf* oncogenes that have been characterized to date, whether identified in retroviruses or by gene transfer experiments, have been found to have sustained deletions resulting in the loss of between 190 and 325 amino acids from the amino terminus of the Raf proteins. The largest of these deletions terminates only ~20 amino acids before the start of the kinase domain. The truncated *raf* genes generated by these deletions are expressed as fusion proteins, either with viral *gag* sequences in the retroviral *raf* oncogenes or with a variety of different cellular coding sequences in the cellular *raf* oncogenes that have been activated during gene transfer. The reproducibility of these amino-terminal deletions suggests that this alteration is critical to *raf* oncogene activation. Moreover, *in vitro* mutagenesis of *raf* proto-oncogenes has established directly that amino-terminal deletion is sufficient to fully activate Raf transforming potential. The mutants with highest transform-

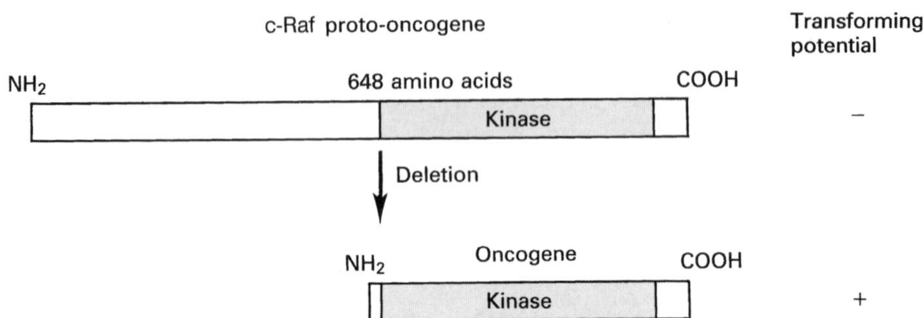

FIGURE 15.2
The c-*raf*-1 proto-oncogene is converted into an active oncogene by deletion of the normal amino-terminal regulatory domain of the Raf-1 protein.

ing activities have sustained deletions of nearly the entire amino-terminal half of the Raf protein, terminating less than 40 amino acids before the start of the kinase domain.

It thus appears that activation of Raf transforming potential requires deletion of an amino-terminal negative regulatory domain, resulting in constitutive activity of the Raf protein kinase. This mode of oncogene activation is reminiscent of the molecular alterations responsible for the activation of several of the receptor and nonreceptor tyrosine kinase oncogenes discussed in chapter 13. Moreover, as discussed in the next section, the functional organization of Raf deduced from these studies is strikingly parallel to that of protein kinase C.

As discussed in the preceding chapter, normal Raf proto-oncogene proteins are activated downstream of Ras in response to growth factor stimulation. This apparently involves a protein–protein interaction in which Ras binds to the amino-terminal regulatory domain of Raf. However, this interaction with Ras has not been found to directly activate the Raf protein kinase. It therefore appears that other events, which will be considered in chapter 17, are also required for Raf activation in response to growth factor stimulation. In any event, the activated Raf kinase then phosphorylates and activates another protein kinase (called mitogen-activated protein kinase kinase), thereby triggering a cascade of protein-serine/threonine kinases that ultimately leads to cell proliferation.

The Protein Kinase C Family

Protein kinase C was first identified in 1977 as a protein kinase with no known physiologic role. Its importance in signal transduction pathways related to cell proliferation emerged from two subsequent lines of investigation. First, it was identified as an effector enzyme of the inositol phospholipid second messenger pathway, being activated by calcium and diacylglycerol. Second, protein kinase C was found to be the intracellular target for activation by the tumor-promoting phorbol esters, which mimic the effect of diacylglycerol. This protein-serine/threonine kinase thus appeared to play a central role in control of cell proliferation, leading to extensive studies of its structure and function.

It is now clear that there are at least ten distinct members of the protein kinase C (PKC) family, which differ both in their catalytic and in their regulatory properties (Table 15.1). The catalytic activity of the classical protein kinase C isotypes (α, ß1, ß2, and γ) requires the binding of both calcium and diacylglycerol. In contrast, several other protein kinase C isotypes (called nonclassical PKCs; δ, ε, η, and θ) are calcium-independent, requiring only diacylglycerol for activity. Moreover, two protein kinase C isotypes (called atypical PKCs; ζ and λ) are activated by neither diacylglycerol nor calcium, instead presumably being regulated by different second messengers or protein interactions.

Early studies demonstrated that classical protein kinase C could also be activated by limited proteolysis, leading to the suggestion that the enzyme contained distinct regulatory and catalytic domains (Fig. 15.3). Limited proteolytic digestion of intact protein kinase C (~80 kd) yielded two fragments of ~50 and ~35 kd. The ~50 kd fragment was catalytically active independent of calcium

TABLE 15.1
The Protein Kinase C Family

Isotype	Molecular Weight	Activators
Classical PKCs		
PKC-α	77,000	Ca^{++}, DAG
PKC-ß1	77,000	Ca^{++}, DAG
PKC-ß2	77,000	Ca^{++}, DAG
PKC-γ	78,000	Ca^{++}, DAG
Nonclassical PKCs		
PKC-δ	77,000	DAG
PKC-ε	83,000	DAG
PKC-η	78,000	DAG
PKC-θ	82,000	DAG
Atypical PKCs		
PKC-ζ	68,000	Unknown
PKC-λ	67,000	Unknown

Note: In addition to diacylglycerol (DAG) and calcium, phospholipid (for example, phosphatidylserine) is required for PKC activity.

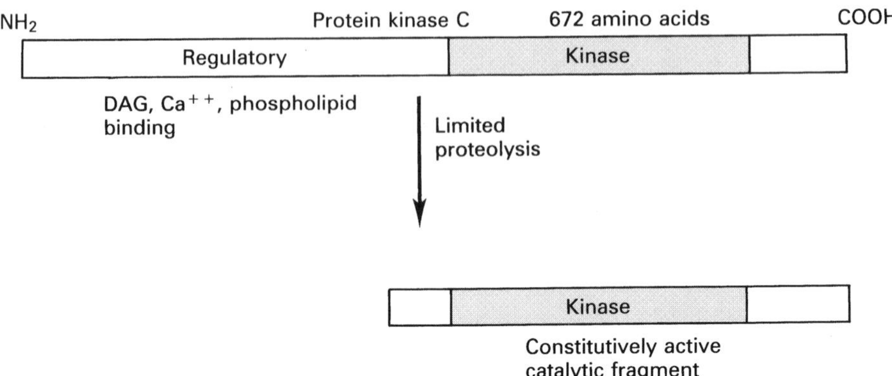

Figure 15.3
Structure of protein kinase C. The classical protein kinase C isotypes (for example, PKC-α) contain amino-terminal regulatory domains that bind diacylglycerol (DAG), calcium, and phospholipid. Removal of the regulatory domain by limited proteolysis results in constitutive activity of the kinase catalytic domain.

and diacylglycerol binding. The smaller fragment retained phorbol ester binding activity. It thus appeared that protein kinase C consisted of two distinct domains that were separated by a hinge region susceptible to proteolytic cleavage: a protein kinase catalytic domain and a regulatory domain containing the sites for calcium and diacylglycerol binding. Because removal of the regulatory domain by proteolysis resulted in constitutive catalytic activity, it appeared that this domain functioned as a negative regulator of kinase function. In the intact molecule, this negative control would be abrogated by binding of calcium and diacylglycerol, perhaps by induction of a conformational change that altered the interaction between regulatory and catalytic domains.

Molecular cloning and sequence analysis of protein kinase C cDNAs have elucidated the structure of the regulatory and catalytic domains inferred from these enzymological studies. All members of the protein kinase C family have similar structures, with their catalytic domains localized to their carboxy-terminal halves. However, their amino-terminal regulatory domains differ according to their requirements for calcium and diacylglycerol for activation. The classical PKC isotypes have binding sites for both diacylglycerol and calcium; the nonclassical PKC isotypes have binding sites only for diacylglycerol; and the atypical PKCs lack both diacylglycerol and calcium binding sites. It is noteworthy that the overall organization of the PKCs is highly analogous to that of the Raf protein kinases, which are also similar to protein kinase C in size. Interestingly, the amino-terminal regulatory domain of Raf most closely resembles that of the atypical PKC isotypes.

Because of the role of protein kinase C in cell proliferation, it was of obvious interest to determine whether protein kinase C could function as an oncogene. This has been demonstrated for four different members of the protein kinase C family (ß1, γ, ε, and ζ), which represent classical, nonclassical, and atypical isotypes. In these cases, PKC overexpression was found to induce cell transformation, although PKC appeared to be less potent as a transforming agent than oncogenes such as *ras* or *raf*. Nevertheless, these studies directly demonstrate the potential of some members of the protein kinase C family to contribute to the transformed phenotype, consistent with the activation of some protein kinase C isotypes as downstream effectors of phosphatidylinositide hydrolysis. Additional studies indicate that some isotypes of protein kinase C can phosphorylate and activate Raf, suggesting their possible participation in Raf activation during mitogenic signaling.

The *mos* Oncogene

The third well-characterized protein-serine/threonine kinase oncogene is *mos*, which was originally identified in Moloney sarcoma virus (chapter 3) and subsequently found to be activated as an oncogene by upstream integration of retroviral-like sequences in some plasmacytomas (chapter 6). In contrast with the *raf* and protein kinase C gene families, *mos* is a single copy gene in several vertebrate species. Its protein-serine/threonine kinase activity has been predicted by sequence analysis and verified by direct biochemical experimentation.

Mos is only 343 amino acids, which is about half the size of Raf or protein kinase C. In fact, the Mos protein encompasses little more than its protein kinase catalytic domain, in contrast with Raf and protein kinase C whose catalytic domains comprise only the carboxy-terminal halves of the protein. This suggests that the Mos protein does not include an extensive amino-terminal regulatory domain, analogous to those present in Raf or protein kinase C, and therefore may be subject to some other form of intracellular regulation.

Consistent with this structural difference, activation of *mos* as a potent oncogene requires only enhanced expression, without any change in its coding sequence. As will be discussed in chapter 19, an interesting feature of the *mos* proto-oncogene is its extremely restricted pattern of expression in normal cells. In contrast to most proto-oncogenes, *mos* is transcriptionally silent in most normal cells and tissues that have been studied, with its major sites of expression limited to the germ cells of both sexes, where it appears to function as a specific regulator of the cell cycle during oocyte meiosis. Taken together, these features of *mos* suggest that the activity of its protein kinase may be constitutive, analogous to truncated Raf or protein kinase C proteins. Rather than being normally regulated at the level of protein catalytic activity, the *mos* proto-oncogene may instead be tightly controlled at the level of gene expression. It is noteworthy that Mos, like Raf, phosphorylates and activates mitogen-activated protein kinase kinase, presumably accounting for its ability to induce transformation when expressed in inappropriate cell types.

Other Protein-Serine/Threonine Kinase Oncogenes

Several additional oncogenes encoding protein-serine/threonine kinases have been isolated from acutely transforming retroviruses (*akt*), as targets for retroviral integration in mouse leukemias (*pim*-1 and *tpl*-2), and by DNA transfection assays (*cot*) (Table 15.2). The proteins encoded by these oncogenes range in size from 313 amino acids (Pim-1) to 481 amino acids (Akt). The *pim*-1 oncogene, like *mos*, is activated by abnormal expression of the structurally normal protein. However, both structural and regulatory changes may activate *akt*, *tpl*-2, and *cot*. The retroviral *akt* oncogene is expressed as a Gag/Akt fusion protein in which approximately 80 amino acids have been deleted from the normal Akt amino terminus. Both the *tpl*-2 and *cot* oncogenes, which are closely related to each other, are activated by DNA rearrangements that both elevate gene expression and result in carboxy-terminal deletions. Although the role of the affected carboxy-terminal sequences is not yet estab-

TABLE 15.2
Protein-Serine/Threonine Kinases with Oncogenic Potential

Proto-oncogene Product	Protein Molecular Weight
Raf Family	
Raf-1	73,000
A-Raf	69,000
B-Raf	95,000
Protein Kinase C Family	
PKC-ß1	77,000
PKC-γ	78,000
PKC-ε	83,000
PKC-ζ	68,000
Mos	37,000
Pim-1	34,000/44,000
Akt	56,000
Cot	52,000
Tpl-2	56,000

Note: The two Pim-1 proteins arise from usage of alternative initiation codons.

lished, their deletion in both the Tpl-2 and Cot oncogene proteins suggests that they may take part in negative regulation of kinase activity, perhaps similar to the carboxy-terminal regulatory sequences of the Src family protein-tyrosine kinases discussed in chapter 13.

Summary

A number of protein-serine/threonine kinases can be activated as oncogenes by either abnormal expression or deletion of regulatory domains. These kinases include Raf and members of the protein kinase C family, which function as downstream effectors in signal transduction pathways initiated by the binding of growth factors to cell surface receptors. Raf in particular plays a central role in signal transduction from protein-tyrosine kinase receptors, being activated downstream of Ras and then serving to activate a protein kinase cascade that ultimately triggers cell proliferation. The activity of protein-serine/threonine kinases as oncogenes thus extends the concept that cell transformation can result from the unregulated activity of enzymes that normally transduce intracellular signals initiated by growth factor binding.

References

General Reference

Taylor, S.S., Knighton, D.R., Zheng, J., Ten Eyck, L.F., and Sowadski, J.M. 1992. Structural framework for the protein kinase family. *Ann. Rev. Cell Biol.* 8:429–462.

The *raf* Oncogene Family

Beck, T.W., Huleihel, M., Gunnel, M., Bonner, T.I., and Rapp, U.R. 1987. The complete coding sequence of the human A-*raf*-1 oncogene and transforming activity of a human A-*raf* carrying retrovirus. *Nucleic Acids Res.* 15:595–609.

Bonner, T.I., Kerby, S.B., Sutrave, P., Gunnell, M.A., Mark, G., and Rapp, U.R. 1985. Structure and biological activity of human homologs of the *raf/mil* oncogene. *Mol. Cell. Biol.* 5:1400–1407.

Dent, P., Haser, W., Haystead, T.A.J., Vincent, L.A., Roberts, T.M., and Sturgill, T.W. 1992. Activation of mitogen-activated protein kinase kinase by v-Raf in NIH 3T3 cells and *in vitro*. *Science* 257:1404–1407.

Force, T., Bonventre, J.V., Heidecker, G., Rapp, U., Avruch, J., and Kyriakis, J.M. 1994. Enzymatic characteristics of the c-Raf-1 protein kinase. *Proc. Natl. Acad. Sci. USA* 91:1270–1274.

Heidecker, G., Huleihel, M., Cleveland, J.L., Kolch, W., Beck, T.W., Lloyd, P., Pawson, T., and Rapp, U.R. 1990. Mutational activation of c-*raf*-1 and definition of the minimal transforming sequence. *Mol. Cell. Biol.* 10:2503–2512.

Howe, L.R., Leevers, S.J., Gomez, N., Nakielny, S., Cohen, P., and Marshall, C.J. 1992. Activation of the MAP kinase pathway by the protein kinase Raf. *Cell* 71:335–342.

Kyriakis, J.M., App, H., Zhang, X., Banerjee, P., Brautigan, D.L., Rapp, U.R., and Avruch, J. 1992. Raf-1 activates MAP kinase-kinase. *Nature* 358:417–421.

Mark, G.E., and Rapp, U.R. 1984. Primary structure of v-*raf*: relatedness to the *src* family of oncogenes. *Science* 224:285–289.

Moodie, S.A., Willumsen, B.M., Weber, M.J., and Wolfman, A. 1993. Complexes of Ras-GTP with Raf-1 and mitogen-activated protein kinase kinase. *Science* 260:1658–1661.

Stanton, V.P., Jr., Nichols, D.W., Laudano, A.P., and Cooper, G.M. 1989. Definition of the human *raf* amino-terminal regulatory region by deletion mutagenesis. *Mol. Cell. Biol.* 9:639–647.

Stephens, R.M., Sithanandam, G., Copeland, T.D., Kaplan, D.R., Rapp, U.R., and Morrison, D.K. 1992. 95-kilodalton B-Raf serine/threonine kinase: identification of the protein and its major autophosphorylation site. *Mol. Cell. Biol.* 12:3733–3742.

Van Aelst, L., Barr, M., Marcus, S., Polverino, A., and Wigler, M. 1993. Complex formation between RAS and RAF and other protein kinases. *Proc. Natl. Acad. Sci. USA* 90:6213–6217.

Vojtek, A.B., Hollenberg, S.M., and Cooper, J.A. 1993. Mammalian Ras interacts directly with the serine/threonine kinase Raf. *Cell* 74:205–214.

Warne, P.H., Viciana, P.R., and Downward, J. 1993. Direct interaction of Ras and the amino-terminal region of Raf-1 *in vitro*. *Nature* 364:352–355.

Zhang, X., Settleman, J., Kyriakis, J.M., Takeuchi-Suzuki, E., Elledge, S.J., Marshall, M.S., Bruder, J.T., Rapp, U.R., and Avruch, J. 1993. Normal and oncogenic $p21^{ras}$ proteins bind to the amino-terminal regulatory domain of c-Raf-1. *Nature* 364:308–313.

The Protein Kinase C Family

Berra, E., Diaz-Meco, M.T., Dominguez, I., Municio, M.M., Sanz, L., Lozano, J., Chapkin, R.S., and Moscat, J. 1993. Protein kinase C ζ isoform is critical for mitogenic signal transduction. *Cell* 74:555–563.

Cacace, A., Guadagno, S.N., Krauss, R.S., Fabbro, D., and Weinstein, I.B. 1993. The epsilon isoform of protein kinase C is an oncogene when overexpressed in rat fibroblasts. *Oncogene* 8:2094–2104.

Housey, G.M., Johnson, M.D., Hsiao, W.-L.W., O'Brian, C.A., Murphy, J.P., Kirschmeier, P., and Weinstein, I.B. 1988. Overproduction of protein kinase C causes disordered growth control in rat fibroblasts. *Cell* 52:343–354.

Kolch, W., Heidecker, G., Kochs, G., Hummel, R., Vahidi, H., Mischak, H., Finkenzeller, G., Marme, D., and Rapp, U.R. 1993. PKC-α activates Raf-1 by direct phosphorylation. *Nature* 364:249–252.

Mischak, H., Goodnight, J., Kolch, W., Martiny-Baron, G., Schaechtle, C., Kazanietz, M.G., Blumberg, P.M., Pierce, J.H., and Mushniski, J.F. 1993. Overexpression of protein kinase C-δ and -ε in NIH 3T3 cells induces opposite effects on growth, morphology, anchorage dependence, and tumorigenicity. *J. Biol. Chem.* 268:6090–6096.

Nishizuka, Y. 1992. Intracellular signaling by hydrolysis of phospholipids and activation of protein kinase C. *Science* 258:607–614.

Persons, D.A., Wilkison, W.O., Bell, R.M., and Finn, O.J. 1988. Altered growth regulation and enhanced tumorigenicity of NIH 3T3 fibroblasts transfected with protein kinase C-I cDNA. *Cell* 52:447–458.

Sozeri, O., Vollmer, K., Liyanage, M., Frith, D., Kour, G., Mark, G.E., III, and Stabel, S. 1992. Activation of the c-Raf protein kinase by protein kinase C phosphorylation. *Oncogene* 7:2259–2262.

The *mos* Oncogene

Barker, W.C., and Dayhoff, M.O. 1982. Viral *src* gene products are related to the catalytic chain of mammalian cAMP-dependent protein kinase. *Proc. Natl. Acad. Sci. USA* 79:2836–2839.

Blair, D.G., Oskarsson, M., Wood, T.G., McClements, W.L., Fischinger, P.J., and Vande Woude, G.F. 1981. Activation of the transforming potential of a normal cell sequence: a molecular model for oncogenesis. *Science* 212:941–943.

Blair, D.G., Wood, T.G., Woodworth, A.M., McGeady, M.L., Oskarsson, M.K., Propst, F., Tainsky, M.A., Cooper, C.S., Watson, R., Baroudy, B.M., and Vande Woude, G.F. 1984. Properties of the mouse and human *mos* oncogene loci. In *Cancer cells,* Vol. 2: *Oncogenes and viral genes,* ed. Vande Woude, G.F., Levine, A.J., Topp, W.C., and Watson, J.D. Cold Spring Harbor, New York, pp. 281–289.

Goldman, D.S., Kiessling, A.A., Millette, C.F., and Cooper, G.M. 1987. Expression of c-*mos* RNA in germ cells of male and female mice. *Proc. Natl. Acad. Sci. USA* 84:4509–4513.

Maxwell, S.A., and Arlinghaus, R.B. 1985. Serine kinase activity associated with Moloney murine sarcoma virus-124-encoded $p37^{mos}$. *Virology* 143:321–333.

Mutter, G.L., and Wolgemuth, D.J. 1987. Distinct developmental patterns of c-*mos* proto-oncogene expression in female and male mouse germ cells. *Proc. Natl. Acad. Sci. USA* 84:5301–5305.

Nebreda, A.R., and Hunt, T. 1993. The c-*mos* proto-oncogene protein kinase turns on and maintains the activity of MAP kinase, but not MPF, in cell-free extracts of *Xenopus* oocytes and eggs. *EMBO J.* 12:1979–1986.

O'Keefe, S.J., Wolfes, H., Kiessling, A.A., and Cooper, G.M. 1989. Microinjection of antisense c-*mos* oligonucleotides prevents meiosis II in the maturing mouse egg. *Proc. Natl. Acad. Sci. USA* 86:7038–7042.

Oskarsson, M., McClements, W.L., Blair, D.G., Maizel, J.V., and Vande Woude, G.F. 1980. Properties of a normal mouse cell DNA sequence (*sarc*) homologous to the *src* sequence of Moloney sarcoma virus. *Science* 207:1222–1224.

Paules, R.S., Buccione, R., Moschel, R.C., Vande Woude, G.F., and Eppig, J.J. 1989. Mouse Mos proto-oncogene product is present and functions during oogenesis. *Proc. Natl. Acad. Sci. USA* 86:5395–5399.

Posada, J., Yew, N., Ahn, N.G., Vande Woude, G.F., and Cooper, J.A. 1993. Mos stimulates MAP kinase in *Xenopus* oocytes and activates a MAP kinase kinase *in vitro*. *Mol. Cell. Biol.* 13:2546–2553.

Propst, F., and Vande Woude, G.F. 1985. Expression of c-*mos* proto-oncogene transcripts in mouse tissues. *Nature* 315:516–518.

Sagata, N., Oskarsson, M., Copeland, T., Brumbaugh, J., and Vande Woude, G.F. 1988.

Function of c-*mos* proto-oncogene in meiotic maturation in *Xenopus* oocytes. *Nature* 335:519–525.

Sagata, N., Watanabe, N., Vande Woude, G.F., and Ikawa, Y. 1989. The c-*mos* proto-oncogene product is a cytostatic factor responsible for meiotic arrest in vertebrate eggs. *Nature* 342:512–518.

Other Protein-Serine/Threonine Kinase Oncogenes

Bellacosa, A., Testa, J.R., Staal, S.P., and Tsichlis, P.N. 1991. A retroviral oncogene, *akt,* encoding a serine-threonine kinase containing an SH2-like region. *Science* 254:274–277.

Miyoshi, J., Higashi, T., Mukai, H., Ohuchi, T., and Kakunaga, T. 1991. Structure and transforming potential of the human *cot* oncogene encoding a putative protein kinase. *Mol. Cell. Biol.* 11:4088–4096.

Patriotis, C., Makris, A., Bear, S.E., and Tsichlis, P.N. 1993. Tumor progression locus 2 (*Tpl-2*) encodes a protein kinase involved in the progression of rodent T-cell lymphomas and in T-cell activation. *Proc. Natl. Acad. Sci. USA* 90:2251–2255.

Saris, C.J.M., Domen, J., and Berns, A. 1991. The *pim*-1 oncogene encodes two related protein-serine/threonine kinases by alternative initiation at AUG and CUG. *EMBO J.* 10:655–664.

Selten, G., Cuypers, H.T., Boelens, W., Robanus-Maandag, E., Verbeek, J., Domen, J., van Beveren, C., and Berns, A. 1986. The primary structure of the putative oncogene *pim*-1 shows extensive homology with protein kinases. *Cell* 46:603–611.

❖ CHAPTER 16

Transcription Factors

The oncogenes discussed so far encode either extracellular growth factors or signal transducing molecules in the plasma membrane or cytoplasm. Ultimately, however, the behavior of a cell is governed by its pattern of gene expression, which is controlled in the nucleus. Not surprisingly, therefore, it has been found that the proteins encoded by a number of oncogenes and tumor suppressor genes are localized to the nucleus, where they function as transcriptional regulatory proteins.

Transcription of eucaryotic genes is controlled by the interaction of proteins, called transcription factors, with specific regulatory DNA sequences, usually located upstream of the gene in question. Because transcription factors are diffusible proteins that affect expression of unlinked genes, they are frequently referred to as *trans*-acting factors. In contrast, regulatory sequences that control transcription of adjacent genes are referred to as *cis*-acting elements. The *cis*-acting regulatory sequences of a number of genes have been identified by gene transfer experiments and found to contain binding sites for factors that either activate or repress transcription. For example (Fig. 16.1), genes that are inducible by glucocorticoid hormones contain an upstream sequence similar to the consensus sequence GGTACANNNTGTTCT. Gene transfer experiments have established that these sequences function as transcriptional enhancer elements that confer glucocorticoid inducibility upon adjacent genes and, therefore, represent the *cis*-acting regulatory sequences responsible for glucocorticoid regulation. In addition, these sequences have been shown to represent the DNA binding sites of glucocorticoid receptors, which are the *trans*-acting factors responsible for mediating increased transcription in re-

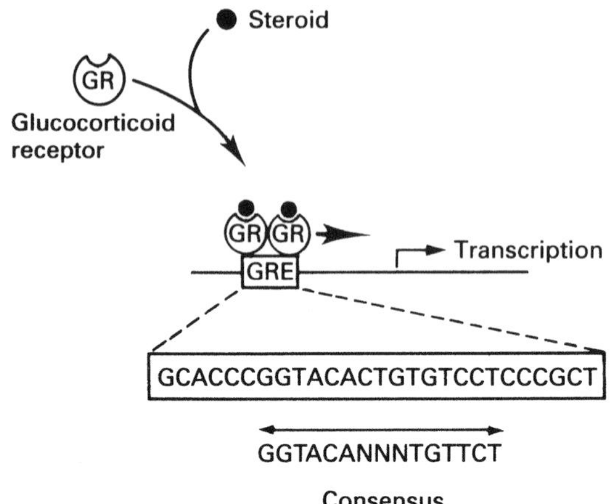

FIGURE 16.1
Regulation of gene expression by glucocorticoid hormones. A dimeric hormone–hormone receptor complex binds to the glucocorticoid response element (GRE) to stimulate transcription of an adjacent gene.

sponse to hormone. The binding sites for glucocorticoid receptor, as well as for a number of other transcription factors, display partial dyad symmetry, reflecting the fact that transcription factors bind to these DNA sequence elements as dimers. Once bound to DNA, transcription factors increase expression of their target genes by interacting with other components of the transcriptional machinery. Thus, transcription factors contain at least two functional domains: (1) a DNA binding domain, responsible for binding of the factor to specific *cis*-acting regulatory sequences and (2) an activation domain, responsible for stimulating transcription through protein–protein interactions.

Regulating the expression and function of transcription factors clearly affords the cell a powerful set of controls on tissue-specific and inducible gene expression. For example, some transcription factors are expressed only in particular differentiated cell types and thus appear to control developmentally programmed patterns of gene expression. In other cases, the expression or activity of transcription factors is modulated in response to growth factors or hormones (for example, the glucocorticoid receptor); so these transcription factors can be viewed as elements in signal transduction pathways that regulate gene expression in response to extracellular stimuli. Considering their importance in the control of cell behavior, it is not unexpected that a variety of oncogenes and tumor suppressor genes have been found to encode transcription factors, thereby extending the spectrum of cell regulatory processes in which oncogenes take part to include direct control of gene expression.

Oncogenes and Steroid Hormone Receptors

The first indications that an oncogene protein could function as a transcription factor came in late 1985 and early 1986 when three groups of researchers reported sequence homologies between steroid receptors and the *erb*A oncogene product (Fig. 16.2). The steroids (including progesterone, estrogen, testosterone, and the glucocorticoids) and related hormones (including thyroid hormone and retinoic acid) exert a variety of biological effects by directly increasing or decreasing transcription of specific genes in their target cells. Steroids enter cells by diffusion through the plasma membrane and bind with high affinity to their intracellular receptors. The conformation of the receptor is altered by hormone binding, which controls the activity of the receptor as a transcriptional regulatory protein. For example, hormone binding to the glucocorticoid receptor results in its translocation to the nucleus, where it can bind DNA and stimulate transcription of glucocorticoid-inducible genes.

The importance of steroid hormone receptors as transcriptional regulatory proteins led a number of laboratories to pursue their purification and structural analysis. Unexpectedly, the first nucleotide sequences reported for glucocorticoid and the estrogen receptor cDNAs revealed a relation between their predicted amino acid sequences and that of the protein encoded by *erb*A, an oncogene initially identified in avian erythroblastosis virus (chapter 3). The sequences of both the glucocorticoid and estrogen receptors indicated that these proteins were composed of three domains (Fig. 16.2): (1) an amino-terminal domain of variable length and sequence (~420 and 180 amino acids in the glu-

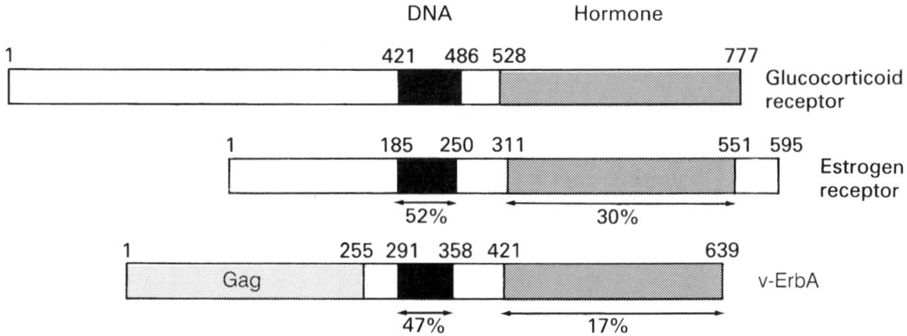

FIGURE 16.2
Relation between steroid receptors and ErbA. The steroid receptors are composed of highly variable amino-terminal domains, central DNA binding domains (labeled DNA), and carboxy-terminal hormone binding domains (labeled hormone). The viral ErbA oncogene protein is a fusion protein with viral Gag sequences. The amino acid sequence identities of the DNA and hormone binding domains of both estrogen receptor and ErbA are presented as percentages with respect to the glucocorticoid receptor.

cocorticoid and estrogen receptors, respectively), (2) a central highly conserved domain ranging from 60 to 70 amino acids that was rich in cysteines and basic amino acids, and (3) a hydrophobic carboxy-terminal domain of ~250 amino acids. Further studies have demonstrated that the highly conserved central domain is responsible for DNA binding, whereas the hydrophobic carboxy-terminal domain is responsible for steroid binding.

Alignment of ErbA with the glucocorticoid and estrogen receptor sequences indicated that the amino acid sequence of ErbA was ~25% identical overall with these steroid receptors. The closest relation (~50% amino acid identities) of ErbA to both receptors was detected in the central DNA binding domain. Significantly, this included conservation of nine cysteine residues that take part in the formation of secondary structures that bind zinc ions and fold into looped structures (called zinc fingers) that bind DNA. These findings indicated that the *erb*A oncogene product is a member of the same gene family as the steroid receptors and led to the hypothesis that the ErbA protein might also function as a transcription factor.

Although apparently a member of the steroid receptor gene family, *erb*A clearly encoded a distinct protein that was only distantly related to the estrogen and glucocorticoid receptors in its carboxy-terminal ligand binding domain. This suggested the possibility that *erb*A might encode a receptor for another class of hormones that affected gene expression in a manner similar to the steroids. Such a notion was substantiated by the end of 1986 when two groups, those of Björn Vennström and Ronald Evans, identified ErbA as the thyroid hormone receptor. The approach taken by both groups of investigators was to clone the *erb*A proto-oncogene cDNA and then to express its encoded protein by *in vitro* transcription and translation. Based on the known similarities between thyroid and steroid hormones, both groups recognized thyroid hormones as plausible candidates for the putative ErbA ligand. Therefore, the binding of radiolabeled thyroid hormone to ErbA was tested, resulting in the direct demonstration that the *erb*A proto-oncogene product is indeed a thyroid hormone receptor. These studies thus provided the first definitive identification of a proto-oncogene that encoded a transcriptional regulatory protein, and they implied that abnormalities in control of gene expression could contribute directly to neoplastic transformation.

The major alteration responsible for the transforming activity of the ErbA oncogene protein is loss of its ability to bind thyroid hormone, presumably as a consequence of multiple mutations in the carboxy-terminal ligand binding domain. This results in the ErbA oncogene protein acting as a dominant negative mutant that interferes with the normal expression of thyroid hormone inducible genes. Unlike the glucocorticoid receptors, the thyroid hormone-receptor resides in the nucleus and binds DNA in either the presence or absence of hormone. However, hormone binding determines whether the receptor ac-

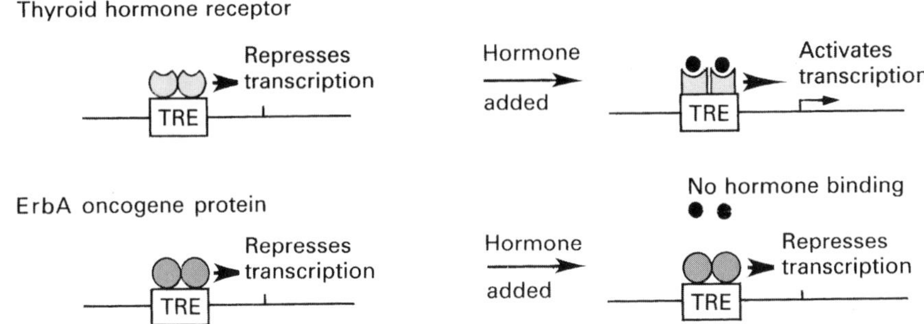

FIGURE 16.3
Regulation of transcription by the thyroid hormone receptor and the ErbA oncogene protein. In the absence of hormone, the thyroid hormone receptor (encoded by the *erb*A proto-oncogene) binds to the thyroid hormone response element (TRE) and represses transcription. Hormone binding converts the normal receptor into a transcriptional activator, resulting in hormonal induction of target gene expression. The ErbA oncogene protein, however, has lost the ability to bind thyroid hormone. Consequently, it acts as a constitutive repressor of thyroid hormone-responsive genes.

tivates or represses transcription (Fig. 16.3). In the absence of hormone, the thyroid hormone receptor acts as a repressor. Hormone binding converts it into an activator, which stimulates transcription of thyroid hormone-inducible genes. The ErbA oncogene protein retains its ability to bind DNA and act as a repressor. However, because it does not bind hormone, the oncogene protein is no longer able to function as a transcriptional activator. Instead, the ErbA oncogene protein is a constitutive repressor that inhibits normal hormone-induced gene expression.

The activity of the ErbA oncogene protein as a constitutive repressor is consistent with its biological activities. As discussed in chapter 3, *erb*A is one of two oncogenes of avian erythroblastosis virus, the other being *erb*B. The *erb*A oncogene is unusual in that it does not appear to directly induce cell transformation by itself but only to enhance the transforming potency of *erb*B in erythroid cells. The effect of *erb*A expression in cells of this lineage is to block the differentiation of erythroblasts to erythrocytes, thereby maintaining the transformed cells in the proliferative state. Rather than directly stimulating cell proliferation, it appears that the ErbA oncogene protein acts to repress transcription of genes that function in erythroid differentiation. It is interesting to note that, based on this mechanism of action, the *erb*A proto-oncogene (thyroid hormone receptor) can be alternatively considered a tumor suppressor gene. In this view, the thyroid hormone receptor normally acts to induce transcription of genes that function in erythroid differentiation, thereby serving as a negative regulator of cell proliferation. Expression of the *erb*A oncogene interferes with the action of the normal gene product, blocking differentiation and contributing to cell transformation.

Another member of the steroid/thyroid hormone receptor superfamily, the retinoic acid receptor α (RARα), is activated as an oncogene by chromosome translocation in human acute promyelocytic leukemias. This translocation fuses RARα with another gene product (called PML), forming a PML-RAR fusion protein. This fusion protein contains most of RARα, including its DNA-binding and hormone-binding domains, in addition to the majority of PML sequences. Because retinoic acid induces differentiation of myeloid cells, one model for the oncogenic activity of the PML-RAR fusion protein is that it interferes with the normal function of RARα, analogous to the action of the ErbA oncogene protein. Consistent with this possibility, the PML-RAR fusion protein exhibits altered transcriptional regulatory properties, although the biological significance of these differences between the PML-RAR and RARα proteins is not yet clear.

Alternatively or in addition, the PML-RAR fusion protein might interfere with the action of PML, which is itself a nuclear protein and putative transcription factor. In this regard, it is noteworthy that administration of retinoic acid is an effective therapy for acute promyelocytic leukemia, which acts by inducing differentiation of the leukemic cells. This observation has led to the suggestion that the PML-RAR fusion protein acts as an oncogenic version of PML whose activity is controlled by retinoic acid binding to the RAR portion of the molecule. Interestingly, recent studies have shown that PML is localized, together with other proteins, in discrete nuclear substructures, called nuclear bodies. In acute promyelocytic leukemia cells, these structures are disrupted, apparently as a result of association of the PML-RAR fusion protein with normal PML. Importantly, treatment with retinoic acid restores the normal organization of PML in nuclear bodies, consistent with its ability to induce differentiation of the leukemic cells. These results thus suggest that the oncogenic activity of the PML-RAR fusion protein results, at least in part, from interfering with the normal localization of PML and associated proteins within the nucleus.

The Jun and Fos Families: Components of Transcription Factor AP-1

Extension of the relation between oncogenes and transcriptional regulatory proteins from the steroid hormone receptors to other oncogenes was not long in coming. The *jun* oncogene of avian sarcoma virus-17 (see chapter 3) was identified and sequenced by Peter Vogt and his colleagues in 1987. A homology search showed a striking relation between the predicted amino acid

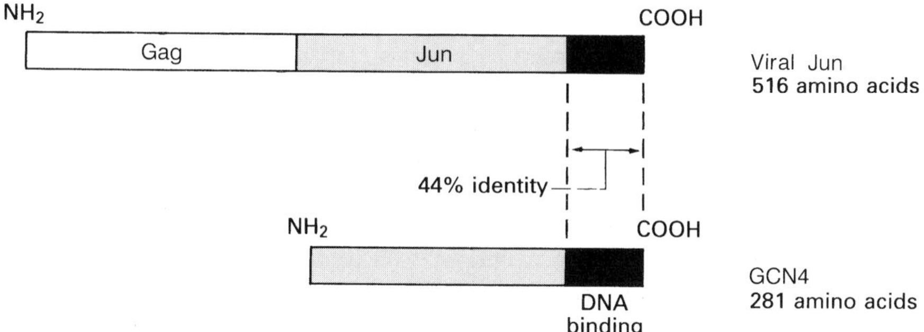

FIGURE 16.4
Relation between Jun and GCN4 proteins. The carboxy-terminal region of Jun displays 44% amino acid sequence identity with the DNA binding domain of the yeast transcriptional activator GCN4.

sequence of Jun and that of the yeast transcriptional activator GCN4 (Fig. 16.4). The relation between Jun and GCN4 was restricted to the DNA binding domain of 60 amino acids located at the carboxy terminus of GCN4. The amino acid sequence of this region of GCN4 was ~45% identical with the carboxy terminus of Jun, leading to the suggestion that Jun might function analogously to GCN4 as a DNA binding transcriptional activator. This suggestion received experimental support in the hands of Kevin Struhl, who generated recombinant molecules in which the DNA binding domain of GCN4 was replaced by the homologous region of Jun. Such a GCN4-Jun hybrid was capable of substituting for normal GCN4 in yeast, indicating that the predicted Jun DNA binding domain is functionally, as well as structurally, related to that of GCN4.

The homology between GCN4 and Jun not only provided the first indication that Jun might function as a transcription factor—it also quickly led to the direct identification of Jun as a component of a known transcription factor in mammalian cells. In 1987, the laboratories of Robert Tjian and Michael Karin identified a human cell transcription factor called AP-1 that was found to interact with sequences in the regulatory elements of genes whose transcription was induced following exposure of cells to growth factors or phorbol ester tumor promoters, which activate protein kinase C. Importantly, the consensus DNA binding site for AP-1 in the regulatory regions of its target genes was identified as the palindrome TGA(C/G)TCA—a sequence that was known to be contained within the yeast GCN4 DNA binding site [ATGA(C/G)TCAT].

This correspondence between the DNA binding sites for AP-1 and GCN4, together with the homology between the DNA binding domains of the GCN4 and Jun proteins, clearly suggested the possibility that Jun was related to AP-1. This was verified by a variety of experiments, including demonstration of immunological cross-reactivity between Jun and AP-1, studies of the DNA-binding properties of Jun expressed in bacteria, and direct sequence analysis of AP-1. Together, these studies demonstrated that the c-*jun* proto-oncogene encodes the major component of the AP-1 transcription factor.

The AP-1 transcription factor, however, consists of more than the single protein encoded by c-*jun*. In fact, purified preparations of AP-1 contain several additional protein species. In this regard, it is important to note that AP-1 has been purified from cells by affinity chromatography procedures based on specific binding to its DNA recognition sequence. Therefore, preparations of AP-1 may in fact consist of multiple proteins that bind to the same DNA target site. Consistent with this notion, AP-1 preparations have been found to contain two additional members of the Jun family, Jun-B and Jun-D, that are highly homologous to c-Jun in their DNA binding domains and also bind to the consensus AP-1 recognition site. Thus, the *jun* gene family encodes a group of related transcription factors that regulate expression of growth factor and phorbol ester inducible genes.

In addition to multiple members of the *jun* family, the proteins encoded by another oncogene family also have been found to function as components of AP-1. The *fos* proto-oncogene has received considerable attention since 1984, when several research groups found that its expression was rapidly and transiently induced following growth factor stimulation of a variety of cell types. Such results led to the hypothesis that Fos might modulate transcription of other genes in response to growth factor stimulation. This first received direct support from studies in the laboratory of Bruce Spiegelman, where it was demonstrated that Fos or a related protein bound specifically to the regulatory sequences of a gene that was transcribed during adipocyte differentiation. Further analysis identified the Fos recognition sequence in this regulatory element as TGACTCA, the same sequence as that recognized by AP-1. Because Fos was known to be present as an intracellular complex with other proteins, these results led to the suggestion that Fos might be associated with Jun/AP-1. Subsequent studies in several laboratories then demonstrated that Jun was present as a complex with Fos in the nucleus and, conversely, that Fos could be detected as a component of purified AP-1 preparations. In addition, several other proteins that were antigenically related to Fos were detected as components of AP-1. These findings raised the possibility that Fos and Jun (and related proteins) might interact to function cooperatively as a complex with transcription-enhancing activity.

Like Jun, Fos contains a putative DNA binding domain related to that of GCN4. It has been established that GCN4 binds DNA as a dimer and that its DNA binding domain includes a leucine repeat sequence responsible for protein dimerization, in addition to a stretch of thirty residues rich in basic amino acids that interacts with DNA directly (Fig. 16.5). The leucine repeat consists of

```
                      Basic Motif                 Leucine Repeat
Jun    ESQERIKAERKRMRNRIAASKCRKRKLERIARLEEKVKTLKAQNSELASTANMLREQVAQL
Fos    SPEEEEKRRIRRERNKHAAAKCRNRRRELTDTLQAETDQLEDEKSALQTEIANLLKEKEKL
GCN4   VPESSDPAALKRARNTEAARRSRARKLQRHKQLEDKVEELLSKNYHLENEVARLKKLVGER
```

FIGURE 16.5
DNA binding domains of Jun, Fos, and GCN4. The DNA binding domains consist of a region of basic amino acids, which interacts with DNA directly, and a leucine repeat region, which is involved in protein dimerization. The highly conserved residues of the basic motif and the leucine residues of the leucine repeat are shaded. Amino acids are designated by single letters: A, alanine; C, cysteine; D, aspartic acid; E, glutamic acid; F, phenylalanine; G, glycine; H, histidine; I, isoleucine; K, lysine; L, leucine; M, methionine; N, asparagine; P, proline; Q, glutamine; R, arginine; S, serine; T, threonine; V, valine; W, tryptophan; Y, tyrosine.

five leucine residues each separated by six other amino acids. It was first suggested by Steve McKnight and his colleagues that such a repeat of leucine residues could stabilize dimer formation through interactions between the hydrophobic leucine side chains in a "leucine zipper" structure. The observation of intracellular complexes between Fos and Jun thus led to the hypothesis that these proteins might bind DNA as a Jun-Fos heterodimer formed by interactions between the leucine repeat regions (Fig. 16.6). This hypothesis was tested by analysis of the ability of Fos and Jun proteins synthesized *in vitro* to form complexes and to bind DNA oligonucleotides containing the AP-1 recog-

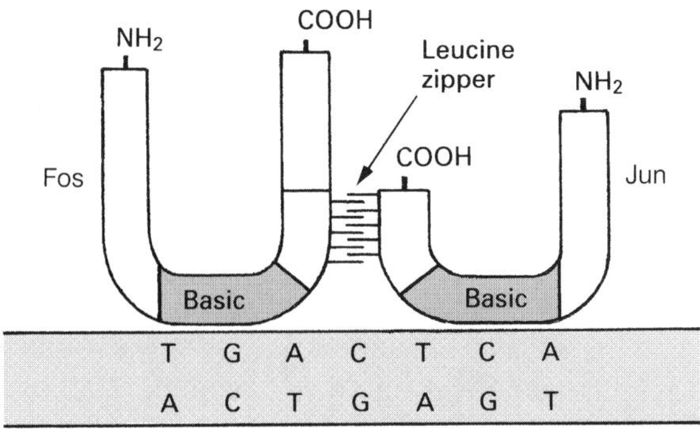

FIGURE 16.6
Binding of a Jun-Fos heterodimer to DNA. Dimerization is accomplished by hydrophobic interactions between leucine side chains to form a "leucine zipper." The basic regions of both Jun and Fos then bind cooperatively to an AP-1 target site, the sequence of which is a symmetrical palindrome.

nition sequence. Experiments of this type demonstrated that Jun could dimerize both with itself and with Fos and that the leucine repeat structure was necessary for dimer formation and subsequent DNA binding. Fos proteins alone neither formed dimers nor bound DNA, but Jun-Fos heterodimers displayed significantly enhanced DNA binding activity compared with homodimers of Jun alone. Thus, Fos and Jun function cooperatively as a complex to activate transcription of AP-1 target genes.

Interestingly, such heterodimer formation and cooperative DNA binding includes multiple members of both the Fos and the Jun gene families. Thus, c-Jun, Jun-B, and Jun-D all form heterodimers with Fos. Likewise, the four members of the Fos family (c-Fos, Fos-B, Fra-1, and Fra-2) all bind cooperatively with Jun to the AP-1 recognition site. Transcription of AP-1 responsive genes therefore appears to be regulated by multiple proteins consisting of complexes between several different members of the Fos and Jun families.

Although mutations within the v-*fos* coding sequence contribute to its transforming activity, constitutive expression of the normal Fos protein is sufficient to activate c-*fos* as an oncogene. This requires both increased transcription and stabilization of *fos* mRNA by deletion of 3' untranslated sequences, including an AU-rich sequence that normally targets rapid degradation of the *fos* message. The combination of these two alterations results in increased deregulated expression of *fos* mRNA and protein, leading to cell transformation. Overexpression of normal Jun also can induce transformation. However, the full activation of its oncogenic potential requires deletion of a negative regulatory domain near its amino terminus, which also increases the activity of Jun as a transcriptional activator. Thus, with both Fos and Jun, transformation results from constitutive expression and function of transcription factors that would normally be activated in response to growth factor stimulation. The identification of Jun and Fos as components of a transcriptional activator regulated by growth factors and protein kinase C clearly defined the role of proto-oncogene transcription factors as elements in signal transduction pathways that function in control of normal cell proliferation.

The Myc Family

Members of the *myc* gene family—c-*myc*, N-*myc*, and L-*myc*—were discussed in chapters 6 through 8 as oncogenes that are activated in a wide variety of neoplasms as a consequence of proviral insertional mutagenesis, chromosome translocation, or DNA amplification. The frequency of activation of members of this gene family in naturally occurring cancers has provided considerable stimulus for investigations of the functions of Myc proteins, which are located in the cell nucleus. In addition, as initially observed in the laboratories of Charles Stiles and Phil Leder in 1983, the c-*myc* gene is rapidly and transiently induced following growth factor stimulation. These results suggested the possibility, as discussed for Fos, that Myc functions to modulate gene expression in response to proliferative stimuli. However, attempts to directly investigate Myc function were frustrated for some years by an inability to detect sequence-specific DNA binding activity of the Myc protein. This obstacle was overcome by experiments based on the sequence relation between Myc and other transcription factors.

The carboxy-terminal region of Myc contains sequences related to the DNA binding domains of two classes of transcription factors: the leucine zipper and helix-loop-helix (HLH) proteins (Fig. 16.7). Like the leucine zipper, the HLH motif is a protein dimerization domain, found in a number of transcription factors, in which two amphipathic α-helical regions are separated by a loop. The Myc HLH domain is preceded by a region of basic amino acids, which is similar to the DNA-binding basic regions of both leucine zipper and HLH proteins. Thus, the carboxy terminus of Myc contains a basic region associated with both leucine zipper and HLH dimerization domains.

The relation of Myc to HLH proteins first led to the demonstration that Myc was capable of sequence-specific DNA binding. It was known that other HLH transcription factors bind to DNA sequences with the consensus CA--TG; so sequences of this type were tested as potential Myc binding sites. Such experiments in the laboratories of Bob Eisenman and Hal Weintraub and that of Ed Ziff first showed, in 1990, that Myc binds specifically to the core sequence CACGTG, similar to that recognized by the DNA binding domains of HLH transcription factors.

The Myc protein dimerizes inefficiently and does not appear to form homodimers under normal physiological conditions. Therefore, its recognition as an HLH-leucine zipper protein implied that Myc normally formed heterodimers with another cellular protein. Such a dimerization partner for Myc was first identified in the laboratory of Bob Eisenman by screening a

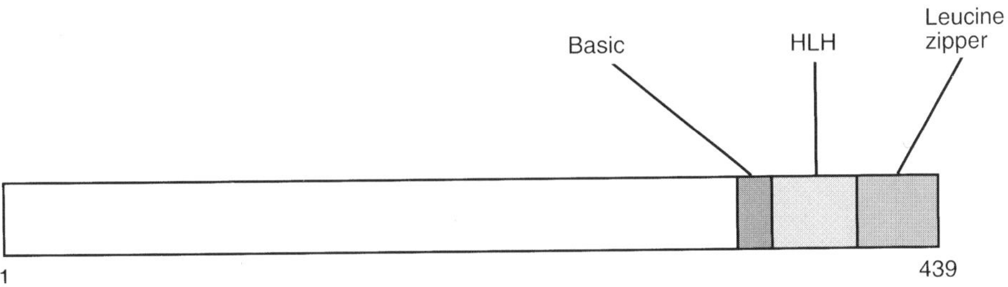

FIGURE 16.7
The Myc DNA-binding domain. The C-terminal region of Myc proteins (human c-Myc is illustrated) has a DNA-binding domain consisting of a region of basic amino acids (designated basic), a helix-loop-helix (HLH) motif, and a leucine zipper.

cDNA expression library for binding to labeled Myc protein. This led to the isolation of a protein, called Max, which contained basic, HLH, and leucine zipper domains similar to those of Myc. An independent approach in Ed Ziff's laboratory also led to the isolation of Max as a protein containing Myc-related DNA binding and dimerization domains. Max is a small protein (~18 kd) that dimerizes with all three members of the Myc family by interactions in both the HLH and the leucine zipper domains. The Myc-Max heterodimers bind to the CACGTG target sequence with high affinity and appear to be the functional form of Myc within cells.

It is important to note that Myc contains a transcriptional activation domain at its amino terminus and stimulates expression of genes containing Myc-Max binding sites (Fig. 16.8). In contrast, Max lacks an activation domain and

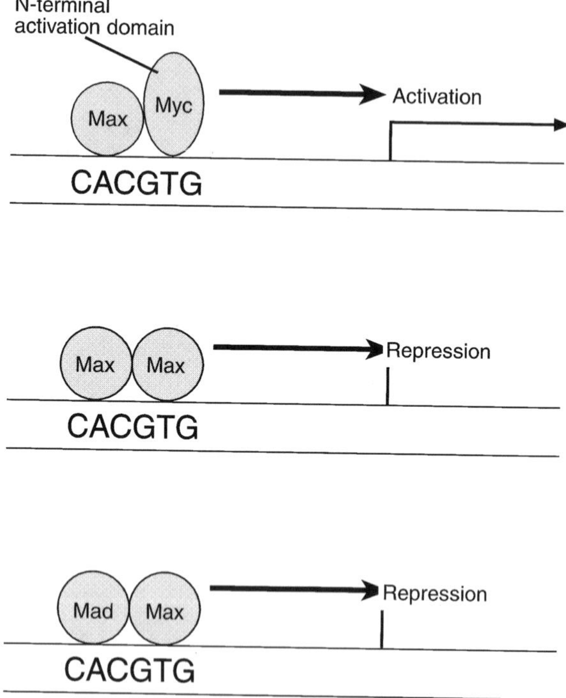

FIGURE 16.8
Activation and repression of Myc target genes. Myc binds to its recognition sequence (CACGTG) as a heterodimer with Max. Myc contains an N-terminal activation domain, so Myc-Max heterodimers activate transcription of target genes. Max can also bind DNA as a Max-Max homodimer, which lacks an activation domain and represses rather than activates transcription. An additional protein (called Mad) also dimerizes with Max, and Mad-Max dimers similarly act as transcriptional repressors.

the binding of Max-Max homodimers serves to repress rather than activate transcription. Consequently, the relative levels of Myc-Max heterodimers versus Max-Max homodimers are an important determinant of activation versus repression of their target genes. Another protein (called Mad) that dimerizes with Max also has been identified and may serve to antagonize Myc function both by competing for Max and by acting directly to repress transcription as a Mad-Max dimer. The details of this potentially complex regulatory network remain to be elucidated. It is clear, however, that elevated Myc expression stimulates transcription of its target genes by increasing the formation of Myc-Max heterodimers, consistent with the frequent activation of *myc* oncogenes by genetic alterations that result in Myc overexpression.

Other Oncogene Transcription Factors

The proteins encoded by a large number of other oncogenes also have been found to function as transcription factors. One well-characterized example is Myb, which is a sequence-specific DNA binding protein that functions as a transcriptional activator. It has an unusual DNA binding domain, characterized by regularly repeated tryptophan residues, and binds DNA as a monomer. Activation of Myb as an oncogene requires deletion of sequences at either the amino or the carboxy terminus, which function as negative regulatory domains.

The *ets* gene family includes a number of transcription factors characterized by DNA binding domains of approximately 85 amino acids that contain conserved tryptophan residues and are thus somewhat similar to the Myb DNA binding domain. The prototype member of this family is the *ets* oncogene of avian erythroblastosis virus E26, but at least twelve *ets* family members have been identified. Ets-1 and Ets-2 are activated by growth factors and phorbol esters and can cooperate with AP-1 (Fos/Jun) to stimulate transcription of some promoters. Consistent with this role in growth factor-mediated activation of gene expression, overexpression of *ets*-1 or *ets*-2 induces cell transformation. Two other members of the *ets* family, *fli*-1 and *spi*-1, are activated as oncogenes by proviral integration in retrovirus-induced murine erythroleukemias (see chapter 6), and *fli*-1 is also activated by chromosome translocation in Ewing's sarcoma (chapter 7). Another *ets*-related gene, *erg*, is translocated in human myeloid leukemias. In addition to these *ets*-related oncogenes, another member of the Ets family, Elk-1, plays a central role in cell proliferation by regulating expression of *fos* in growth factor-stimulated cells, as discussed in chapter 17.

The *rel* oncogene is a member of another family of transcription factors, the NF-κB family. The activity of these factors, of which NF-κB is the prototype, is regulated by translocation from the cytoplasm to the nucleus in response to a variety of stimuli, including phorbol esters and cytokines (Fig. 16.9). In its in-

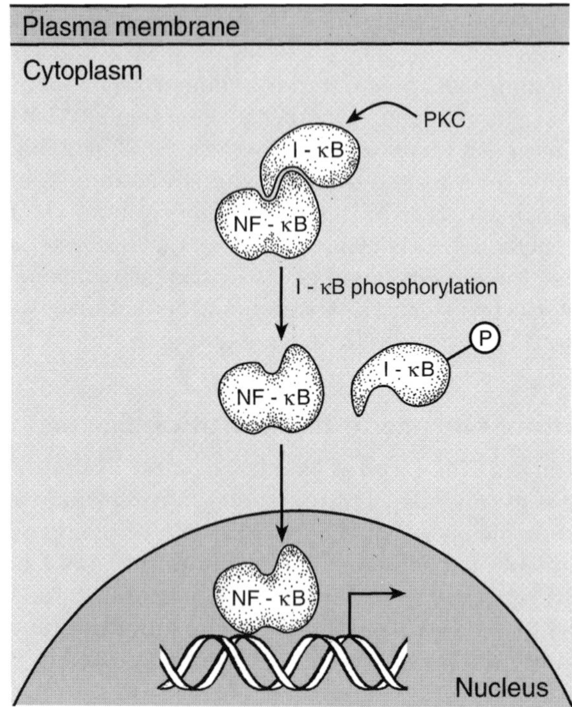

FIGURE 16.9
Regulation of NF-κB. The NF-κB transcription factor consists of two subunits that, in the inactive state, are complexed with an inhibitory subunit (I-κB) in the cytoplasm. Phosphorylation of I-κB by protein kinase C (PKC) results in dissociation of the complex. The released NF-κB then translocates to the nucleus, where it activates transcription from κB sites.

active state, NF-κB is complexed with an inhibitory subunit (I-κB) and retained in the cytoplasm. Following cell stimulation, the NF-κB complex (consisting of 50 and 65 kd subunits) is released from I-κB, allowing its translocation to the nucleus, where it binds to κB DNA sites and activates expression of target genes. Cloning and sequencing of NF-κB revealed that it was related to the *rel* oncogene product, which was subsequently shown to also bind κB DNA sequences. Like NF-κB, the Rel proto-oncogene protein is a transcriptional activator. Interestingly, however, the Rel oncogene protein lacks the transcriptional activation domain of normal Rel and represses rather than activates transcription from κB sites. It thus appears that cell transformation results from action of the Rel oncogene protein as a dominant negative repressor, analogous to the action of the ErbA oncogene protein. It is noteworthy that the products of two other oncogenes, *lyt*-10 and *bcl*-3, also are related to NF-κB and I-κB, respectively, suggesting that abnormalities affecting multiple members of the Rel/NF-κB family can result in cell transformation.

A variety of additional oncogenes encode proteins with characteristic DNA binding domains and are thus likely also to represent transcription factors. As noted in chapter 7, the activation of transcription factor oncogenes by chromosome translocations is particularly frequent, especially in leukemias and lymphomas. In some cases, such as c-*myc* in Burkitt's lymphomas, translocations result in abnormal gene expression: the oncogenes activated by this mechanism include *bcl*-6, *hox*11, *lyl*-1, *rhom*-1, *rhom*-2, *tal*-1, and *tal*-2. In other cases, however, translocations generate fusion proteins that contain distinct functional domains derived from two different transcription factor proto-oncogenes. The prototypical example is provided by the *E2A-pbx*1 oncogene, which is generated by a translocation between chromosomes 1 and 19 in pre-B-cell leukemias. The *E2A* gene on chromosome 19 encodes a member of the HLH family of transcription factors, and the *pbx*1 gene on chromosome 1 encodes a protein with a DNA-binding homeodomain. The result of the chromosome translocation is a chimeric protein in which the transcriptional activation domain of E2A is fused to the DNA-binding domain of Pbx1 (Fig. 16.10). Pbx1 does not appear to activate transcription of its target genes and may instead function as a repressor. In contrast, the E2A-Pbx1 fusion protein is a strong transcriptional activator, suggesting that its activity as an oncogene results from abnormal expression of Pbx1 target genes. Other examples of chimeric transcription factors include the PML-RAR fusion protein discussed earlier in

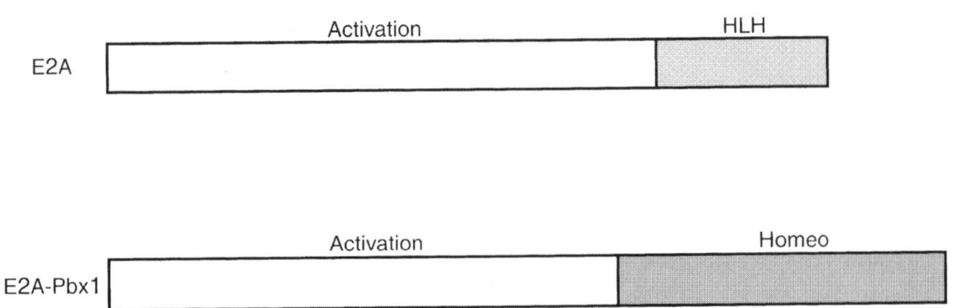

FIGURE 16.10
The E2A-Pbx1 oncogene protein is a chimeric transcription factor. The *E2A* gene encodes a member of the HLH transcription factor family with an N-terminal activation domain and a C-terminal HLH DNA binding domain. It is converted to an oncogene by a chromosome translocation that generates the E2A-Pbx1 fusion protein, in which the activation domain of E2A is joined to the DNA-binding homeodomain (homeo) of Pbx1. The chimeric E2A-Pbx1 protein therefore activates transcription of Pbx1 target genes.

this chapter, the Ews-Fli1 fusion protein formed by translocation of the *ets*-related *fli*-1 gene in Ewing's sarcomas, and possibly some of the fusion proteins discussed in chapter 7 (Aml1-Mtg8, Hrx-AF4, and Hrx-Enl).

Tumor Suppressor Gene Products

Tumor suppressor gene products, as well as oncogene-encoded proteins, can also regulate cell growth by controlling transcription. As discussed earlier in this chapter, the *erb*A proto-oncogene might alternatively be considered a tumor suppressor gene because leukemia results from interference with the normal function of its gene product (thyroid hormone receptor) in promoting differentiation and inhibiting cell proliferation. Moreover, three of the characterized tumor suppressor genes—*WT1*, *p53*, and *Rb*—have been found to encode transcriptional regulatory proteins.

WT1, encoded by a tumor suppressor gene associated with Wilms tumors, was first recognized as a candidate transcription factor because it contained a zinc finger DNA-binding domain. Subsequent experiments then demonstrated that WT1 was a sequence-specific DNA binding protein that recognized the same target sequence as a family of proteins that activate transcription of a number of growth factor-induced genes. However, WT1 was found to act as a repressor rather than a transcriptional activator, suggesting that the action of WT1 as a tumor suppressor gene product results from repressing the transcription of genes involved in cell proliferation. Interestingly, the targets for repression by WT1 include the gene encoding insulin-like growth factor II (IGF-II), which is characteristically overexpressed in Wilms tumors and is thought to play a role as an autocrine growth factor in the development of these neoplasms.

The product of the *p53* tumor suppressor gene, which is the single gene most frequently involved in human cancers, appears to act as a transcriptional activator rather than a repressor. The initial insight into p53 function came from the observation that it contained an activation domain related to those of other transcription factors. The p53 activation domain was then shown to stimulate gene expression when fused to a heterologous DNA binding domain. Subsequent studies then showed that p53 was itself a sequence-specific DNA binding protein that activated transcription of adjacent genes. Mutant p53 proteins were defective in transcriptional activation, and the activity of wild-type p53 was blocked by the binding of both the viral (SV40 T antigen, adenovirus E1B, and human papillomavirus E6) and the cellular (Mdm-2) oncogene proteins that interfere with the function of p53 as a tumor suppressor (see chapter 10). Although p53 can also repress some promoters, its function as an activator appears to correlate with its ability to act as a tumor suppressor. This implies that target genes whose expression is stimulated by p53 will themselves act to inhibit cell proliferation. In confirmation of this prediction, one such p53-induc-

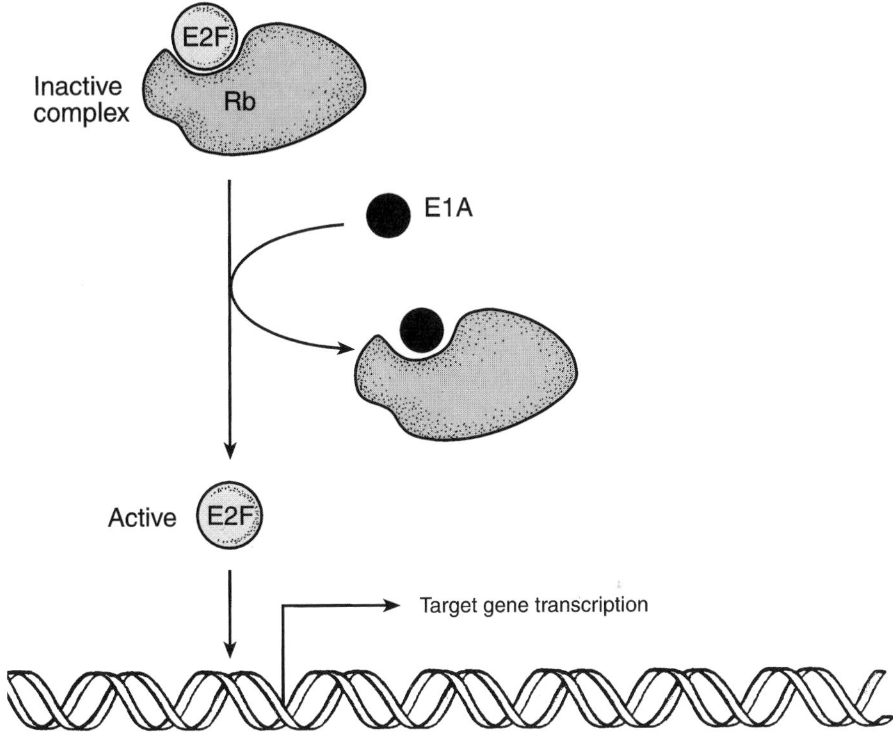

FIGURE 16.11
Interaction of Rb with the E2F transcription factor. Rb binds to E2F, blocking its function as a transcriptional activator. Oncogene proteins of several DNA tumor viruses (for example, adenovirus E1A) bind to Rb and disrupt the Rb-E2F complex, releasing E2F, which then activates transcription of its target genes.

ible gene has recently been identified and found to suppress cell proliferation, as discussed further in chapter 18.

The *Rb* gene product also functions as a transcriptional regulator, although Rb does not itself bind to target DNA sequences. Rather, Rb interacts with other cellular transcription factors, in particular members of the E2F family (Fig. 16.11). The link between Rb and E2F came from studies of the adenovirus *E1A* oncogene product, which (like SV40 T antigen and human papillomavirus E7) acts by inhibiting Rb function (see chapter 10). The E2F transcription factor was first recognized as a regulator of the adenovirus *E2* gene, which is activated in response to the adenovirus E1A protein. Subsequent studies indicated that E2F also regulates transcription of a number of cellular genes involved in cell proliferation and DNA synthesis, including the *myc* and *myb* proto-oncogenes and genes encoding dihydrofolate reductase, thymidine kinase, and DNA polymerase. Further experiments then demonstrated that E2F was present in normal cells as an inactive complex with Rb. Binding to Rb blocks the function of E2F as a transcriptional activator, and the Rb-E2F complex may instead act as a repressor. E1A and other viral oncogene proteins bind to the same region of Rb as E2F; the binding of these proteins disrupts the Rb-E2F complex, releasing E2F to activate transcription of its target genes. Overexpression of E2F can stimulate cell proliferation, and the activity of Rb as a tumor suppressor correlates with E2F binding. It thus appears that inhibiting transcriptional activation by E2F is at least one mechanism of Rb action.

It is important to note that E2F denotes a family of transcription factors, members of which might differ in their interactions with Rb. Moreover, E2F may not be the only target for Rb or Rb-related proteins. In this regard, a distinct Rb-related protein (called p107) is of interest. Like Rb, p107 binds to viral oncogene proteins and to E2F. In addition, however, recent experiments indicate that p107 binds to the transcriptional activation domain of Myc and suppresses Myc activity, suggesting that Myc may be a critical target for growth inhibition by p107.

Summary

The proteins encoded by a variety of oncogenes and tumor suppressor genes function as transcriptional regulatory molecules. The identification of Jun and Fos as components of the AP-1 transcription factor provided the first clear demonstration of oncogene proteins that normally functioned to activate gene expression in response to signals that stimulate cell proliferation. This has been extended to include the Myc family and a variety of other oncogene products. Proteins encoded by other oncogenes, such as *erb*A and *rel*, act as constitutive repressors rather than activators of their target genes. The WT1 and p53 tumor suppressors similarly act to repress or activate transcription of specific target genes, respectively. Rb also acts as a transcriptional regulator, although it functions by inhibiting the transcriptional activator E2F rather than by binding directly to target DNA sequences. Regulation of transcription is thus a central mechanism by which both oncogenes and tumor suppressor genes act to control cell proliferation.

References

General References

McKnight, S.L., and Yamamoto, K.R., eds. 1992. *Transcriptional regulation.* Cold Spring Harbor Laboratory Press, Plainview, NY.

Mitchell, P.J., and Tjian, R. 1989. Transcriptional regulation in mammalian cells by sequence-specific DNA binding proteins. *Science* 245:371–378.

Oncogenes and Steroid Hormone Receptors

Damm, K., and Evans, R.M. 1993. Identification of a domain required for oncogenic activity and transcriptional suppression by v-*erb*A and thyroid-hormone receptor α. *Proc. Natl. Acad. Sci. USA* 90:10668–10672.

Damm, K., Thompson, C.C., and Evans, R.M. 1989. Protein encoded by v-*erb*A functions as a thyroid-hormone receptor antagonist. *Nature* 339:593–597.

De Thé, H., Lavau, C., Marchio, A., Chomienne, C., Degos, L., and Dejean, A. 1991. The PML-RARα fusion mRNA generated by the t(15;17) translocation in acute promyelocytic leukemia encodes a functionally altered RAR. *Cell* 66:675–684.

Dyck, J.A., Maul, G.G., Miller, W.H., Jr., Chen, J.D., Kakizuka, A., and Evans, R.M. 1994. A novel macromolecular structure is a target of the promyelocyte-retinoic acid receptor oncoprotein. *Cell* 76:333–343.

Goddard, A.D., Borrow, J., Freemont, P.S., and Solomon, E. 1991. Characterization of a zinc finger gene disrupted by the t(15;17) in acute promyelocytic leukemia. *Science* 254:1371–1374.

Green, S., Walter, P., Kumar, V., Krust, A., Bornert, J.-M., Argos, P., and Chambon, P. 1986. Human oestrogen receptor cDNA: sequence, expression and homology to v-*erb*-A. *Nature* 320:134–139.

Greene, G.L., Gilna, P., Waterfield, M., Baker, A., Hort, Y., and Shine, J. 1986. Sequence and expression of human estrogen receptor complementary DNA. *Science* 231:1150–1154.

Kakizuka, A., Miller, W.H., Jr., Umesono, K., Warrell, R.P., Jr., Frankel, S.R., Murty, V.V.V.S., Dmitrovsky, E., and Evans, R.M. 1991. Chromosomal translocation t(15;17) in human acute promyelocytic leukemia fuses RARα with a novel putative transcription factor, PML. *Cell* 66:663–674.

Koken, M.H.M., Puvion-Dutilleul, F., Guillemin, M.C., Viron, A., Linares-Cruz, G., Stuurman, N., de Jong, L., Szostecki, C., Calvo, F., Chomienne, C., Degos, L., Puvion, E., and de Thé, H. 1994. The t(15;17) translocation alters a nuclear body in a retinoic acid-reversible fashion. *EMBO J.* 13:1073–1083.

Sap, J., Muñoz, A., Damm, K., Goldberg, Y., Ghysdael, J., Leutz, A., Beug, H., and Vennström, B. 1986. The c-*erb*A protein is a high-affinity receptor for thyroid hormone. *Nature* 324:635–640.

Sap, J., Muñoz, A., Schmitt, J., Stunnenberg, H., and Vennström, B. 1989. Repression of transcription mediated at a thyroid hormone response element by the v-*erb*-A oncogene product. *Nature* 340:242–244.

Weinberger, C., Hollenberg, S.M., Rosenfeld, M.G., and Evans, R.M. 1985. Domain structure of human glucocorticoid receptor and its relationship to the v-*erb*-A oncogene product. *Nature* 318:670–672.

Weinberger, C., Thompson, C.C., Ong, E.S., Lebo, R., Gruol, D.J., and Evans, R.M. 1986. The c-*erb*-A gene encodes a thyroid hormone receptor. *Nature* 324:641–646.

Weis, K., Rambaud, S., Lavau, C., Jansen, J., Carvalho, T., Carmo-Fonseca, M., Lamond, A., and Dejean, A. 1994. Retinoic acid regulates aberrant nuclear localization of PML-RARα in acute promyelocytic leukemia cells. *Cell* 76:345–356.

The Jun and Fos Families: Components of Transcription Factor AP-1

Angel, P., Allegretto, E.A., Okino, S.T., Hattori, K., Boyle, W.J., Hunter, T., and Karin, M. 1988. Oncogene *jun* encodes a sequence-specific *trans*-activator similar to AP-1. *Nature* 332:166–171.

Bohmann, D., Bos, T.J., Admon, A., Nishimura, T., Vogt, P.K., and Tjian, R. 1987. Human proto-oncogene c-*jun* encodes a DNA binding protein with structural and functional properties of transcription factor AP-1. *Science* 238:1386–1392.

Bohmann, D., and Tjian, R. 1989. Biochemical analysis of transcriptional activation by Jun: differential activity of c- and v-Jun. *Cell* 59:709–717.

Bos, T.J., Monteclaro, F.S., Mitsunobu, F., Ball, A.R., Chang, C.H.W., Nishimura, T., and Vogt, P.K. 1990. Efficient transformation of chicken embryo fibroblasts by c-*jun* requires structural modification in coding and noncoding sequences. *Genes Dev.* 4:1677–1687.

Chiu, R., Boyle, W.J., Meek, J., Smeal, T., Hunter, T., and Karin, M. 1988. The c-fos protein interacts with c-*jun*/AP-1 to stimulate transcription of AP-1 responsive genes. *Cell* 54:541–552.

Distel, R.J., Ro, H.-S., Rosen, B.S., Groves, D.L., and Spiegelman, B.M. 1987. Nucleoprotein complexes that regulate gene expression in adipocyte differentiation: direct participation of c-*fos*. *Cell* 49:835–844.

Franza, B.R., Jr., Rauscher, F.J., III, Josephs, S.F., and Curran, T. 1988. The *fos* complex and *fos*-related antigens recognize sequence elements that contain AP-1 binding sites. *Science* 239:1150–1153.

Halazonetis, T.D., Georgopoulos, K., Greenberg, M.E., and Leder, P. 1988. c-Jun dimerizes with itself and with c-Fos, forming complexes of different DNA binding affinities. *Cell* 55:917–924.

Kouzarides, T., and Ziff, E. 1988. The role of the leucine zipper in the Fos-Jun interaction. *Nature* 336:646–651.

Landschulz, W.H., Johnson, P.F., and McKnight, S.L. 1988. The leucine zipper: a hypothetical structure common to a new class of DNA binding proteins. *Science* 240:1759–1764.

Nakabeppu, Y., Ryder, K., and Nathans, D. 1988. DNA binding activities of three murine Jun proteins: stimulation by Fos. *Cell* 55:907–915.

Nishina, H., Sato, H., Suzuki, T., Sato, M., and Iba, H. 1990. Isolation and characterization of *fra*-2, an additional member of the *fos* gene family. *Proc. Natl. Acad. Sci. USA* 87:3619–3623.

Ransome, L.J., and Verma, I.M. 1990. Nuclear proto-oncogenes *FOS* and *JUN*. *Ann. Rev. Cell Biol.* 6:539–557.

Rauscher, F.J., III, Cohen, D.R., Curran, T., Bos, T.J., Vogt, P.K., Bohmann, D., Tjian, R., and Franza, B.R., Jr. 1988. Fos-associated protein p39 is the product of the *jun* proto-oncogene. *Science* 240:1010–1016.

Rauscher, F.J., III, Sambucetti, L.C., Curran, T., Distel, R.J., and Spiegelman, B.M. 1988. Common DNA binding site for *fos* protein complexes and transcription factor AP-1. *Cell* 52:471–480.

Sassone-Corsi, P., Lamph, W.W., Kamps, M., and Verma, I.M. 1988. *fos*-associated cellular p39 is related to nuclear transcription factor AP-1. *Cell* 54:553–560.

Sassone-Corsi, P., Ransone, L.J., Lamph, W.W., and Verma, I.M. 1988. Direct interaction between fos and jun nuclear oncoproteins: role of the "leucine zipper" domain. *Nature* 336:692–695.

Struhl, K. 1987. The DNA-binding domains of the *jun* oncoprotein and the yeast GCN4 transcriptional activator protein are functionally homologous. *Cell* 50:841–846.

Turner, R., and Tjian, R. 1989. Leucine repeats and an adjacent DNA binding domain mediate the formation of functional cFos-cJun heterodimers. *Science* 243:1689–1694.

Vinson, C.R., Sigler, P.B., and McKnight, S.L. 1989. Scissors-grip model for DNA recognition by a family of leucine zipper proteins. *Science* 246:911–916.

Vogt, P.K., Bos, T.J., and Doolittle, R.F. 1987. Homology between the DNA-binding domain of the GCN4 regulatory protein of yeast and the carboxyl-terminal region of a protein coded for by the oncogene *jun*. *Proc. Natl. Acad. Sci. USA* 84:3316–3319.

The Myc Family

Amati, B., Brooks, M.W., Levy, N., Littlewood, T.D., Evan, G.I., and Land, H. 1993. Oncogenic activity of the c-Myc protein requires dimerization with Max. *Cell* 72:233–245.

Ayer, D.E., Kretzner, L., and Eisenman, R.N. 1993. Mad: a heterodimeric partner for Max that antagonizes Myc transcriptional activity. *Cell* 72:211–222.

Blackwell, T.K., Kretzner, L., Blackwood, E.M., Eisenman, R.N., and Weintraub, H. 1990. Sequence-specific DNA binding by the c-Myc protein. *Science* 250:1149–1151.

Blackwood, E.M., and Eisenman, R.N. 1991. Max: a helix-loop-helix zipper protein that forms a sequence-specific DNA-binding complex with Myc. *Science* 251:1211–1217.

Kato, G.J., Barrett, J., Villa, G.M., and Dang, C.V. 1990. An amino-terminal c-*myc* domain required for neoplastic transformation activates transcription. *Mol. Cell. Biol.* 10:5914–5920.

Kretzner, L., Blackwood, E.M., and Eisenman, R.N. 1992. Myc and Max possess distinct transcriptional activities. *Nature* 359:426–429.

Marcu, K.B., Bossone, S.A., and Petel, A.J. 1992. *Myc* function and regulation. *Ann. Rev. Biochem.* 61:809–860.

Prendergast, G.C., Lawe, D., and Ziff, E.B. 1991. Association of Myn, the murine homolog of Max, with c-Myc stimulates methylation-sensitive DNA binding and Ras cotransformation. *Cell* 65:395–407.

Prendergast, G.C., and Ziff, E.B. 1991. Methylation-sensitive sequence-specific DNA binding by the c-Myc basic region. *Science* 251:186–189.

Zervos, A.S., Gyuris, J., and Brent, R. 1993. Mxi1, a protein that specifically interacts with Max to bind Myc-Max recognition sites. *Cell* 72:223–232.

Other Oncogene Transcription Factors

Ballard, D.W., Walker, W.H., Doerre, S., Sista, P., Molitor, J.A., Dixon, E.P., Peffer, N.J., Hannink, M., and Greene, W.C. 1990. The v-*rel* oncogene encodes a κB enhancer binding protein that inhibits NF-κB function. *Cell* 63:803–814.

Biedenkapp, H., Borgmeyer, U., Sippel, A.E., and Klempnauer, K.-H. 1988. Viral *myb* oncogene encodes a sequence-specific DNA-binding activity. *Nature* 335:835–837.

Cleary, M.L. 1991. Oncogenic conversion of transcription factors by chromosomal translocations. *Cell* 66:619–622.

Franzoso, G., Bours, V., Park, S., Tomita-Yamaguchi, M., Kelly, K., and Siebenlist, U. 1992. The candidate oncoprotein Bcl-3 is an antagonist of p50/NF-κB-mediated inhibition. *Nature* 359:339–342.

Gabrielson, O.S., Sentenac, A., and Fromageot, P. 1991. Specific DNA binding by c-Myb: evidence for a double helix-turn-helix-related motif. *Science* 253:1140–1143.

Ghosh, S., Gifford, A.M., Rivieve, L.R., Tempst, P., Nolan, G.P., and Baltimore, D. 1990. Cloning of the p50 DNA binding subunit of NF-κB: homology to *rel* and *dorsal*. *Cell* 62:1019–1029.

Gu, Y., Nakamura, T., Alder, H., Prasad, R., Canaani, O., Cimino, G., Croce, C.M., and Canaani, E. 1992. The t(4;11) chromosome translocation of human acute leukemias fuses the *ALL-1* gene, related to *Drosophila trithorax*, to the *AF-4* gene. *Cell* 71:701–708.

Hipskind, R.A., Rao, V.N., Mueller, C.G.F., Reddy, E.S.P., and Nordheim, A. 1991. Ets-related protein Elk-1 is homologous to the c-*fos* regulatory factor $p62^{TCF}$. *Nature* 354:531–534.

Inoue, J., Kerr, L.D., Ransone, L.J., Bengal, E., Hunter, T., and Verma, I.M. 1991. c-rel activates but v-rel suppresses transcription from κB sites. *Proc. Natl. Acad. Sci. USA* 88:3715–3719.

Kamps, M.P., Murre, C., Sun, X., and Baltimore, D. 1990. A new homeobox gene contributes the DNA binding domain of the t(1;19) translocation protein in pre-B ALL. *Cell* 60:547–555.

Kerr, L.D., Duckett, C.S., Wamsley, P., Zhang, Q., Chiao, P., Nabel, G., McKeithan, T.W., Baeuerle, P.A., and Verma, I.M. 1992. The proto-oncogene *BCL-3* encodes an IκB protein. *Genes Dev.* 6:2352–2363.

Kieran, M., Blank, V., Logeat, F., Vandekerckhove, J., Lottspeich, F., Le Bail, O., Urban, M.B., Kourilsky, P., Baeuerle, P.A., and Israel, A. 1990. The DNA binding subunit of NF-κB is identical to factor KBF1 and homologous to the *rel* oncogene product. *Cell* 62:1007–1018.

Luscher, B., and Eisenman, R.N. 1990. New light on Myc and Myb. Part II. Myb. *Genes Dev.* 4:2235–2241.

Macleod, K., Leprince, D., and Stehelin, D. 1992. The *ets* gene family. *Trends Biochem. Sci.* 17:251–256.

May, W.A., Lessnick, S.L., Braun, B.S., Klemsz, M., Lewis, B.C., Lunsford, L.B., Hromas, R., and Denny, C.T. 1993. The Ewing's sarcoma *EWS/FLI-1* fusion gene encodes a more potent transcriptional activator and is a more powerful transforming gene than *FLI-1*. *Mol. Cell. Biol.* 13:7393–7398.

Miyoshi, H., Kozu, T., Shimizu, K., Enomoto, K., Maseki, N., Kaneko, Y., Kamada, N., and Ohki, M. 1993. The t(8;21) translocation in acute myeloid leukemia results in production of an *AML1-MTG8* fusion transcript. *EMBO J.* 12:2715–2721.

Neri, A., Chang, C.-C., Lombardi, L., Salina, M., Corradini, P., Maiolo, A.T., Chaganti, R.S.K., and Dalla-Favera, R. 1991. B cell lymphoma-associated chromosomal translocation involves candidate oncogene *lyt*-10, homologous to NF-κB p50. *Cell* 67:1075–1087.

Nourse, J., Mellentin, J.D., Galili, N., Wilkinson, J., Stanbridge, E., Smith, S.D., and Cleary, M.L. 1990. Chromosomal translocation t(1;19) results in synthesis of a homeobox fusion mRNA that codes for a potential chimeric transcription factor. *Cell* 60:535–545.

Rabbitts, T.H. 1991. Translocations, master genes, and differences between the origins of acute and chronic leukemias. *Cell* 67:641–644.

Tkachuk, D.C., Kohler, S., and Cleary, M.L. 1992. Involvement of a homolog of *Drosophila trithorax* by 11q23 chromosomal translocations in acute leukemias. *Cell* 71:691–700.

Van Dijk, M.A., Voorhoeve, P.M., and Murre, C. 1993. Pbx1 is converted into a transcriptional activator upon acquiring the N-terminal region of E2A in pre-B-cell acute lymphoblastoid leukemia. *Proc. Natl. Acad. Sci. USA* 90:6061–6065.

Wasylyk, B., Wasylyk, C., Flores, P., Begue, A., Leprince, D., and Stehelin, D. 1990. The c-*ets* proto-oncogenes encode transcription factors that cooperate with c-Fos and c-Jun for transcriptional activation. *Nature* 346:191–193.

Weston, K., and Bishop, J.M. 1989. Transcriptional activation by the v-*myb* oncogene and its cellular progenitor, c-*myb*. *Cell* 58:85–93.

Wulczyn, F.G., Naumann, M., and Scheidereit, C. 1992. Candidate proto-oncogene *bcl*-3 encodes a subunit-specific inhibitor of transcription factor NF-κB. *Nature* 358:597–599.

Ye, B.H., Lista, F., Lo Coco, F., Knowles, D.M., Offit, K., Chaganti, R.S.K., and Dalla-Favera, R. 1993. Alterations of a zinc finger-encoding gene, *BCL-6*, in diffuse large-cell lymphoma. *Science* 262:747–750.

Tumor Suppressor Gene Products

Bandara, L.R., and La Thangue, N.B. 1991. Adenovirus E1A prevents the retinoblastoma gene product from complexing with a common transcription factor. *Nature* 351:494–497.

Chellapan, S., Hiebert, S., Mudryj, M., Horowitz, J.M., and Nevins, J.R. 1991. The E2F transcription factor is a cellular target for the RB protein. *Cell* 58:1193–1198.

Chittenden, T., Livingston, D.M., and Kaelin, W.G., Jr. 1991. The T/E1A-binding domain of the retinoblastoma product can interact selectively with a sequence-specific DNA-binding protein. *Cell* 65:1073–1082.

Drummond, I.A., Madden, S.L., Rohwer-Nutter, P., Bell, G.I., Sukhatme, V.P., and Rauscher, F.J., III. 1992. Repression of the insulin-like growth factor II gene by the Wilms tumor suppressor WT1. *Science* 257:674–678.

El-Deiry, W., Tokino, T., Velculescu, V.E., Levy, D.B., Parsons, R., Trent, J.M., Lin, D., Mercer, W.E., Kinzler, K.W., and Vogelstein, B. 1993. *WAF1*, a potential mediator of p53 tumor suppression. *Cell* 75:817–825.

Fields, S., and Jang, S.K. 1990. Presence of a potent transcription activating sequence in the p53 protein. *Science* 249:1046–1049.

Flemington, E.K., Speck, S.H., and Kaelin, W.G., Jr. 1993. E2F-1-mediated transactivation is inhibited by complex formation with the retinoblastoma susceptibility gene product. *Proc. Natl. Acad. Sci. USA* 90:6914–6918.

Gu, W., Bhatia, K., Magrath, I.T., Dang, C.V., and Dalla-Favera, R. 1994. Binding and suppression of the Myc transcriptional activation domain by p107. *Science* 264:251–254.

Johnson, D.G., Schwarz, J.K., Cress, W.D., and Nevins, J.R. 1993. Expression of transcription factor E2F1 induces quiescent cells to enter S phase. *Nature* 365:349–352.

Kern, S.E., Kinzler, K.W., Bruskin, A., Jarosz, D., Friedman, P., Prives, C., and Vogelstein, B. 1991. Identification of p53 as a sequence specific DNA binding protein. *Science* 256:827–830.

Levine, A.J. 1993. The tumor suppressor genes. *Ann. Rev. Biochem.* 62:623–651.

Madden, S.L., Cook, D.M., Morris, J.F., Gashler, A., Sukhatme, V.P., and Rauscher, F.J., III. 1991. Transcriptional repression mediated by the *WT1* Wilms tumor gene product. *Science* 253:1550–1553.

Nevins, J.R. 1992. E2F: a link between the Rb tumor suppressor protein and viral oncoproteins. *Science* 258:424–429.

Pietenpol, J.A., Tokino, T., Thiagalingam, S., El-Deiry, W.S., Kinzler, K.W., and Vogelstein, B. 1994. Sequence-specific transcriptional activation is essential for growth suppression by p53. *Proc. Natl. Acad. Sci. USA* 91:1998–2002.

Rauscher, F.J., III, Morris, J.F., Tournay, O.E., Cook, D.M., and Curran, T. 1990. Binding of the Wilms tumor locus zinc finger protein to the EGR-1 consensus sequence. *Science* 250:1259–1262.

Raycroft, L., Wu, H., and Lozano, G. 1990. Transcriptional activation by wild-type but not transforming mutants of the p53 anti-oncogene. *Science* 249:1049–1051.

Vogelstein, B., and Kinzler, K.W. 1992. p53 function and dysfunction. *Cell* 70:523–526.

Weintraub, S.J., Prater, C.A., and Dean, D.C. 1992. Retinoblastoma protein switches the E2F site from positive to negative element. *Nature* 358:259–261.

❖ CHAPTER 17

Pathways of Mitogenic Signal Transduction

The preceding five chapters dealt with the biochemical activities of a number of oncogene and tumor suppressor gene products, most of which can be viewed as elements in signal transduction pathways. Many of the proto-oncogenes thus appear to encode signal transducing proteins with a normal role in control of cell proliferation by transmission of mitogenic stimuli. This chapter will consider the links between the different functional groups of oncogene proteins in signal transduction pathways. How do these gene products integrate into a network by which a mitogenic signal, generated at the plasma membrane by binding of an extracellular growth factor, is transmitted to the nucleus, ultimately leading to cell division?

As discussed in chapter 1, the proliferation of most normal cells is controlled by extracellular growth factors that bind to specific receptors on the cell surface. In the absence of growth factor stimulation, cells cease proliferation in the G_1 phase of the cell cycle and enter a quiescent state called G_0. Exposure to appropriate mitogens induces quiescent cells to reenter the cell cycle and initiate DNA synthesis (S phase) several hours later. The stimulus provided by growth factor binding at the cell surface must therefore be transmitted through the cytoplasm to the nucleus, resulting in a long-term commitment of the cell to proliferation.

The first evidence linking different proto-oncogenes in a physiological pathway of mitogenic signal transduction was reported in 1983 from the laboratories of Charles Stiles and Phil Leder (Fig. 17.1). These investigators observed that increased transcription of the c-*myc* proto-oncogene was an early event in the response of both lymphocytes and fibroblasts to stimulation with

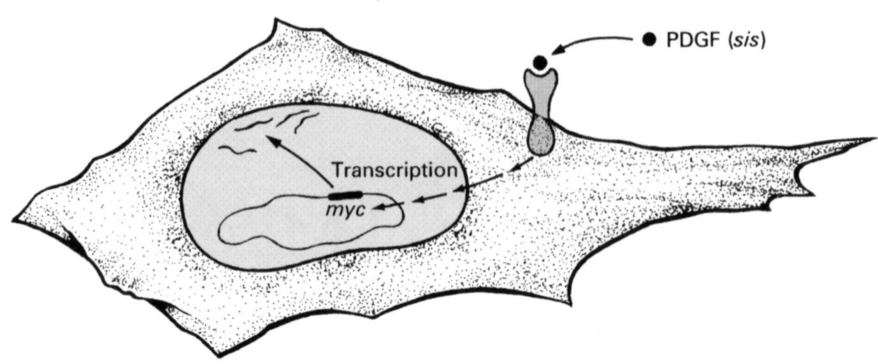

FIGURE 17.1
Linkage of proto-oncogenes in mitogenic signal transduction. Transcription of c-*myc* is induced following mitogenic stimulation of quiescent fibroblasts with PDGF (the *sis* proto-oncogene product).

appropriate mitogens (lipopolysaccharide for B lymphocytes, concanavalin A for T lymphocytes, and PDGF for fibroblasts). Thus, a mitogenic signal initiated at the cell surface of fibroblasts by the product of one proto-oncogene (PDGF/*sis*) was transmitted to the nucleus, where the expression of a second proto-oncogene was increased. The implication of these findings was that different proto-oncogenes could constitute elements of signal transduction pathways linking cell surface events to alterations in gene expression, ultimately leading to cell proliferation. The roles of oncogenes as individual components of such pathways has already been discussed. This chapter will focus on the connections by which different groups of proto-oncogenes act in series to transmit signals from the cell surface to the nucleus.

Receptor Tyrosine Kinases and Their Targets

Chapters 12 and 13 dealt with the relation of oncogenes to the initial events in mitogenic signal transduction pathways—namely, the interaction of extracellular growth factors with their cell surface receptors. The prototypical examples of oncogenes encoding growth factors and growth factor receptors were provided by the platelet-derived growth factor (PDGF) and epidermal growth factor (EGF) systems. How is the proliferative signal normally generated by growth factor binding transmitted from the cell surface receptor to the interior of the cell? The PDGF and EGF receptors, like a number of other cell surface receptors, are protein-tyrosine kinases. Growth factor binding stimulates tyrosine kinase activity, implicating this activity in signal transduction. In addition, mutants of these receptors that are defective in tyrosine kinase fail to induce cell proliferation after growth factor stimulation, indicating that protein-tyrosine kinase activity is necessary for receptor function.

Growth factor binding induces dimerization of the receptor protein-tyrosine kinases, which then cross-phosphorylate each other on tyrosine residues (Fig. 17.2). Dimerization of some receptors is induced directly by the binding of a dimeric growth factor, such as PDGF, which consists of two polypeptide chains. Other growth factors (for example, EGF) are monomeric, but their binding appears to induce conformational alterations that also result in receptor dimerization. Interactions between the cytoplasmic domains of receptor dimers are then thought to stimulate their protein kinase activities, and the receptors phosphorylate both themselves (autophosphorylation) and other substrates. Receptor autophosphorylation, which occurs by cross-phosphorylation of adjacent polypeptides, plays a critical role in signal transduction by mediating the association of other signaling molecules with the activated receptor tyrosine kinases. In particular, autophosphorylation results in the phosphorylation of specific tyrosine residues in the receptor cytoplasmic domains. These phosphotyrosines then serve as binding sites for the SH2 domains of other signaling molecules. As discussed in chapter 13, SH2 domains bind specific

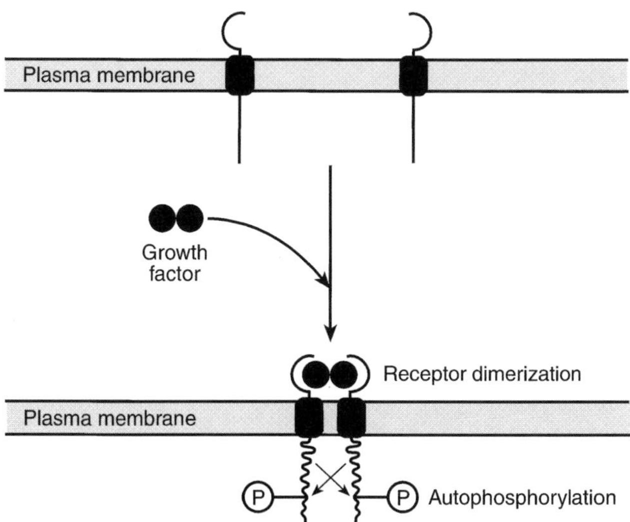

FIGURE 17.2
Dimerization and autophosphorylation of receptor tyrosine kinases. Growth factor binding induces receptor dimerization, in some cases because the growth factor is itself a dimeric polypeptide. The activated receptor kinase domains then cross-phosphorylate each other on tyrosine residues.

phosphotyrosine-containing peptides. The presence of SH2 domains in a variety of signal transducing molecules promotes their association with activated tyrosine kinase receptors, thereby mediating further transmission of the signals initiated by growth factor binding.

One of the targets of receptor tyrosine kinases is an isotype of phospholipase C (PLC-γ), which catalyzes the hydrolysis of phosphatidylinositol 4,5-bisphosphate (PIP_2) to generate the second messengers diacylglycerol (DAG) and inositol 1,4,5- triphosphate (IP_3) (Fig. 17.3). The role of this second messenger system in activation of protein kinase C and stimulation of cell proliferation was discussed in chapter 14 with respect to the activation of a different isotype of phospholipase C (PLC-ß) downstream of some G protein-coupled receptors. Whereas PLC-ß is controlled by G proteins, PLC-γ contains SH2 domains that lead to its activation by receptor tyrosine kinases. In particular, the PLC-γ SH2 domains bind to specific autophosphorylated tyrosine residues of the activated receptors. PLC-γ is then itself phosphorylated by the receptor, and this tyrosine phosphorylation increases PLC-γ catalytic activity, thereby stimulating PIP_2 hydrolysis.

Despite the common activation of PIP_2 hydrolysis by both tyrosine kinase and G protein-coupled receptors, it should be noted that activation of PLC-γ does not appear to be required for mitogenic signaling by receptor tyrosine ki-

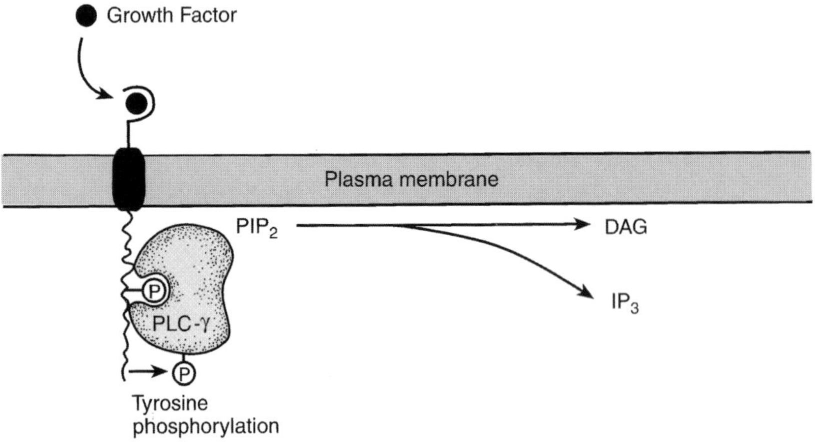

FIGURE 17.3
Stimulation of phospholipase C-γ by receptor tyrosine kinases. Phospholipase C-γ (PLC-γ) binds to phosphotyrosine-containing peptides of activated receptors via its SH2 domains. Tyrosine phosphorylation then increases the activity of PLC-γ, leading to increased hydrolysis of phosphatidylinositol 4,5-bisphosphate (PIP_2) and production of the second messengers diacylglycerol (DAG) and inositol 1,4,5-triphosphate (IP_3).

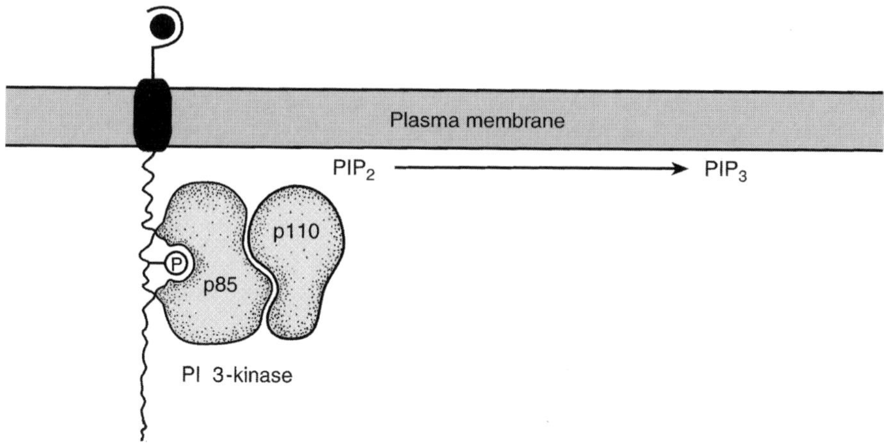

FIGURE 17.4
Receptor association of PI 3-kinase. PI 3-kinase consists of 85 kd and 110 kd subunits, designated p85 and p110, respectively. p85 binds to activated receptors through its SH2 domain and the catalytic subunit (p110) then catalyzes phosphorylation of phosphatidylinositides on the 3 position of the inositol ring (see Fig. 14.2 for structures). The principal intracellular reaction is thought to be the conversion of PI 4,5-bisphosphate (PIP_2) into PI 3,4,5-trisphosphate (PIP_3).

nases. The binding of PLC-γ to specific receptor phosphotyrosine residues has allowed its role to be investigated by the construction of mutant receptors in which the tyrosines responsible for PLC-γ binding have been changed to phenylalanines. Perhaps surprisingly, such mutant receptors are still able to stimulate cell proliferation, suggesting that activation of other pathways is sufficient to signal mitogenesis in the absence of PLC-γ activation.

Stimulation of PLC-γ is not the only interaction between the receptor tyrosine kinases and phosphatidylinositide metabolism. A phosphatidylinositol kinase (PI 3-kinase) that phosphorylates phosphatidylinositides at the 3 position of the inositol ring is also associated with activated tyrosine kinases (Fig. 17.4). PI 3-kinase is a heterodimer consisting of 110 kd and 85 kd subunits. The 110 kd subunit is responsible for catalytic activity, whereas the 85 kd subunit contains SH2 domains and mediates the association of PI 3-kinase with activated receptor tyrosine kinases. This translocates PI 3-kinase to the plasma membrane where it presumably has improved access to its substrates. In addition, binding of the 85 kd subunit SH2 domains to phosphotyrosine residues appears to activate the 110 kd catalytic subunit by an allosteric mechanism, resulting in elevated PI 3-kinase activity.

The principal reaction catalyzed by PI 3-kinase is the phosphorylation of PIP_2 (PI 4,5-bisphosphate) to PIP_3 (PI 3,4,5-trisphosphate). PIP_3 is not cleaved by phospholipase C and is thought to be a novel second messenger, but its functions are not yet clear. One target may be the ζ isotype of protein kinase C (PKC-ζ). As discussed in chapter 15, PKC-ζ is an atypical PKC that is not activated by either calcium or diacylglycerol. However, activation of PKC-ζ by PIP_3 has been reported, as has a role for PKC-ζ in mitogenic signal transduction downstream of growth factor receptors. It is also noteworthy that two genes encoding proteins related to PI 3-kinase have been identified in yeast. One of them (TOR2) may function in a signal transduction pathway in the regulation of cell proliferation, but the other (VPS34) functions in protein sorting and transport to the yeast vacuole. Consistent with a similar role for PI 3-kinase in mammalian cells, it has been reported that a mutant PDGF receptor lacking its PI 3-kinase binding sites is not normally internalized by endocytosis after ligand binding, raising the possibility that PI 3-kinase functions analogously to VPS34 to regulate receptor trafficking.

As discussed in chapter 13, members of the Src family of nonreceptor tyrosine kinases are associated with a variety of receptors that lack intrinsic protein kinase activity. In this context, the Src family kinases function as noncovalently linked receptor subunits; they are activated by ligand binding and couple nontyrosine kinase receptors to PLC-γ, PI 3-kinase, and other tyrosine kinase targets discussed below. In addition, however, members of the Src family are associated, through their SH2 domains, with activated receptor tyrosine kinases, where they may function to amplify the intrinsic kinase activity of the receptors. It is noteworthy that inhibiting the function of Src with antibodies or dominant interfering mutants blocks signaling from receptor tyrosine kinases, including the PDGF receptor, indicating that the Src nonreceptor kinases are functionally important components of the receptor-initiated mitogenic signaling cascade.

Activated receptor tyrosine kinases are also associated with a protein-tyrosine phosphatase, called Syp, that contains SH2 domains. Phosphorylation of Syp by the receptor tyrosine kinases has further been shown to enhance its catalytic activity. The role of Syp in signal transduction, however, is unclear. One straightforward possibility is that the Syp tyrosine phosphatase acts to downregulate the receptor tyrosine kinases, providing a negative feedback control mechanism. However, studies in *Drosophila* indicate that Syp plays a positive role in signaling from receptor tyrosine kinases. In particular, the *Drosophila* gene *corkscrew* (*csw*) encodes a protein-tyrosine phosphatase that is closely related to Syp, and *csw* mutants have been found to be defective in signal transduction from a receptor tyrosine kinase that functions in embryonic development (see chapter 19). One potential positive role for Syp might be activation of Src family kinases by dephosphorylation of their inhibitory phosphotyrosine residues (for example, Src tyr-527), as discussed in chapter 13. In addition, Syp may function not only as a phosphatase, but also as an adapter molecule that couples some receptor tyrosine kinases to Ras, as discussed further in the next section.

Another group of proteins associated with receptor tyrosine kinases are members of a family of transcription factors that contain SH2 domains and are regulated by phosphorylation on tyrosine residues (Fig. 17.5). These proteins, called STATs (*s*ignal *t*ransducers and *a*ctivators of *t*ranscription), were first identified in association with cytokine receptors (for example, the interferon receptor) and subsequently found to also be activated by receptor tyrosine kinases. The STAT proteins bind to activated receptors through their SH2 domains and are then phosphorylated by receptor (or associated nonreceptor) tyrosine kinases. Phosphorylation activates the STAT transcription factors and promotes their translocation to the nucleus, where they function as direct activators of gene expression. Although the role of STAT proteins in mitogenic signaling is not yet clear, it is noteworthy that one of their actions is stimulation of c-*fos* transcription.

The SH2-containing proteins discussed above thus link the receptor tyrosine kinases to a variety of intracellular signaling systems, including metabolism of phosphatidylinositides, nonreceptor tyrosine kinases, tyrosine phosphatases, and transcription factors. In addition, several other SH2-containing proteins serve as adapters that couple the receptor tyrosine kinases to acti-

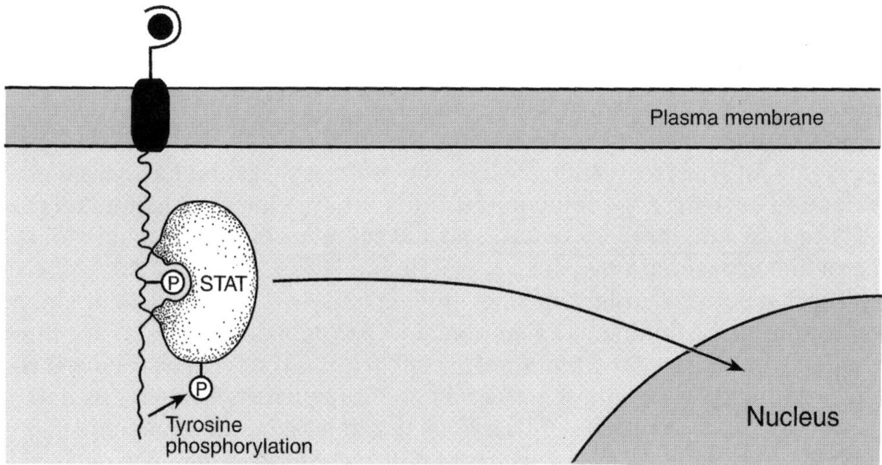

FIGURE 17.5
Activation of STAT transcription factors. The STAT transcription factors contain SH2 domains that bind to activated receptor tyrosine kinases. Tyrosine phosphorylation then activates the STAT factors, which translocate to the nucleus and stimulate target gene expression.

vation of the Ras proteins. Because Ras plays a central role in the transmission of mitogenic signals from both receptor and nonreceptor tyrosine kinases, its regulation and function are discussed in the following section.

Ras Activation

The importance of Ras in mitogenic signaling was first indicated by microinjection experiments, in which the introduction of Ras oncogene proteins into cultured cells was found to induce cell proliferation in the absence of growth factor stimulation. Conversely, inhibiting the activity of endogenous Ras, either by microinjection of anti-Ras monoclonal antibody or by expression of a dominant negative Ras mutant, was found to block cell proliferation in response to a variety of mitogenic growth factors. Taken together, these findings strongly implicated the *ras* gene products as central elements in mitogenic signaling pathways. Not only could exogenously introduced Ras protein directly stimulate cell proliferation, but endogenous Ras function was also required for normal cell proliferation and growth factor-induced mitogenesis. Furthermore, Ras activity was found to be necessary for cell transformation by oncogenes encoding both receptor and nonreceptor tyrosine kinases, suggesting that Ras functioned downstream of the tyrosine kinases in a signal transduction pathway.

As discussed in chapter 14, Ras proteins alternate between inactive GDP-bound and active GTP-bound forms. Consistent with a role for Ras downstream of receptor tyrosine kinases, growth factor stimulation increases the proportion of intracellular Ras bound to GTP. In principle, this could be accomplished either by increasing the rate of exchange of GTP for bound GDP or by decreasing the rate of inactivation of the Ras-GTP complex by the Ras GTPase (see Fig. 14.5). Thus, regulation of either the proteins that stimulate Ras guanine nucleotide exchange (guanine nucleotide exchange factors) or Ras GTPase (GTPase activating proteins or GAPs) could serve to couple the activity of Ras to growth factor receptors.

The first potential insight into the coupling of Ras to receptor tyrosine kinases stemmed from the finding that GAP contains SH2 domains, which lead to its association with activated receptors and phosphorylation on tyrosine residues. This suggested that tyrosine phosphorylation might regulate GAP, perhaps linking tyrosine kinases to activation of Ras. However, at least in mitogen-stimulated fibroblasts, activation of Ras results from an increased rate of guanine nucleotide exchange, rather than from a decreased rate of GTP hydrolysis. It thus appears that guanine nucleotide exchange factors, rather than GAP, play the critical role in Ras activation by growth factors. Rather than mediating Ras activation, the association of GAP with receptor tyrosine kinases may serve to down-regulate Ras following growth factor stimulation. In addition, as reviewed in chapter 14, GAP may also act as a Ras effector, although a role for GAP in signal transduction downstream of Ras has not been directly established.

The key to activation of Ras in response to mitogenic growth factors appears to be the guanine nucleotide exchange factor Sos, which becomes associated with receptor tyrosine kinases through adapter molecules that contain SH2 and SH3 domains. This association serves to activate Ras downstream of tyrosine kinases, not only in mammalian cells, but also in *Drosophila* and *C. elegans*, where genetic studies have clearly defined the critical role of both Sos and adapter molecules in Ras activation. Indeed, genetic studies in *Drosophila* first identified Sos as a factor that was required for activation of Ras by a receptor tyrosine kinase. Likewise, genetic analysis in *C. elegans* led to the identification of Sem5— an adapter molecule containing only SH2 and SH3 domains that functioned upstream of Ras in signal transduction from a receptor tyrosine kinase required for vulval development. Mammalian homologs of both Sos and Sem5 were subsequently cloned and shown to provide the biochemical link between tyrosine kinases and Ras (Fig. 17.6). Sem5 and its mammalian homolog (Grb2) contain two SH3 domains that bind to proline-rich peptides of Sos. The Grb2-Sos complex is then recruited to activated tyrosine kinase receptors through the SH2 domain of Grb2. Hence, Grb2 acts as an SH2/SH3 adapter to mediate the association of Sos with tyrosine kinases. This does not appear to alter the catalytic activity of Sos, but instead brings Sos to the plasma membrane, where it can interact with Ras and stimulate formation of the active Ras-GTP complex.

It is noteworthy that multiple mechanisms may be responsible for association of the Grb2-Sos complex with different receptors. The simplest case is the EGF receptor, in which the Grb2 SH2 domain binds directly to a receptor phosphotyrosine-containing peptide (Fig. 17.6A). Other receptor and nonreceptor tyrosine kinases, however, do not interact with Grb2 directly, but employ additional adapters to bind to the Grb2-Sos complex. One such adapter is Shc, an SH2-containing protein that is phosphorylated by both receptor and nonreceptor tyrosine kinases (Fig. 17.6B). The SH2 domain of Grb2 binds to phosphotyrosine-containing peptides of Shc, and the binding of Shc to activated growth factor receptors (through its SH2 domain) results in receptor association of the Grb2-Sos complex. As noted earlier, a similar adapter role may be played by the Syp protein-tyrosine phosphatase, which is associated (through its SH2 domain) with the PDGF receptor. Tyrosine phosphorylation of Syp creates a binding site for the SH2 domain of Grb2, thereby recruiting the Grb2-Sos complex to the receptor. The complexity of the system may be further increased by the potential association of other SH2/SH3-containing adapters (for example, Crk and Nck) with Sos or with other guanine nucleotide exchange factors. A variety of different mechanisms may thus serve to couple tyrosine kinases to guanine nucleotide exchange factors, providing the potential for differential regulation of multiple pathways leading to Ras activation.

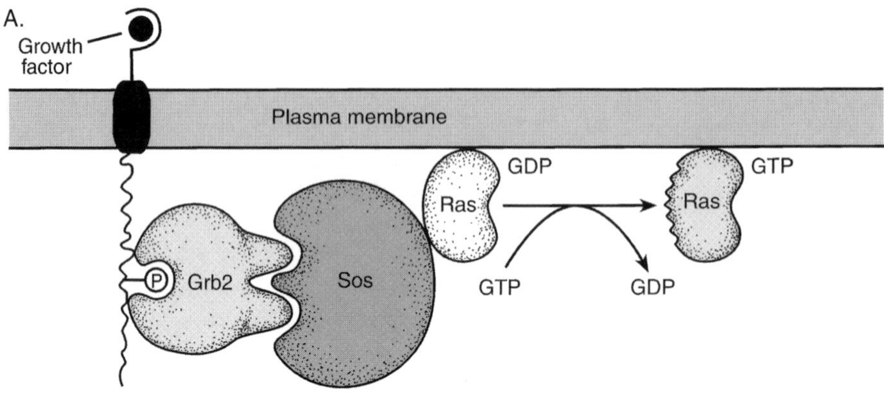

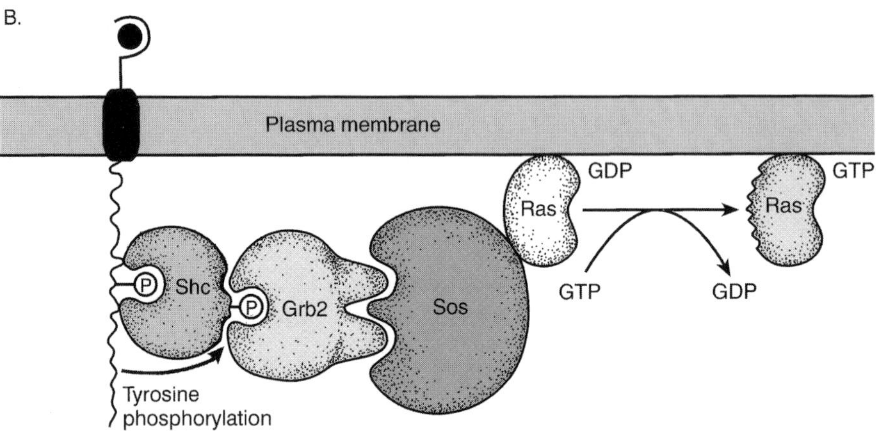

FIGURE 17.6
Activation of Ras proteins. The Sos guanine nucleotide exchange factor is bound to the SH3 domains of Grb2. The Grb2 SH2 domains then bind to phosphotyrosine-containing peptides of activated receptor tyrosine kinases (A) or of other receptor-associated adapter molecules, such as Shc (B). In either case, this recruits Sos to the plasma membrane, where it can activate Ras by stimulating GDP/GTP exchange.

The Raf Protein-Serine/Threonine Kinase

As discussed in chapter 14, the activated Ras-GTP complex then interacts with effector proteins to further transmit the mitogenic signal initiated by growth factor binding. In mammalian cells, as well as in *Drosophila* and *C. elegans*, the Raf protein-serine/threonine kinase appears to be the critical Ras target. First, Raf oncogene proteins are able to induce cell proliferation even when Ras function is blocked by antibody or a dominant inhibitory Ras mutant, suggesting that Raf acts downstream of Ras. Consistent with this hypothesis, dominant inhibitory mutants of Raf block the action of both tyrosine kinases and Ras. Moreover, the protein kinase activity of Raf is stimulated by treatment of cells with growth factors, and this activation of Raf is blocked by expression of dominant inhibitory Ras.

Further experiments have demonstrated a direct physical association between the Ras and Raf proteins (Fig. 17.7). Formation of the Ras-Raf complex requires the active GTP-bound form of Ras and is mediated by a protein-protein interaction between the effector domain of Ras and the amino-terminal regulatory domain of Raf. These findings indicate that Raf is a direct effector of Ras and that a critical function of Ras is to couple the Raf protein-serine/threonine kinase to growth factor stimulation of receptor protein-tyrosine kinases. It is important to note, however, that this does not exclude the possibility that Ras also interacts with other effectors, possibly including GAP or NF1 (see chapter 14).

Although the Ras-Raf interaction identifies Raf as a direct target of Ras, this interaction does not appear to directly activate the Raf protein kinase. Treatment of cells with mitogenic growth factors increases both the level of Raf phosphorylation (primarily on serine residues) and its protein kinase activity. However, interaction with Ras has not been found to stimulate Raf kinase ac-

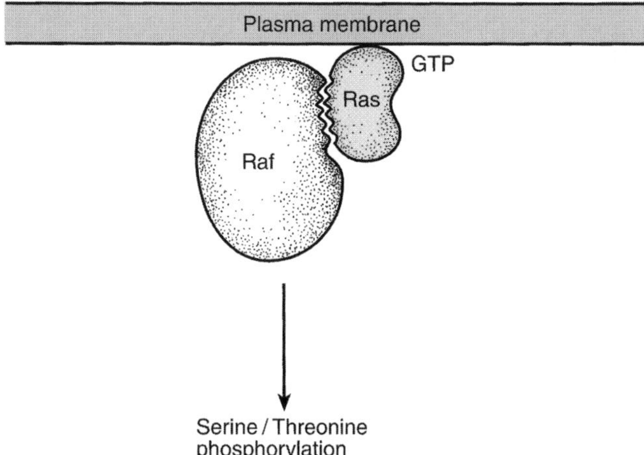

FIGURE 17.7
Interaction of Ras and Raf. The Ras and Raf proteins form a complex, in which the effector domain of GTP-bound Ras interacts with the amino-terminal regulatory domain of the Raf protein-serine/threonine kinase. Because the Ras proteins are membrane associated, this interaction recruits Raf to the plasma membrane.

tivity, suggesting that other factors are required for Raf activation. One hypothesis suggested by these results is that the Ras-Raf interaction serves to bring Raf to the plasma membrane, where it can be activated by membrane lipids or other protein kinases. Consistent with this possibility, addition of a membrane-targeting signal to Raf results in its localization to the plasma membrane and constitutive activation, independent of Ras function. It thus appears that the major role of the Ras-Raf interaction is simply to recruit Raf to the plasma membrane, where it can be activated by other signals.

The nature of the signals that activate Raf remains to be determined. One possibility is the direct activation of Raf by membrane lipids, such as diacylglycerol or PIP_3, although such lipid activators of Raf have not so far been identified. Alternatively, Raf activation may result from its phosphorylation by other membrane-associated protein kinases. Although Raf is primarily phosphorylated on serine residues, tyrosine phosphorylation of Raf also occurs and may play a role. In addition, several isotypes of protein kinase C have been found to phosphorylate and activate Raf. Both protein-tyrosine and protein-serine/threonine kinases thus appear to be plausible candidates for mediators of Raf activation.

Additional experiments further implicate second messengers derived from phosphatidylcholine in Raf activation. A variety of growth factors stimulate hydrolysis of phosphatidylcholine (PC), in addition to the hydrolysis of PIP_2 discussed earlier. PC hydrolysis yields diacylglycerol but not inositol triphosphate; so it is thought to activate nonclassical calcium-independent protein kinase C isotypes. Treatment of cells with exogenous PC-phospholipase C (PC-PLC) is itself mitogenic, and overexpression of a cloned bacterial PC-PLC gene induces cell transformation, indicating that PC hydrolysis is sufficient to signal cell proliferation. Although the mechanism by which growth factors stimulate PC hydrolysis is not known, it appears to be activated downstream of Ras, because it is inhibited by expression of a dominant negative Ras mutant. Interestingly, however, PC hydrolysis appears to function upstream of Raf, because overexpression of PC-PLC overcomes the inhibitory effects of dominant negative Ras but not dominant negative Raf mutants. In addition, treatment of cells with exogenous PC-PLC activates the Raf protein kinase. It thus appears that growth factor stimulation of PC hydrolysis provides at least one pathway for the generation of second messengers (such as diacylglycerol) that can induce Raf activation, possibly mediated by nonclassical isotypes of protein kinase C.

Further studies are clearly required to understand the mechanisms that regulate Raf kinase activity. Nonetheless, it is apparent that Raf is a critical target of Ras and plays a central role in the transmission of signals from activated growth factor receptors. As discussed in the next section, Raf initiates a protein kinase cascade that ultimately results in phosphorylation of transcription factors and consequent alterations in gene expression.

The MAP Kinase Pathway

Activation of Raf leads to activation of mitogen-activated protein kinases (MAP kinases) that, as their name implies, are stimulated by a variety of mitogenic growth factors and other extracellular stimuli. The MAP kinases are a family of protein-serine/threonine kinases that are highly conserved in evolution and participate in signal transduction pathways in yeast, insects, and amphibians, as well as in mammalian cells. In mammals, the two best-characterized forms of MAP kinase (also called ERK1 and ERK2 for *e*xtracellular-signal *r*egulated *k*inase) are related proteins of 42 and 44 kd.

MAP kinases are themselves activated by dual phosphorylation on both threonine and tyrosine residues. For example, activation of the 42 kd MAP kinase (ERK2) requires phosphorylation of both threonine-183 and tyrosine-185. Such activating phosphorylations are catalyzed by another protein kinase, called MAP kinase kinase (MKK) or MEK (for MAPK/Erk kinase). MEK is a dual specificity kinase that phosphorylates MAP kinase on both threonine and tyrosine residues. Two closely related forms of MEK have been identified in mammalian cells. Importantly, MEK is also activated by phosphorylation—in this case, on serine residues. Thus, MEK appears to be an intermediate component of a protein kinase cascade leading to MAP kinase activation.

The MAP kinase pathway was first linked to Ras by the finding that activation of MAP kinase in response to growth factors was blocked by expression of a dominant inhibitory Ras mutant. The cloning and sequencing of MEK further revealed its similarity to yeast protein kinases known to function downstream of Ras in yeast signal transduction pathways. Moreover, a direct biochemical link between Ras and MAP kinase has been provided by the finding that Raf phosphorylates and activates MEK. Thus, a direct protein kinase cascade, initiated by activation of Raf, leads to the activation of MAP kinase in response to mitogenic growth factors (Fig. 17.8).

A variety of mechanisms also regulate the MAP kinase pathway downstream of G protein-coupled receptors. For example, stimulation of protein kinase C leads to activation of MAP kinase downstream of receptors that activate PLC-ß. In addition, the MAP kinase pathway can be activated by the ßγ subunits of some G proteins. Although the details remain to be elucidated, it appears that activation of MAP kinase by these ßγ subunits is mediated by Ras, because it is blocked by expression of a dominant negative Ras mutant. The MAP kinase pathway is also regulated by cAMP, which stimulates MAP kinase in some cells and inhibits its activation in others. For example, phosphorylation by cAMP-dependent protein kinase inhibits the activation of Raf-1 in mitogen-stimulated rat fibroblasts, thereby blocking MAP kinase activation in this cell type. Because cAMP also inhibits the proliferation of these cells, this intersection between the cAMP system and Raf may be an important crossover point

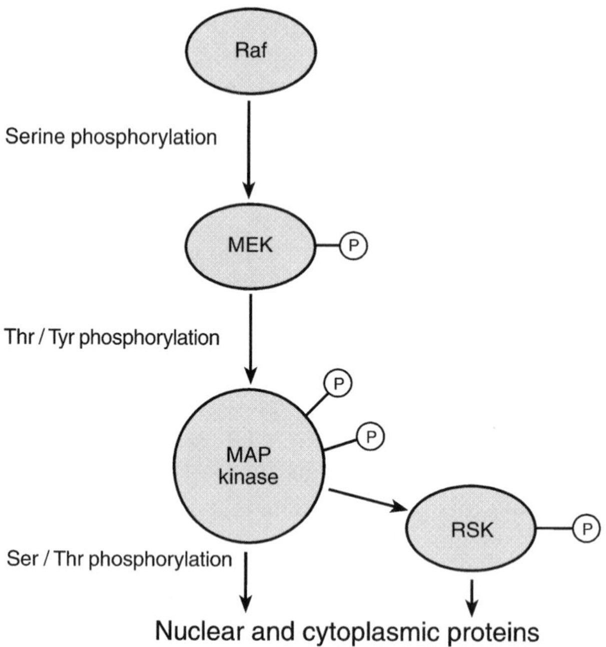

FIGURE 17.8
The MAP kinase pathway. Raf phosphorylates the MEK protein kinase on serine residues. MEK is a dual specificity kinase that catalyzes phosphorylation of MAP kinase on threonine-183 and tyrosine-185. MAP kinase then phosphorylates a variety of substrates, including the RSK protein-serine/threonine kinase. Both MAP kinase and RSK translocate between the cytoplasm and nucleus, and their substrates include a number of transcription factors.

in the control of cell proliferation by different signaling systems. In other cells, where cAMP stimulates MAP kinase, inhibition of Raf-1 by cAMP-mediated phosphorylation may be bypassed by the expression of other protein kinases that provide alternative routes to MEK activation. These might include the other members of the Raf family, A-Raf and B-Raf, both of which lack sites for phosphorylation by cAMP-dependent protein kinase. In addition, MEK can be phosphorylated by at least one distinct protein kinase (called MEKK for MEK kinase) that is unrelated to Raf and might represent an independent route to MEK activation.

The MAP kinase pathway thus appears to play a central role in signaling from both tyrosine kinase and G protein coupled receptors. Its importance is directly supported by experiments indicating that dominant negative mutants of MAP kinase inhibit mitogenic signaling in mammalian cells and by genetic studies demonstrating that MAP kinase is required for signal transduction downstream of tyrosine kinase receptors in *Drosophila.* Thus, although activation of the MAP kinase pathway is not the only target of growth factor receptors, MAP kinase is clearly a central element in pathways of signal transduction that have been conserved throughout the evolution of eucaryotes.

Once activated, MAP kinase phosphorylates a variety of proteins, including phospholipase A_2 and another protein-serine/threonine kinase, called RSK, which phosphorylates ribosomal protein S6. Importantly, both MAP kinase and RSK translocate between the nucleus and the cytoplasm. Furthermore, their substrates include a number of transcription factors, as discussed in the next section. The MAP kinase pathway thus provides at least one clear link between the cytosol and the nucleus, coupling the stimulation of growth factor receptors to changes in gene expression.

Transcription Factors and Alterations in Gene Expression

As noted at the beginning of this chapter, the first evidence that implicated oncogenes in intracellular signaling pathways was the finding that expression of c-*myc* was rapidly induced in response to PDGF stimulation. Increased c-*myc* transcription was detected within an hour after exposure to PDGF and did not require new protein synthesis, indicating that c-*myc* induction was a rapid and direct effect of growth factor stimulation. Further studies have shown that c-*myc* is only one of a family of fifty to one hundred genes, called immediate-early genes, that are rapidly and transiently induced following mitogenic stimulation. The immediate-early genes include members of the *fos* and *jun* families, as well as a number of other genes encoding transcriptional regulatory proteins. As with c-*myc,* induced transcription of these genes is a primary response to growth factor stimulation for which protein synthesis is not required. Because many of the immediate-early genes encode transcription factors, their rapid induction in response to growth factor stimulation is expected to alter transcription of still other cellular genes, leading to an altered program of gene expression during the transition from the quiescent to the proliferative state. The transcription factors encoded by these genes thus serve as signal transducing elements that convert a short-term signal at the cell surface into a longer-term alteration in gene expression and cell behavior.

The mechanisms that control expression of the immediate-early genes are complex and appear to include regulation at both the transcriptional and the post-transcriptional levels. First, transcription of these genes increases rapidly in response to mitogenic stimulation. Following their initial induction by growth factors, the transcription of immediate-early genes is then repressed, and intracellular levels of both mRNA and the protein return to basal levels.

Unlike gene induction, the subsequent repression of immediate-early genes requires protein synthesis, suggesting that synthesis of a repressor is required to down-regulate their expression. Both the mRNAs and the proteins encoded by immediate-early genes are rapidly degraded; so these changes in transcription effectively modulate the levels of protein within the cell. In addition, the activities of the transcription factors encoded by these genes are frequently regulated by phosphorylation.

The c-*fos* gene provides a particularly well studied example of immediate-early gene induction. Induction of c-*fos* is extremely rapid, being detectable within 15 minutes and clearly preceding the induction of c-*myc* in growth factor-stimulated cells. This transcriptional activation of c-*fos* is mediated by two major regulatory elements: the serum response element (SRE) and the Sis inducible element (SIE) (Fig. 17.9). It is noteworthy that SREs not only are respon-

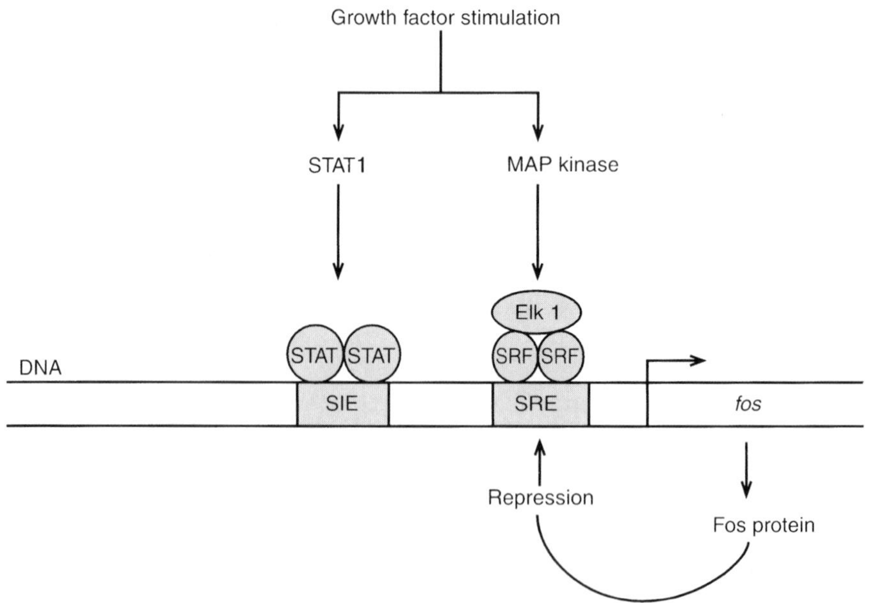

FIGURE 17.9
Regulation of c-*fos* transcription. Induction of c-*fos* in response to mitogenic stimulation is mediated by proteins that bind to the serum response element (SRE) and to the Sis inducible element (SIE). The SRE-binding proteins are a complex between the SRF and Elk-1, which is activated by MAP kinase. Transcription through the SIE is induced by a member of the STAT family of transcription factors, which is activated directly by tyrosine phosphorylation. The Fos protein subsequently acts through the SRE to repress its own transcription.

sible for induction of c-*fos*, but also mediate induction of a number of other immediate-early genes. The SRE is recognized by the serum response factor (SRF), which forms a complex with a second protein of 62 kd, originally called ternary complex factor (TCF). Subsequent experiments identified TCF as the product of the *ets*-related gene, Elk1, which was further found to be phosphorylated by MAP kinase in response to growth factor stimulation. This phosphorylation appears to enhance the transcriptional activity of Elk1, providing a direct link between the MAP kinase pathway and c-*fos* induction. In addition, the SRF is phosphorylated by RSK, which may increase binding of the SRF to DNA and also contribute to c-*fos* induction. A second, independent pathway activates c-*fos* transcription through the SIE, which is the binding site of a protein called SIF (Sis-inducible factor). SIF has been found to belong to the STAT family of transcription factors that, as discussed earlier in this chapter, are associated with receptor tyrosine kinases and activated by direct tyrosine phosphorylation. Growth factor induction of c-*fos* is thus mediated both by the MAP kinase pathway and by direct tyrosine phosphorylation of a transcription factor, which then translocates to the nucleus.

The induced rate of c-*fos* transcription and the increased levels of c-*fos* mRNA are present very transiently in stimulated cells, returning to basal levels within 2 hours after growth factor stimulation. Although protein synthesis is not required for c-*fos* induction, the synthesis of new proteins is required for the subsequent "turning off" of c-*fos* transcription, suggesting that synthesis of a transcriptional repressor is required for this programmed down-regulation. Interestingly, it appears that synthesis of the Fos protein itself results in repression of transcription from the c-*fos* promoter (Fig. 17.9). The down-regulation of c-*fos* transcription is thus due to a negative autoregulatory mechanism, in which Fos represses its own transcription. Although the mechanism of Fos repression is not established, it is mediated through the SRE. Interestingly, Fos is phosphorylated by both MAP kinase and RSK at sites in its carboxy-terminal repression domain, suggesting that the MAP kinase pathway may regulate both induction and repression of c-*fos* transcription.

In contrast to c-*fos*, induction of c-*jun* in response to mitogenic growth factors is mediated by an AP-1 site and Jun appears to function as a positive autoregulatory protein that induces its own transcription (Fig. 17.10). Mitogenic stimulation results in phosphorylation of preexisting Jun at sites in its amino-terminal activation domain. This enhances the transcriptional activity of c-Jun, resulting in increased expression of AP-1 target genes, including c-*jun* itself. A kinase (called JNK for Jun N-terminal kinase) that phosphorylates the amino-terminal activating sites of c-Jun has recently been identified and found to be a distant member of the MAP kinase family. Although c-Jun is also phosphorylated by MAP kinase, the role of MAP kinase in c-Jun activation is unclear because it also phosphorylates carboxy-terminal sites that inhibit c-Jun

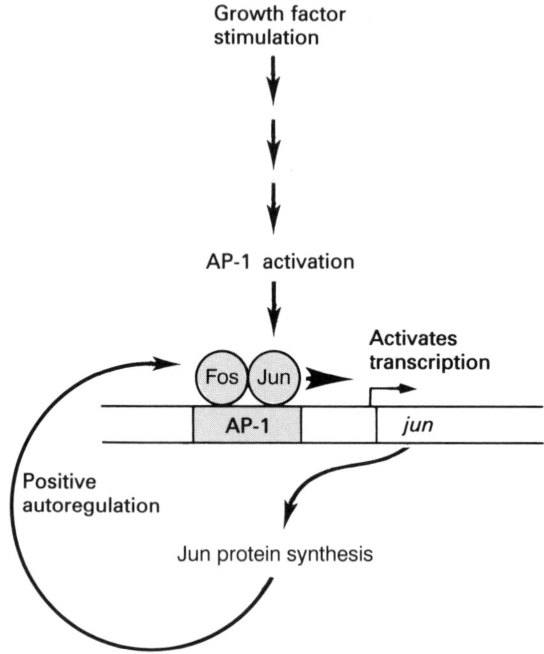

FIGURE 17.10
Regulation of c-*jun* transcription. Increased AP-1 activity is induced by phosphorylation of preexisting Jun. This stimulates c-*jun* transcription, which is regulated by an AP-1 site in the c-*jun* promoter. Synthesis of Jun protein therefore results in further induction of c-*jun* by a positive autoregulatory mechanism.

DNA binding. In any event, activation of the Ras-Raf pathway leads to phosphorylation of preexisting c-Jun, increasing AP-1 activity and stimulating further c-*jun* transcription. The positive autoregulatory activity of c-Jun prolongs its expression after growth factor stimulation; so the increase in c-*jun* transcription persists longer than that of c-*fos* before returning to basal levels.

The mechanisms responsible for induction of Fos and Jun in response to mitogenic growth factors link the expression of these immediate-early genes to stimulation of growth factor receptors. Because c-*fos*, c-*jun*, and other immediate-early genes (including c-*myc*) encode transcription factors, their induction is expected to regulate the expression of additional target genes. Consistent with this expectation, several genes inducible by either AP-1 or Myc are expressed following mitogenic stimulation. However, identifying the critical targets of the immediate-early genes and understanding their relation to cell proliferation remain challenges for further research.

Summary

Mitogenic signals, initiated by the binding of a growth factor to its cell surface receptor, are transmitted to the nucleus, leading to changes in gene expression, DNA synthesis, and cell proliferation. Members of several groups of proto-oncogene products, including growth factors, receptor and nonreceptor tyrosine kinases, GTP binding proteins, serine/threonine kinases, and transcriptional regulatory proteins, clearly function in signal transduction pathways that regulate proliferation of normal cells. There are likely to be multiple parallel and intersecting pathways for transduction of mitogenic signals in which different proto-oncogenes take part. However, and in spite of the fact that some intervening regulatory steps remain to be elucidated, at least one clear pathway exists linking the action of growth factors at the cell surface with the transient induction of transcriptional regulatory proteins (Fig. 17.11). Growth factor binding activates receptor tyrosine kinases, resulting in activation of Ras, Raf, and MAP kinase. Phosphorylation then activates transcription factors, leading to transient induction of c-*fos*, c-*jun*, and other immediate-early genes. The transcription factor proto-oncogenes are thus terminal elements of signal transduction pathways

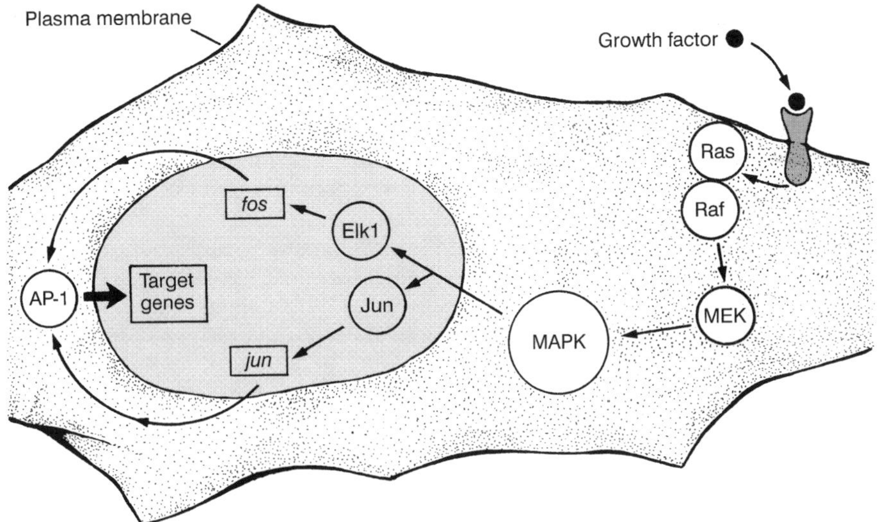

FIGURE 17.11
The Ras-Raf-MAP kinase pathway of mitogenic signal transduction. Growth factor binding activates Ras and Raf, leading to activation of MEK and MAP kinase. Phosphorylation of transcription factors (for example, Elk1 and Jun) by MAP kinase or related enzymes then leads to induction of immediate-early genes (for example, c-*fos* and c-*jun*), followed by induction of the target genes for these transcription factors.

that activate expression of critical target genes, thereby converting the transient action of extracellular growth factors into a longer-term alteration in gene expression, ultimately resulting in cell proliferation.

References

General References

Aaronson, S.A. 1991. Growth factors and cancer. *Science* 254:1146–1153.

Cantley, L.C., Auger, K.R., Carpenter, C., Duckworth, B., Graziani, A., Kapeller, R., and Soltoff, S. 1991. Oncogenes and signal transduction. *Cell* 64:281–302.

Receptor Tyrosine Kinases and Their Targets

Auger, K.R., Serunian, L.A., Soltoff, S.P., Libby, P., and Cantley, L.C. 1989. PDGF-dependent tyrosine phosphorylation stimulates production of novel polyphosphoinositides in intact cells. *Cell* 57:167–175.

Berra, E., Diaz-Meco, M.T., Dominguez, I., Municio, M.M., Sanz, L., Lozano, J., Chapkin, R.S., and Moscat, J. 1993. Protein kinase C ζ isoform is critical for mitogenic signal transduction. *Cell* 74:555–563.

Escobedo, J.A., Navankasattusas, S., Kavanaugh, W.M., Milfay, D., Fried, V.A., and Williams, L.T. 1991. cDNA cloning of a novel 85 kd protein that has SH2 domains and regulates binding of PI3-kinase to the PDGF ß-receptor. *Cell* 65:75–82.

Fantl, W.J., Escobedo, J.A., Martin, G.A., Turck, C.W., del Rosario, M., McCormick, F., and Williams, L.T. 1992. Distinct phosphotyrosines on a growth factor receptor bind to specific molecules that mediate different signaling pathways. *Cell* 69:414–423.

Fantl, W.J., Johnson, D.E., and Williams, L.T. 1993. Signalling by receptor tyrosine kinases. *Ann. Rev. Biochem.* 62:453–481.

Feng, G.-S., Hui, C.-C., and Pawson, T. 1993. SH2-containing phosphotyrosine phosphatase as a target of protein-tyrosine kinases. *Science* 259:1607–1611.

Fu, X-Y., and Zhang, J.-J. 1993. Transcription factor p91 interacts with the epidermal growth factor receptor and mediates activation of the c-*fos* gene promoter. *Cell* 74:1135–1145.

Hawkins, P.T., Jackson, T.R., and Stephens, L.R. 1992. Platelet-derived growth factor stimulates synthesis of PtdIns(3,4,5)P_3 by activating a PtdIns(4,5)P_2 3-OH kinase. *Nature* 358:157–159.

Hiles, I.D., Otsu, M., Volinia, S., Fry, M.J., Gout, I., Dhand, R., Panayotou, G., Ruiz-Larrea, F., Thompson, A., Totty, N.F., Hsuan, J.J., Courtneidge, S.A., Parker, P.J., and Waterfield, M.D. 1992. Phosphatidylinositol 3-kinase: structure and expression of the 110 kd catalytic subunit. *Cell* 70:419–429.

Joly, M., Kazlauskas, A., Fay, F.S., and Corvera, S. 1994. Disruption of PDGF receptor trafficking by mutation of its PI-3 kinase binding sites. *Science* 263:684–687.

Kazlauskas, A., and Cooper, J.A. 1989. Autophosphorylation of the PDGF receptor in the kinase insert region regulates interactions with cell proteins. *Cell* 58:1121–1133.

Krypta, R.M., Goldberg, Y., Ulug, E.T., and Courtneidge, S.A. 1990. Association between the PDGF receptor and members of the *src* family of tyrosine kinases. *Cell* 62:481–492.

Kunz, J., Henriquez, R., Schneider, U., Deuter-Reinhard, M., Movva, N.R., and Hall, M.N. 1993. Target of rapamycin in yeast, TOR2, is an essential phosphatidylinositol kinase homolog required for G_1 progression. *Cell* 73:585–596.

Li, W., Nishimura, R., Kashishian, A., Batzer, A.G., Kim, W.J.H., Cooper, J.A., and Schlessinger, J. 1994. A new function for a phosphotyrosine phosphatase: linking GRB2-Sos to a receptor tyrosine kinase. *Mol. Cell. Biol.* 14:509–517.

Mohammadi, M., Dionne, C.A., Li, W., Spivak, T., Honegger, A.M., Jaye, M., and Schlessinger, J. 1992. Point mutation in FGF receptor eliminates phosphatidylinositol hydrolysis without affecting mitogenesis. *Nature* 358:681–684.

Nakanishi, H., Brewer, K.A., and Exton, J.H. 1993. Activation of the ζ isozyme of protein kinase C by phosphatidylinositol 3,4,5-trisphosphate. *J. Biol. Chem.* 268:13–16.

Nishibe, S., Wahl, M.I., Hernandez-Sotomayor, S.M., Tonks, N.K., Rhee, S.G., and Carpenter, G. 1990. Increase of the catalytic activity of phospholipase C-γ1 by tyrosine phosphorylation. *Science* 250:1253–1256.

Otsu, M., Hiles, I., Gout, I., Fry, M.J., Ruiz-Larrea, F., Panayotou, G., Thompson, A., Dhand, R., Hsuan, J., Totty, N., Smith, A.D., Morgan, S.J., Courtneidge, S.A., Parker, P.J., and Waterfield, M.D. 1991. Characterization of two 85 kd proteins that associate with receptor tyrosine kinases, middle-T/pp60$^{c\text{-}src}$ complexes, and PI3-kinase. Cell 65:91-104.

Perkins, L.A., Larsen, I., and Perrimon, N. 1992. *Corkscrew* encodes a putative protein tyrosine phosphatase that functions to transduce the terminal signal from the receptor tyrosine kinase torso. *Cell* 70:225–236.

Peters, K.G., Marie, J., Wilson, E., Ives, H.E., Escobedo, J., Del Rosario, M., Mirda, D., and Williams, L.T. 1992. Point mutation of an FGF receptor abolishes phosphatidylinositol turnover and Ca^{2+} flux but not mitogenesis. *Nature* 358:678–681.

Ruff-Jamison, S., Chen, K., and Cohen, S. 1993. Induction by EGF and interferon-γ of tyrosine phosphorylated DNA binding proteins in mouse liver nuclei. *Science* 261:1733–1736.

Sadowski, H.B., Shuai, K., Darnell, J.E., Jr., and Gilman, M.Z. 1993. A common nuclear signal transduction pathway activated by growth factor and cytokine receptors. *Science* 261:1739–1744.

Schu, P.V., Takegawa, K., Fry, M.J., Stack, J.H., Waterfield, M.D., and Emr, S.D. 1993. Phosphatidylinositol 3-kinase encoded by yeast *VPS34* gene essential for protein sorting. *Science* 260:88–91.

Shoelson, S.E., Sivaraja, M., Williams, K.P., Hu, P., Schlessinger, J., and Weiss, M.A. 1993. Specific phosphopeptide binding regulates a conformational change in the PI 3-kinase SH2 domain associated with enzyme activation. *EMBO J.* 12:795–802.

Silvennoinen, O., Schindler, C., Schlessinger, J., and Levy, D.E. 1993. Ras-independent growth factor signaling by transcription factor tyrosine phosphorylation. *Science* 261:1736–1739.

Skolnik, E.Y., Margolis, B., Mohammadi, M., Lowenstein, E., Fischer, R., Drepps, A., Ullrich, A., and Schlessinger, J. 1991. Cloning of PI3 kinase-associated p85 utilizing a novel method for expression/cloning of target proteins for receptor tyrosine kinases. *Cell* 65:83–90.

Twamley-Stein, G.M., Pepperkok, R., Ansorge, W., and Courtneidge, S.A. 1993. The Src family tyrosine kinases are required for platelet-derived growth factor-mediated signal transduction in NIH 3T3 cells. *Proc. Natl. Acad. Sci. USA* 90:7696–7700.

Ullrich, A., and Schlessinger, J. 1990. Signal transduction by receptors with tyrosine kinase activity. *Cell* 61:203–212.

Vogel, W., Lammers, R., Huang, J., and Ullrich, A. 1993. Activation of a phosphotyrosine phosphatase by tyrosine phosphorylation. *Science* 259:1611–1614.

Zhong, Z., Wen, Z., and Darnell, J.E., Jr. 1994. Stat3: a STAT family member activated by tyrosine phosphorylation in response to epidermal growth factor and interleukin-6. *Science* 264:95–98.

Ras Activation

Buday, L., and Downward, J. 1993. Epidermal growth factor regulates $p21^{ras}$ through the formation of a complex of receptor, Grb2 adapter protein, and Sos nucleotide exchange factor. *Cell* 73:611–620.

Cai, H., Szeberenyi, J., and Cooper, G.M. 1990. Effect of a dominant inhibitory Ha-*ras* mutation on mitogenic signal transduction in NIH 3T3 cells. *Mol. Cell. Biol.* 12:5329–5335.

Chardin, P., Camonis, J.H., Gale, N.W., Van Aelst, L., Schlessinger, J., Wigler, M.H., and Bar-Sagi, D. 1993. Human Sos-1: a guanine nucleotide exchange factor for Ras that binds to Grb2. *Science* 260:1338–1343.

Clark, S.G., Stern, M.J., and Horvitz, H.R. 1992. *C. elegans* cell-signalling gene *sem*-5 encodes a protein with SH2 and SH3 domains. *Nature* 356:340–344.

Egan, S.E., Giddings, B.W., Brooks, M.W., Buday, L., Sizeland, A.M., and Weinberg, R.A. 1993. Association of Sos Ras exchange protein with Grb2 is implicated in tyrosine kinase signal transduction and transformation. *Nature* 363:45–51.

Ellis, C., Moran, M., McCormick, F., and Pawson, T. 1990. Phosphorylation of GAP and GAP-associated proteins by transforming and mitogenic tyrosine kinases. *Nature* 343:377–381.

Feig, L.A. 1993. The many roads that lead to Ras. *Science* 260:767–768.

Feig, L.A., and Cooper, G.M. 1988. Inhibition of NIH 3T3 cell proliferation by a mutant *ras* protein with preferential affinity for GDP. *Mol. Cell. Biol.* 8:3235–3243.

Feramisco, J.R., Gross, M., Kamata, T., Rosenberg, M., and Sweet, R.W. 1984. Microinjection of the oncogene form of the human H-*ras* (T-24) protein results in rapid proliferation of quiescent cells. *Cell* 38:109–117.

Gale, N.W., Kaplan, S., Lowenstein, E.J., Schlessinger, J., and Bar-Sagi, D. 1993. Grb2 mediates the EGF-dependent activation of guanine nucleotide exchange on Ras. *Nature* 363:88–92.

Kaplan, D.R., Morrison, D.K., Wong, G., McCormick, F., and Williams, L.T. 1990. PDGF ß-receptor stimulates tyrosine phosphorylation of GAP and association of GAP with a signaling complex. *Cell* 61:125–133.

Li, B.-Q., Kaplan, D., Kung, H., and Kamata, T. 1992. Nerve growth factor stimulation of the Ras-guanine nucleotide exchange factor and GAP activities. *Science* 256:1456–1459.

Li, N., Batzer, A., Daly, R., Yajnik, V., Skolnik, E., Chardin, P., Bar-Sagi, D., Margolis, B., and Schlessinger, J. 1993. Guanine-nucleotide-releasing factor hSos1 binds to Grb2 and links receptor tyrosine kinases to Ras signalling. *Nature* 363:85–88.

Li, W., Nishimura, R., Kashishian, A., Batzer, A.G., Kim, W.J.H., Cooper, J.A., and Schlessinger, J. 1994. A new function for a phosphotyrosine phosphatase: linking GRB2-Sos to a receptor tyrosine kinase. *Mol. Cell. Biol.* 14:509–517.

Lowy, D.R., and Willumsen, B.M. 1993. Function and regulation of Ras. *Ann. Rev. Biochem.* 62:851–891.

Molloy, C.J., Bottaro, D.P., Fleming, T.P., Marshall, M.S., Gibbs, J.B., and Aaronson, S.A. 1989. PDGF induction of tyrosine phosphorylation of GTPase activating protein. *Nature* 342:711–714.

Mulcahy, L.S., Smith, M.R., and Stacey, D.W. 1985. Requirement for *ras* proto-oncogene function during serum-stimulated growth of NIH 3T3 cells. *Nature* 313:241–243.

Rozakis-Adcock, M., Fernley, R., Wade, J., Pawson, T., and Bowtell, D. 1993. The SH2 and SH3 domains of mammalian Grb2 couple the EGF receptor to the Ras activator mSos1. *Nature* 363:83–85.

Rozakis-Adcock, M., McGlade, J., Mbamalu, G., Pelicci, G., Daly, R., Li, W., Batzer, A., Thomas, S., Brugge, J., Pelicci, P.G., Schlessinger, J., and Pawson, T. 1992. Association of the Shc and Grb2/Sem5 SH2-containing proteins is implicated in activation of the Ras pathway by tyrosine kinases. *Nature* 360:689–692.

Satoh, T., Endo, M., Nakafuku, M., Nakamura, S., and Kaziro, Y. 1990. Platelet-derived growth factor stimulates formation of active $p21^{ras}\cdot$GTP complex in Swiss mouse 3T3 cells. *Proc. Natl. Acad. Sci. USA* 87:5993–5997.

Schlessinger, J. 1993. How receptor tyrosine kinases activate Ras. *Trends Biochem. Sci.* 18:273–275.

Simon, M.A., Bowtell, D.D., Dodson, G.S., Laverty, T.R., and Rubin, G.M. 1991. Ras1 and a putative guanine nucleotide exchange factor perform crucial steps in signaling by the sevenless protein tyrosine kinase. *Cell* 67: 701–716.

Smith, M.R., DeGudicibus, S.J., and Stacey, D.W. 1986. Requirement for c-*ras* proteins during viral oncogene transformation. *Nature* 320:540–543.

Stacey, D.W., and Kung, H.-F. 1984. Transformation of NIH 3T3 cells by microinjection of Ha-*ras* p21 protein. *Nature* 310:508–511.

Tanaka, S., Morishita, T., Hashimoto, Y., Hattori, S., Nakamura, S., Shibuya, M., Matuoka, K., Takenawa, T., Kurata, T., Nagashima, K., and Matsuda, M. 1994. C3G, a guanine nucleotide-releasing protein expressed ubiquitously, binds to the Src homology 3 domains of CRK and GRB2/ASH proteins. *Proc. Natl. Acad. Sci. USA* 91:3443–3447.

Zhang, K., Papageorge, A.G., and Lowy, D.R. 1992. Mechanistic aspects of signaling through Ras in NIH 3T3 cells. *Science* 257:671–674.

The Raf Protein-Serine/Threonine Kinase

Bruder, J.T., Heidecker, G., and Rapp, U.R. 1992. Serum, TPA, and Ras-induced expression from AP-1/Ets-driven promoters requires Raf-1 kinase. *Genes Dev.* 6:545–556.

Cai, H., Erhardt, P., Szeberenyi, J., Diaz-Meco, M.T., Moscat, J., and Cooper, G.M. 1992.

Hydrolysis of phosphatidylcholine is stimulated by Ras proteins during mitogenic signal transduction. *Mol. Cell. Biol.* 12:5329–5335.

Cai, H., Erhardt, P., Troppmair, J., Diaz-Meco, M.T., Sithanandam, G., Rapp, U.R., Moscat, J., and Cooper, G.M. 1993. Hydrolysis of phosphatidylcholine couples Ras to activation of Raf protein kinase during mitogenic signal transduction. *Mol. Cell. Biol.* 13:7645–7651.

Fabian, J.R., Daar, I.O., and Morrison, D.K. 1993. Critical tyrosine residues regulate the enzymatic and biological activity of Raf-1 kinase. *Mol. Cell. Biol.* 13:7170–7179.

Force, T., Bonventre, J.V., Heidecker, G., Rapp, U., Avruch, J., and Kyriakis, J.M. 1994. Enzymatic characteristics of the c-Raf-1 protein kinase. *Proc. Natl. Acad. Sci. USA* 91:1270–1274.

Johansen, T., Bjorkoy, G., Overvatn, A., Diaz-Meco, M.T., Traavik, T., and Moscat, J. 1994. NIH 3T3 cells stably transfected with the gene encoding phosphatidylcholine-hydrolyzing phospholipase C from *Bacillus cereus* acquire a transformed phenotype. *Mol. Cell. Biol.* 14:646–654.

Kolch, W., Heidecker, G., Kochs, G., Hummel, R., Vahidi, H., Mischak, H., Finkenzeller, G., Marme, D., and Rapp, U.R. 1993. PKC-α activates Raf-1 by direct phosphorylation. *Nature* 364:426–428.

Kolch, W., Heidecker, G., Lloyd, P., and Rapp, U.R. 1991. Raf-1 protein kinase is required for growth of induced NIH/3T3 cells. *Nature* 349:426–428.

Larrodera, P., Cornet, M.E., Diaz-Meco, M.T., Lopez-Barrahona, M., Diaz-Laviada, I., Guddal, P.H., Johansen, T., and Moscat, J. 1990. Phospholipase C-mediated hydrolysis of phosphatidylcholine is an important step in PDGF-stimulated DNA synthesis. *Cell* 61:1113–1120.

Leevers, S.J., Paterson, H.F., and Marshall, C.J. 1994. Requirement for Ras in Raf activation is overcome by targeting Raf to the plasma membrane. *Nature* 369:411–414.

Moodie, S.A., Willumsen, B.M., Weber, M.J., and Wolfman, A. 1993. Complexes of Ras-GTP with Raf-1 and mitogen-activated protein kinase kinase. *Science* 260:1658–1661.

Morrison, D.K., Kaplan, D.R., Rapp, U., and Roberts, T.M. 1988. Signal transduction from membrane to cytoplasm: growth factors and membrane-bound oncogene products increase Raf-1 phosphorylation and associated protein kinase activity. *Proc. Natl. Acad. Sci. USA* 85:8855–8859.

Nishizuka, Y. 1992. Intracellular signaling by hydrolysis of phospholipids and activation of protein kinase C. *Science* 258:607–614.

Sozeri, O., Vollmer, K., Liyanage, M., Frith, D., Kour,G., Mark, G.E., III, and Stabel, S. 1992. Activation of the c-Raf protein kinase by protein kinase C phosphorylation. *Oncogene* 7:2259–2262.

Stokoe, D., Macdonald, S.G., Cadwallader, K., Symons, M., and Hancock, J.F. 1994. Activation of Raf as a result of recruitment to the plasma membrane. *Science* 264:1463–1467.

Troppmair, J., Bruder, J.T., App, H., Cai, H., Liptak, L., Szeberenyi, J., Cooper, G.M., and Rapp, U.R. 1992. Ras controls coupling of growth factor receptors and protein kinase C in the membrane to Raf-1 and B-Raf protein serine kinases in the cytosol. *Oncogene* 7:1867–1873.

Van Aelst, L., Barr, M., Marcus, S., Polverino, A., and Wigler, M. 1993. Complex formation between RAS and RAF and other protein kinases. *Proc. Natl. Acad. Sci. USA* 90:6213–6217.

Vojtek, A.B., Hollenberg, S.M., and Cooper, J.A. 1993. Mammalian Ras interacts directly with the serine/threonine kinase Raf. *Cell* 74:205–214.

Warne, P.H., Viciana, P.R., and Downward, J. 1993. Direct interaction of Ras and the amino-terminal region of Raf-1 *in vitro*. *Nature* 364:352–355.

Wood, K.W., Sarnecki, C., Roberts, T.M., and Blenis, J. 1992. Ras mediates nerve growth factor receptor modulation of three signal-transducing protein kinases: MAP kinase, Raf-1, and RSK. *Cell* 68:1041–1050.

Zhang, X., Settleman, J., Kyriakis, J.M., Takeuchi-Suzuki, E., Elledge, S.J., Marshall, M.S., Bruder, J.T., Rapp, U.R., and Avruch, J. 1993. Normal and oncogenic p21^{ras} proteins bind to the amino-terminal regulatory domain of c-Raf-1. *Nature* 364:308–313.

The MAP Kinase Pathway

Biggs, W.H., III, Zavitz, K.H., Dickson, B., van der Straten, A., Brunner, D., Hafen, E., and Zipursky, S.L. 1994. The *Drosophila rolled* locus encodes a MAP kinase required in the sevenless signal transduction pathway. *EMBO J.* 13:1628–1635.

Blenis, J. 1993. Signal transduction via the MAP kinases: proceed at your own RSK. *Proc. Natl. Acad. Sci. USA* 90:5889–5992.

Chen, R.-H., Sarnecki, C., and Blenis, J. 1992. Nuclear localization and regulation of *erk*- and *rsk*-encoded protein kinases. *Mol. Cell. Biol.* 12:915–927.

Cook, S.J., and McCormick, F. 1993. Inhibition by cAMP of Ras-dependent activation of Raf. *Science* 262:1069–1072.

Crespo, P., Xu, N., Simonds, W.F., and Gutkind, J.S. 1994. Ras-dependent activation of MAP kinase pathway mediated by G-protein ßγ subunits. *Nature* 369:418–420.

Crews, C.M., Alessandrini, A., and Erikson, R.L. 1992. The primary structure of MEK, a protein kinase that phosphorylates the *ERK* gene product. *Science* 258:478–480.

Dent, P., Haser, W., Haystead, T.A.J., Vincent, L.A., Roberts, T.M., and Sturgill, T.W. 1992. Activation of mitogen-activated protein kinase kinase by v-Raf in NIH 3T3 cells and *in vitro*. *Science* 257:1404–1407.

De Vries-Smits, A.M.M., Burgering, B.M.Th., Leevers, S.J., Marshall, C.J., and Bos, J.L. 1992. Involvement of p21^{ras} in activation of extracellular signal-regulated kinase 2. *Nature* 357:602–604.

Faure, M., Voyno-Yasenetskaya, T.A., and Bourne, H.R. 1994. cAMP and ßγ subunits of heterotrimeric G proteins stimulate the mitogen-activated protein kinase pathway in COS-7 cells. *J. Biol. Chem.* 269:7851–7854.

Frodin, M., Peraldi, P., and van Obberghen, E. 1994. Cyclic AMP activates the mitogen-activated protein kinase cascade in PC12 cells. *J. Biol. Chem.* 269:6207–6214.

Frost, J.A., Geppert, T.D., Cobb, M.H., and Feramisco, J.R. 1994. A requirement for extracellular signal-regulated kinase (ERK) function in the activation of AP-1 by Ha-Ras, phorbol 12-myristate 13-acetate, and serum. *Proc. Natl. Acad. Sci. USA* 91:3844–3848.

Gomez, N., and Cohen, P. 1991. Dissection of the protein kinase cascade by which nerve growth factor activates MAP kinases. *Nature* 353:170–173.

Graves, L.M., Bornfeldt, K.E., Raines, E.W., Potts, B.C., Macdonald, S.G., Ross, R., and Krebs, E.G. 1993. Protein kinase A antagonizes platelet-derived growth factor-induced signaling by mitogen-activated protein kinase in human arterial smooth muscle cells. *Proc. Natl. Acad. Sci. USA* 90:10300–10304.

Howe, L.R., Leevers, S.J., Gomez, N., Nakielny, S., Cohen, P., and Marshall, C.J. 1992. Activation of the MAP kinase pathway by the protein kinase raf. *Cell* 71:335–342.

Kyriakis, J.M., App, H., Zhang, X., Banerjee, P., Brautigan, D.L., Rapp, U.R., and Avruch, J. 1992. Raf-1 activates MAP kinase-kinase. *Nature* 358:417–421.

Lange-Carter, C.A., Pleiman, C.M., Gardner, A.M., Blumer, K.J., and Johnson, G.L. 1993. A divergence in the MAP kinase regulatory network defined by MEK kinase and Raf. *Science* 260:315–319.

Lin, L.-L., Wartmann, M., Lin, A.Y., Knopf, J.L., Seth, A., and Davis, R.J. 1993. $cPLA_2$ is phosphorylated and activated by MAP kinase. *Cell* 72:269–278.

Pagès, G., Lenormand, P., L'Allemain, G., Chambard, J.-C., Meloche, S., and Pouysségur, J. 1993. Mitogen-activated protein kinases $p42^{mapk}$ and $p44^{mapk}$ are required for fibroblast proliferation. *Proc. Natl. Acad. Sci. USA* 90:8319–8323.

Rossomando, A., Wu, J., Weber, M.J., and Sturgill, T.W. 1992. The phorbol ester-dependent activator of the mitogen-activated protein kinase $p42^{mapk}$ is a kinase with specificity for the threonine and tyrosine regulatory sites. *Proc. Natl. Acad. Sci. USA* 89:5221–5225.

Sevetson, B.R., Kong, X., and Lawrence, J.C., Jr. 1993. Increasing cAMP attenuates activation of mitogen-activated protein kinase. *Proc. Natl. Acad. Sci. USA* 90:10305–10309.

Sturgill, T.W., Ray, L.B., Erikson, E., and Maller, J.L. 1988. Insulin-stimulated MAP-2 kinase phosphorylates and activates ribosomal protein S6 kinase II. *Nature* 334:715–718.

Thomas, G. 1992. MAP kinase by any other name smells just as sweet. *Cell* 68:3–6.

Thomas, S.M., DeMarco, M., D'Arcangelo, G., Halegoua, S., and Brugge, J.S. 1992. Ras is essential for nerve growth factor- and phorbol ester-induced tyrosine phosphorylation of MAP kinases. *Cell* 68:1031–1040.

Wood, K.W., Sarnecki, C., Roberts, T.M., and Blenis, J. 1992. Ras mediates nerve growth factor receptor modulation of three signal-transducing protein kinases: MAP kinase, Raf-1, and RSK. *Cell* 68:1041–1050.

Wu, J., Dent, P., Jelinek, T., Wolfman, A., Weber, M.J., and Sturgill, T.W. 1993. Inhibition of the EGF-activated MAP kinase signaling pathway by adenosine 3',5'-monophosphate. *Science* 262:1065–1069.

Wu, J., Harrison, J.K., Vincent, L.A., Haystead, C., Haystead, T.A.J., Michel, H., Hunt, D.F., Lynch, K.R., and Sturgill, T.W. 1993. Molecular structure of a protein-tyrosine/threonine kinase activating p42 mitogen-activated protein (MAP) kinase: MAP kinase kinase. *Proc. Natl. Acad. Sci. USA* 90:173–177.

Transcription Factors and Alterations in Gene Expression

Angel, P., Hattori, K., Smeal, T., and Karin, M. 1988. The *jun* proto-oncogene is positively autoregulated by its product, jun/AP-1. *Cell* 55:875–885.

Bello-Fernandez, C., Packham, G., and Cleveland, J.L. 1993. The ornithine decarboxylase gene is a transcriptional target of c-Myc. *Proc. Natl. Acad. Sci. USA* 90:7804–7808.

Chen, R.-H., Abate, C., and Blenis, J. 1993. Phosphorylation of the c-Fos transrepression domain by mitogen-activated protein kinase and 90-kDa ribosomal S6 kinase. *Proc. Natl. Acad. Sci. USA* 90:10952–10956.

Derijard, B., Hibi, M., Wu, I.-H., Barrett, T., Su, B., Deng, T., Karin, M., and Davis, R.J.

1994. JNK1: a protein kinase stimulated by UV light and Ha-Ras that binds and phosphorylates the c-Jun activation domain. *Cell* 76:1025–1037.

Fu, X-Y., and Zhang, J.-J. 1993. Transcription factor p91 interacts with the epidermal growth factor receptor and mediates activation of the c-*fos* gene promoter. *Cell* 74:1135–1145.

Gille, H., Sharrocks, A.D., and Shaw, P.E. 1992. Phosphorylation of transcription factor $p62^{TCF}$ by MAP kinase stimulates ternary complex formation at c-*fos* promoter. *Nature* 358:414–417.

Greenberg, M.E., and Ziff, E.B. 1984. Stimulation of 3T3 cells induces transcription of the c-*fos* proto-oncogene. *Nature* 311:433–438.

Hill, C.S., Marais, R., John, S., Wynne, J., Dalton, S., and Treisman, R. 1993. Functional analysis of a growth factor-responsive transcription factor complex. *Cell* 73:395–406.

Hipskind, R.A., Rao, V.N., Mueller, C.G.F., Reddy, E.S.P., and Nordheim, A. 1991. Ets-related protein Elk-1 is homologous to the c-*fos* regulatory factor $p62^{TCF}$. *Nature* 354:531–534.

Hunter, T., and Karin, M. 1992. The regulation of transcription by phosphorylation. *Cell* 70:375–387.

Karin, M., and Smeal, T. 1992. Control of transcription factors by signal transduction pathways: the beginning of the end. *Trends Biochem. Sci.* 17:418–422.

Kelly, K., Cochran, B.H., Stiles, C.D., and Leder, P. 1983. Cell-specific regulation of the c-*myc* gene by lymphocyte mitogens and platelet-derived growth factor. *Cell* 35:603–610.

Konig, H., Ponta, H., Rahmsdorf, U., Buscher, M., Schonthal, A., Rahmsdorf, H.J., and Herrlich, P. 1989. Autoregulation of *fos:* the dyad symmetry element as the major target of repression. *EMBO J.* 8:2559–2566.

Kruijer, W., Cooper, J.A., Hunter, T., and Verma, I.M. 1984. Platelet-derived growth factor induces rapid but transient expression of the c-*fos* gene and protein. *Nature* 312:711–716.

Marais, R., Wynne, J., and Treisman, R. 1993. The SRF accessory protein Elk-1 contains a growth factor-regulated transcriptional activation domain. *Cell* 73:381–393.

Marcu, K.B., Bossone, S.A., and Patel, A.J. 1992. *myc* function and regulation. *Ann. Rev. Biochem.* 61:809–860.

Muller, R., Bravo, R., Burckhardt, J., and Curran, T. 1984. Induction of c-*fos* gene and protein by growth factors precedes activation of c-*myc*. Nature 312:716–720.

Pulverer, B.J., Kyriakis, J.M., Avruch, J., Nikolakaki, E., and Woodgett, J.R. 1991. Phosphorylation of c-jun mediated by MAP kinases. *Nature* 353:670–674.

Ransone, L.J., and Verma, I.B. 1990. Nuclear proto-oncogenes *FOS* and *JUN*. *Ann. Rev. Biochem.* 6:539–557.

Rivera, V.M., Miranti, C.K., Misra, R.P., Ginty, D.D., Chen, R.-H., Blenis, J., and Greenberg, M.E. 1993. A growth factor-induced kinase phosphorylates the serum response factor at a site that regulates its DNA-binding activity. *Mol. Cell. Biol.* 13:6260–6273.

Ruff-Jamison, S., Chen, K., and Cohen, S. 1993. Induction by EGF and interferon-γ of tyrosine phosphorylated DNA binding proteins in mouse liver nuclei. *Science* 261:1733–1736.

Sadowski, H.B., Shuai, K., Darnell, J.E., Jr., and Gilman, M.Z. 1993. A common nuclear signal transduction pathway activated by growth factor and cytokine receptors. *Science* 261:1739–1744.

Sassone-Corsi, P., Sisson, J.C., and Verma, I.M. 1988. Transcriptional autoregulation of the proto-oncogene *fos*. *Nature* 334:314–319.

Schonthal, A., Herrlich, P., Rahmsdorf, H.J., and Ponta, H. 1988. Requirement for *fos* gene expression in the transcriptional activation of collagenase by other oncogenes and phorbol esters. *Cell* 54:325–334.

Shaw, P.E., Frasch, S., and Nordheim, A. 1989. Repression of c-*fos* transcription is mediated through $p67^{SRF}$ bound to the SRE. *EMBO J.* 8:2567–2574.

Silvennoinen, O., Schindler, C., Schlessinger, J., and Levy, D.E. 1993. Ras-independent growth factor signaling by transcription factor tyrosine phosphorylation. *Science* 261:1736–1739.

Smeal, T., Binetruy, B., Mercola, D., Grover-Bardwick, A., Heidecker, G., Rapp, U.R., and Karin, M. 1992. Oncoprotein-mediated signalling cascade stimulates c-Jun activity by phosphorylation of serines 63 and 73. *Mol. Cell. Biol.* 12:3507–3513.

Treisman, R. 1992. The serum response element. *Trends Biochem. Sci.* 17:423–426.

❖ CHAPTER 18

Regulation of the Cell Cycle

Growth factors stimulate quiescent cells to resume proliferation and progress through the G_1 phase of the cell cycle, which is followed by DNA synthesis (S phase), G_2, and mitosis (M phase). This progression through the cell cycle is controlled by a family of protein-serine/threonine kinases that are highly conserved in evolution and regulate the cell cycle of all eucaryotes. In particular, these cell cycle regulatory proteins function at two major points, controlling the progression of cells from G_1 to S and the transition from G_2 to M.

Because extracellular growth factors stimulate the proliferation of cells in G_1, it may be expected that at least some growth factor-induced responses interface with components of the cell cycle machinery that controls G_1 to S progression. This is indeed the case, and components of the cell cycle machinery have been found to be regulated by both mitogenic growth factors and extracellular factors that inhibit cell proliferation. Moreover, abnormalities in cell cycle regulation can lead directly to abnormal cell proliferation and tumor development, as indicated by the fact that both oncogenes and tumor suppressor genes function as regulators of cell cycle progression.

Cyclins and Cyclin-dependent Kinases

Our basic understanding of cell cycle regulation comes from studies of the G_2/M transition in amphibian oocytes, which are arrested in G_2 until hormonal stimulation triggers their entry into the M phase of meiosis (a process also called oocyte maturation and characterized by breakdown of the oocyte nucleus or germinal vesicle). In 1971, experiments in two independent laboratories demonstrated the activity of a cytoplasmic factor that could drive the G_2/M transition. These investigators injected oocytes with cytoplasm from eggs that had already undergone germinal vesicle breakdown and were in the process of meiosis. Such injections directly induced the G_2/M transition in recipient oocytes without the normal requirement for hormonal stimulation, indicating that the donor eggs contained a cytoplasmic factor with M phase-inducing activity. The factor identified by these experiments was called maturation promoting factor (MPF), which has since proved to be a universal regulator of the G_2/M transition in both mitosis and meiosis of all eucaryotic cells.

The biochemical purification of MPF, accomplished in the laboratory of James Maller in 1988, led to our current understanding of the cell cycle machinery through a convergence with yeast genetics. MPF was found to be a protein-serine/threonine kinase composed of both catalytic and regulatory subunits. The catalytic subunit was the vertebrate homolog of a protein kinase encoded by a yeast gene called *cdc2,* which had been isolated as a yeast cell cycle mutant in the early 1970s. The other subunit of MPF was found to be cyclin B, which had been initially identified in sea urchin eggs as a protein that was periodically synthesized throughout interphase and degraded at mitosis.

Cyclin B is a regulatory subunit, that controls the protein kinase activity of Cdc2 during the cell cycle (Fig. 18.1). In mammalian cells, synthesis of cyclin B begins in S phase. Complexes of cyclin B and Cdc2 then accumulate as the cell progresses through S and G_2. As these complexes form, Cdc2 is phosphorylated on threonine-161, which is required for its activation. In addition, it is phosphorylated on threonine-14 and tyrosine-15, which inhibits Cdc2 protein kinase activity until the G_2/M transition. Dephosphorylation of Thr-14 and Tyr-15 then activates the Cdc2 kinase, which triggers entry into M phase. Cdc2 phosphorylates a number of proteins, including nuclear lamins and histone H1, leading to nuclear envelope breakdown, chromosome condensation, and assembly of the mitotic spindle. In addition, Cdc2 triggers the degradation of cyclin B by activating a ubiquitin-dependent protease. Proteolytic degradation of cyclin B then inactivates MPF, leading the cell to exit mitosis and undergo cytokinesis.

Both Cdc2 and cyclin B are members of larger families of proteins that function to regulate additional cell cycle transitions. In yeast, Cdc2 is responsible not only for regulating the G_2/M transition, but also for controlling progression through a critical regulatory point in G_1 (called START). However, the activity of Cdc2 in G_1 is controlled by a distinct family of cyclins, called G_1 cyclins in contrast with the mitotic cyclins that control the G_2/M transition.

Animal cell cycles are regulated not only by multiple cyclins but also by multiple Cdc2-related kinases. In addition to the mitotic (B-type) cyclins, animal cells have distinct A-type cyclins, which appear to play a role in both the G_2/M transition and DNA replication, as well as several types of cyclins (C-, D-, and

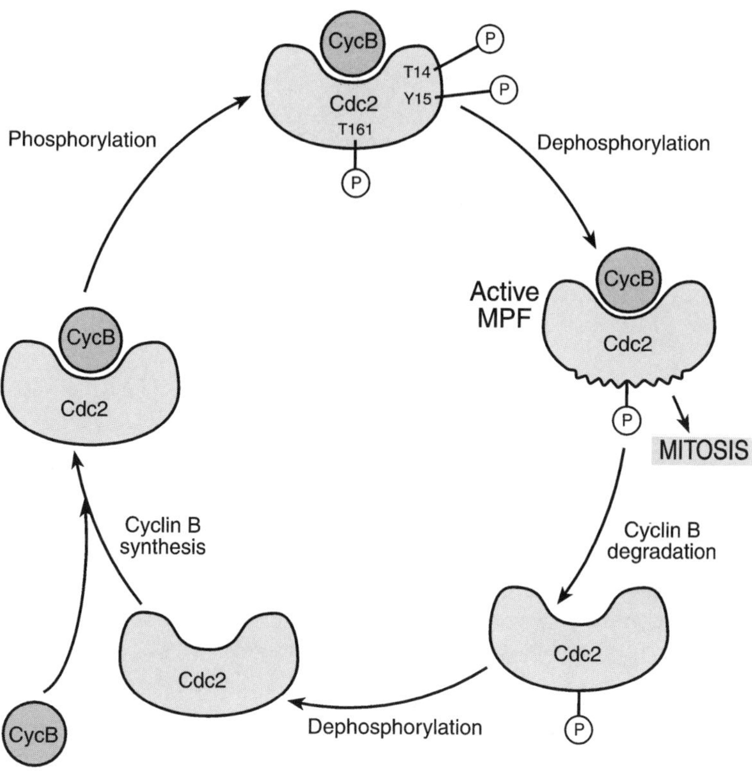

FIGURE 18.1
Regulation of Cdc2 kinase activity. Cdc2 forms complexes with cyclin B, which is synthesized during late S and G_2. These complexes are then phosphorylated on Cdc2 residues threonine-161 (T161), which is required for activation, as well as on threonine-14 (T14) and tyrosine-15 (Y15), which inhibit Cdc2 kinase activity. The complex is then activated by dephosphorylation of T14 and Y15 at the G_2/M transition, forming active MPF which induces mitosis. The activity of MPF is then terminated by proteolytic degradation of cyclin B, which signals the cell to exit from mitosis and undergo cytokinesis. Cdc2 is then dephosphorylated at T161 in preparation for reforming a new complex with cyclin B in the next cell cycle.

E-type) that are expressed in G_1. Members of a family of Cdc2-related kinases (called Cdks, for *c*yclin-*d*ependent *k*inases) associate with these different cyclins at different stages of the cell cycle (Fig. 18.2). For example, Cdc2 acts specifically at the G_2/M transition in association with cyclin B, whereas Cdk2, Cdk4, and Cdk6 associate with D- and E-type cyclins to regulate progression through G_1 and entry into S phase. Cdk2 is also associated with cyclin A during S phase, and Cdc2 is associated with cyclin A as well as cyclin B during G_2.

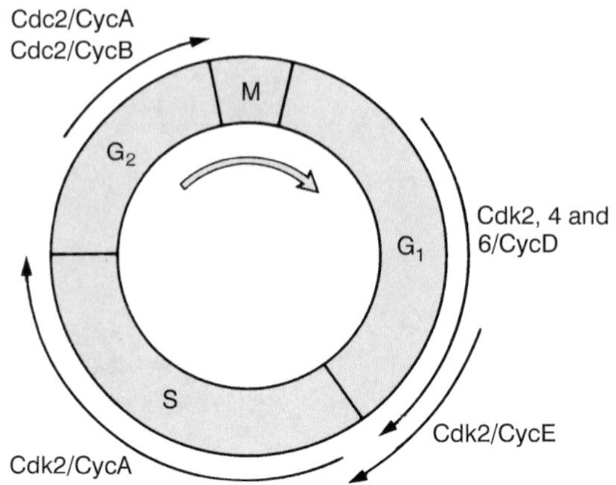

FIGURE 18.2
Cyclin/Cdk complexes in the animal cell cycle. Different stages of the cell cycle are regulated by distinct combinations of cyclins and Cdks. The G_2 to M transition is regulated by Cdc2 complexed with cyclins A and B. Progression through G_1 is controlled by cyclin D complexed with Cdks 2, 4, and 6. Cyclin E complexed with Cdk2 appears to act later in G_1 to regulate the G_1 to S transition. Cdk2 is also associated with cyclin A during S phase.

It thus appears that multiple combinations of cyclins and cyclin-dependent kinases regulate the progression of vertebrate cells through the cell cycle. Because the proliferation of most cells is controlled in G_1, the G_1 cyclins and their associated Cdks may be expected to provide a critical intersection between cell cycle regulation and the signal transduction pathways that govern the response of cells to growth factors. The role of cyclin C is unclear, but both the D-type and the E-type cyclins appear to regulate progression through G_1 and entry into S phase. The D-type cyclins appear to be expressed and to function at an earlier stage of G_1 than cyclin E, suggesting that the D-type cyclins may be primary regulators of G_1 progression, whereas cyclin E may act later to regulate the G_1/S transition. This suggested importance of D-type cyclins as regulators of cell proliferation has been supported both by their response to growth factors and by their ability to act as oncogenes.

Cyclin D: A Proto-oncogene Inducible by Growth Factors

Mammalian D-type cyclins were initially isolated by three independent approaches, two of which provided immediate indications that the D-type cyclins played a critical role in control of cell proliferation and provided a link between the action of mitogenic growth factors and the cell cycle machinery. In the most direct approach to isolating G_1 cyclins from mammalian cells, cyclin D (as well as cyclins C and E) was functionally identified by its ability to complement G_1 cyclin-deficient yeast mutants. In independent approaches, however, cyclin D was also identified as (1) a growth factor-inducible member of the cyclin family and (2) an oncogene-encoded protein.

The first of these routes to isolation of cyclin D included the screening of cDNA libraries to isolate growth factor-induced genes that were expressed late in G_1, subsequent to the initial transient induction of immediate-early genes such as c-*fos*. Such genes, called delayed response genes, are presumably induced downstream of the immediate-early genes and may play direct roles in the subsequent proliferation of growth factor-stimulated cells. The cDNA cloning of delayed response genes from mRNAs expressed three hours after growth factor stimulation led to the isolation of three genes encoding D-type cyclins (cyclins D1, D2, and D3). As expected for delayed response genes, induced transcription of the cyclin D genes required new protein synthesis, in contrast with the immediate-early genes. Moreover, induced transcription of cyclin D persisted as long as growth factors were present, rather than following the pattern of transient induction characteristic of the immediate-early genes. It therefore appears that induction of cyclin D is part of the secondary response of cells to growth factor stimulation, suggesting that cyclin D expression plays a key role in coupling the action of mitogenic growth factors to the subsequent progression of stimulated cells through G_1. Consistent with this hypothesis, interfering with cyclin D expression or function by antibodies or antisense RNAs has been shown to block DNA synthesis in growth factor-stimulated cells, indicating that cyclin D is required for their progression to S phase. In addition, overexpression of cyclin D shortens the duration of G_1, indicating that it is rate limiting for G_1 progression. As noted in the preceding section, cyclin E is induced at a later time in G_1 than cyclin D and is thought to take part in the G_1/S transition.

The importance of cyclin D in control of cell proliferation was further evidenced by its independent identification as an oncogene product. In particular, the *PRAD*-1 oncogene was initially isolated by molecular cloning of a site of chromosomal translocation in parathyroid adenomas. Sequence analysis of *PRAD*-1 revealed that it encoded a novel member of the cyclin family, which was designated cyclin D1. As discussed in chapters 7 and 8, *PRAD*-1 is also activated by chromosomal translocation in some B-cell lymphomas and by amplification in breast and squamous cell carcinomas, indicating that abnormal expression of cyclin D can contribute to development of a variety of neoplasms. Moreover, the oncogenic potential of cyclin D1 has been directly demonstrated by its ability to cooperate with *ras* to induce cell transformation in gene transfer experiments and with *myc* to induce lymphomas in transgenic mice. The D-type cyclins are thus clearly critical regulators of cell proliferation and provide a direct link between cell transformation and perturbations of the cell cycle regulatory machinery.

Cell Cycle Regulation of Rb

A further link between the cell cycle machinery and cell transformation derives from studies of the *Rb* tumor suppressor gene product. Phosphorylation of Rb is regulated during the cell cycle, such that it is underphosphorylated in G_0 and early G_1, but becomes more highly phosphorylated in late G_1, S, G_2, and M. The underphosphorylated form of Rb is specifically recognized by the DNA tumor virus oncogene proteins (for example, SV40 T antigen), which do not interact with the hyperphosphorylated forms of Rb. This suggests that underphosphorylated Rb is the active form of the molecule, which acts to suppress cell proliferation in G_1. Phosphorylation would then inactivate the growth suppressive function of Rb, allowing cells to progress through G_1 and the rest of the cell cycle.

The Rb protein is a substrate for Cdc2, suggesting that its regulated phosphorylation during the cell cycle may be mediated by Cdc2 or other members of the Cdk family. Because the relevant phosphorylation appears to inactivate Rb in G_1, one would predict the involvement of Cdks associated with the G_1 cyclins in this process. This prediction has been substantiated by the findings that at least some of the D-type cyclins can bind Rb and promote its phosphorylation by Cdk2 and Cdk4. Moreover, overexpression of D-type cyclins has been shown to reverse the growth suppressive activity of Rb. It thus appears that D-type cyclins can interact with Rb and promote its phosphorylation, correlated with the loss of Rb growth suppressive activity in G_1. Overexpression of cyclins A and E also can overcome Rb suppression of cell proliferation and lead to hyperphosphorylation of Rb, suggesting that Cdks associated with these cyclins may mediate Rb phosphorylation at later stages of the cell cycle.

As discussed in chapter 16, Rb forms complexes with members of the E2F family of transcription factors, inhibiting transcription of E2F regulated genes. Phosphorylation of Rb results in its dissociation from E2F, leading to activation of E2F target genes (Fig. 18.3). These include a number of genes required for cell cycle progression and DNA synthesis, such as c-*myc*, B-*myb*, *cdc*2, and those encoding thymidine kinase, dihydrofolate reductase, and DNA polymerase. Thus, cyclin-Cdk–mediated phosphorylation of Rb activates the transcription of a series of genes that function in the progression of cells through G_1 to S phase. During S phase, E2F is also associated with the Rb-related protein p107 in a complex with cyclin A and Cdk2. Although the function of this complex has not been established, it is possible that it plays a role in regulating DNA replication or other S phase events.

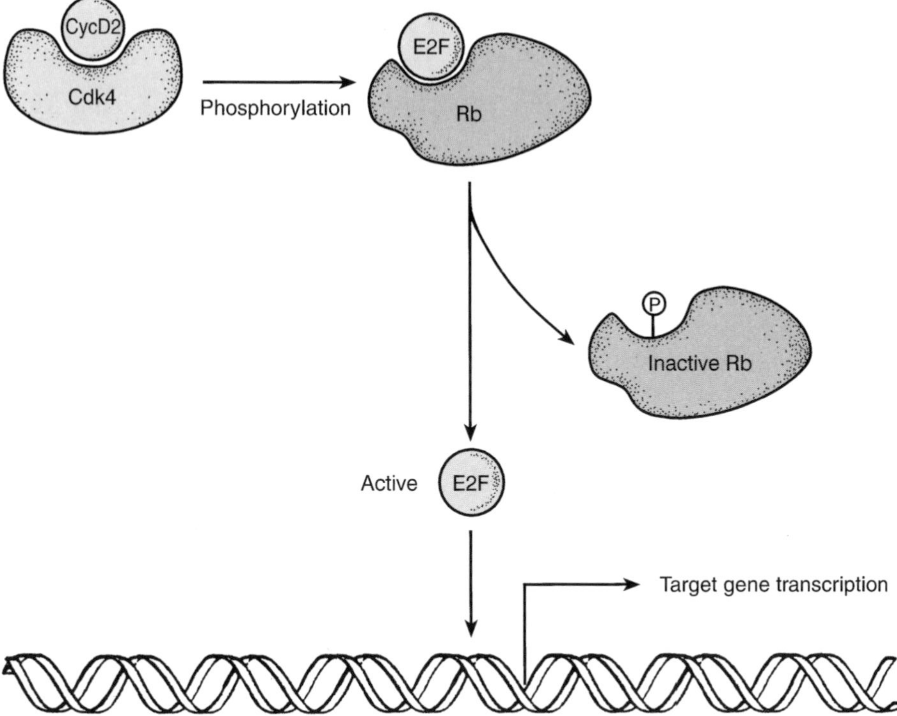

FIGURE 18.3
Cell cycle regulation of Rb. Some D-type cyclins (for example, cyclin D2) specifically bind to Rb and direct its phosphorylation by Cdks. Phosphorylation inactivates Rb and results in its dissociation from the E2F transcription factor. Active E2F then stimulates the transcription of target genes required for cell cycle progression and DNA synthesis.

The phosphorylation of Rb by Cdk/cyclin D complexes provides a direct link between the Rb tumor suppressor and the action of mitogenic growth factors, which induce expression of the D-type cyclins. In addition, it is important to note that cyclins and Cdks also regulate the activity of Rb in response to TGF-ß, a growth factor that acts to inhibit proliferation of epithelial cells. Treatment with TGF-ß arrests cell proliferation in G_1 and results in accumulation of Rb in the underphosphorylated, active form. This appears to result from inhibition of the synthesis of Cdk4, as well as from inducing the synthesis of a protein ($p27^{Kip1}$) that inhibits the activity of Cdk2/cyclin E complexes. Although the pathways remain to be fully elucidated, it is apparent that cyclins and Cdks regulate the activity of Rb in response to extracellular factors that act to inhibit as well as to stimulate cell proliferation.

p53 and MTS1: Tumor Suppressors That Regulate Cdks

The regulation of Rb by cyclin-Cdk complexes provided the first indication of a direct intersection between cell cycle regulation and the function of a tumor suppressor protein, as well as linking Rb to the action of extracellular growth factors. Part of this linkage is the induction of a Cdk inhibitor (p27) by the growth inhibitory factor TGF-ß. Further links between tumor suppressor genes and the cell cycle machinery have been elucidated by the identification of two additional cellular proteins that inhibit the activity of Cdks and thereby suppress cell cycle progression.

One negative regulator of the Cdks is a protein of 21 kd (called p21), which was isolated and characterized because of its association with Cdk/cyclin complexes. This protein (which is related to but not identical with p27) was then found to inhibit the protein kinase activity of all members of the Cdk family, indicating that p21 is a general Cdk inhibitor. Importantly, p21 was also found to be absent from cells lacking functional p53, as a result either of transformation by DNA tumor viruses or of mutational inactivation of the *p53* gene. As discussed in chapter 16, p53 is a transcriptional activator; so its activity as a tumor suppressor is thought to result from stimulating the expression of target genes that suppress cell proliferation. The absence of p21 from p53-deficient cells therefore suggested the possibility that expression of p21 was induced downstream of p53. This prediction was supported by the independent isolation of a p21 cDNA by directly screening for p53-inducible genes. Further analysis then showed that the p21 gene contains a regulatory site to which p53 binds and directly activates transcription. Moreover, overexpression of p21 itself inhibits cell proliferation. Thus, p21 appears to be a critical link between the *p53* tumor suppressor gene and regulation of the cell cycle (Fig. 18.4). Normally, p53 induces expression of p21, which suppresses cell proliferation by inhibiting Cdk activity. Inactivation of p53 results in a failure to express p21, leading to a loss of Cdk regulation and uncontrolled cell cycle progression.

The roles of p21 and p27 as negative regulators of Cdks suggest that these inhibitors of cell cycle progression might also function as tumor suppressors. Although this prediction has not yet been verified for p21 or p27, it has been shown to be the case for another Cdk inhibitor, p16. Like p21, p16 was isolated as a molecule that formed complexes with Cdks. In contrast with the general inhibition of Cdks by p21, p16 specifically binds to Cdk4 and inhibits the activity of Cdk4/cyclin D complexes. Thus, inactivation of p16 would be expected to specifically result in unregulated phosphorylation of Rb and loss of cell cycle control during G_1. The importance of p16 in growth regulation is confirmed by the finding that it is encoded by the *MTS*1 tumor suppressor gene, which is a target for deletions in melanomas and a variety of other tumors. This finding

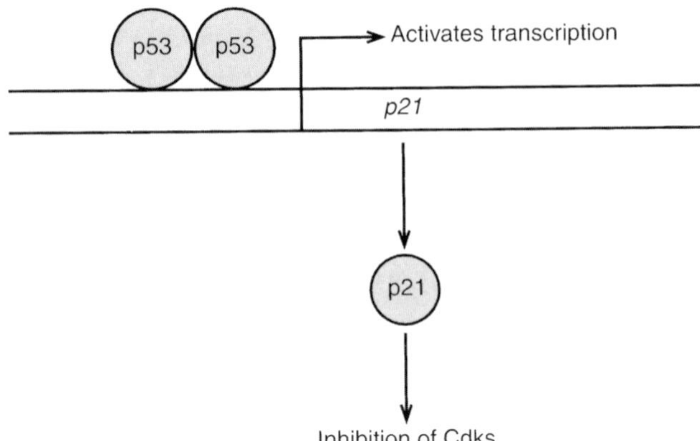

FIGURE 18.4
The *p53* tumor suppressor gene and cell cycle regulation. p53 activates transcription of the gene encoding p21, which then suppresses cell proliferation by inhibiting Cdks.

is especially noteworthy in light of the fact that mutations of *MTS1*, like those of *p53*, may be extremely common in human cancers. It thus appears that two tumor suppressor genes, both of which are implicated in a broad array of human malignancies, function by negatively regulating Cdks and inhibiting cell cycle progression.

Role of p53 in Cell Cycle Arrest and Apoptosis Induced by DNA Damage

All cells possess a variety of mechanisms for repairing DNA damage, which can result from exposure to chemicals or irradiation. In most cells, part of the response to irradiation is to delay the progression of irradiated cells through both the G_1 and the G_2 stages of the cell cycle. Delays at these two points in the cell cycle (called "checkpoints") presumably allow time for DNA repair to take place before S phase and mitosis, respectively, thereby preventing both the replication of damaged DNA and its transmission to daughter cells. The *p53* tumor suppressor gene product plays a central role in arresting cells at the G_1 checkpoint; so it is a major determinant of genetic stability and cell survival following exposure of cells to irradiation or other agents that damage DNA.

The involvement of p53 in the G_1 checkpoint was first suggested by the finding that levels of intracellular p53 substantially increased in irradiated cells. A direct functional role for p53 was then indicated by the fact that cells lacking p53 failed to arrest in G_1 following irradiation, although the G_2 checkpoint continued to function normally. It thus appeared not only that irradiation induced expression of p53, but also that p53 was required for G_1 arrest. This proposal is further supported by the finding that p21, the p53-inducible Cdk inhibitor, accumulates in irradiated cells, blocking the activity of cyclin-Cdk complexes required for G_1 progression. Thus, induction of p53, presumably by some mechanism that senses the presence of damaged DNA, leads to the arrest of irradiated cells in G_1 and thereby facilitates the repair of DNA damage before replication (Fig. 18.5).

For many cells, however, DNA damage is not successfully repaired and instead leads to cell death by apoptosis—a process in which p53 also plays a critical role. Apoptosis (also known as programmed cell death) is an active process of cell death characterized by cell shrinkage, chromatin condensation, and DNA degradation. It is a normal physiological process leading to the elimination of some cells in the course of the development of multicellular organisms. For example, the development of fingers or toes requires the programmed death of cells in the tissue between the digits. In addition, apoptosis can be trig-

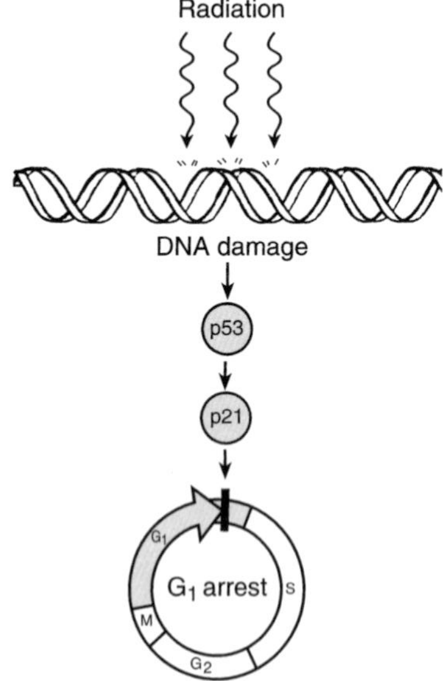

FIGURE 18.5
Role of p53 in radiation-induced G_1 arrest. p53 is induced by DNA damage resulting from irradiation. This results in induction of the Cdk inhibitor p21 and the arrest of irradiated cells in G_1.

gered by DNA damage, which is presumably advantageous to a multicellular organism because it eliminates cells carrying potentially deleterious mutations. Importantly, however, p53-deficient cells fail to normally undergo apoptosis in response to either irradiation or treatment with other agents that damage DNA, including a number of cancer chemotherapeutic drugs.

It thus appears that p53 plays a dual role in the response of cells to DNA damage, being required for apoptosis as well as for G_1 arrest. Because they fail to undergo apoptosis, p53-deficient cells display substantially increased resistance to cell killing by irradiation and other DNA damaging agents. Such inhibition of apoptosis may contribute to tumor development, as discussed for the *bcl*-2 oncogene in the next chapter, as well as affecting the susceptibility of cancer cells to irradiation and chemotherapy. Moreover, the absence of both the G_1 checkpoint and radiation-induced apoptosis in p53-deficient cells would be expected to allow the replication and transmission of damaged DNA to daughter cells, leading to increased mutation frequencies and a general instability of the cellular genome. It is noteworthy that such genetic instability is a common property of cancer cells, which may contribute to further alterations in oncogenes and tumor suppressor genes during tumor progression as well as to the continuing development of resistance to chemotherapeutic agents.

Summary

The progression of animal cells through the cell cycle is controlled by a family of protein kinases (Cdks) and their regulatory subunits (cyclins). The D-type cyclins, which control progression through G_1, play a critical role in regulating cell proliferation and serve to couple the cell cycle machinery to extracellular growth factors. This is indicated by their induction as delayed response genes following mitogenic stimulation and by the fact that cyclin D can act as an oncogene. Moreover, a critical action of cyclin D/Cdk complexes is phosphorylation of the Rb tumor suppressor protein, which inactivates Rb and allows progression through G_1. Two other tumor suppressor genes, *p53* and *MTS1*, also are linked to cell cycle regulation by inhibitors of Cdks. The *p53* tumor suppressor induces transcription of a gene encoding a general Cdk inhibitor (p21), and the *MTS1* gene encodes a specific inhibitor (p16) of Cdk4. Thus, two tumor suppressor genes implicated in a variety of human cancers act by negatively regulating cell cycle progression. It is also noteworthy that p53 is required for G_1 arrest and apoptosis of cells following DNA damage by irradiation or chemicals, including cancer chemotherapeutic drugs. Consequently, p53-deficient cells are resistant to cell killing by these agents and may display increased genetic instability, potentially contributing to continuing tumor progression and development of drug resistance.

References

General References

Murray, A., and Hunt, T. 1993. *The cell cycle*. W.H. Freeman, New York.

Pardee, A.B. 1989. G_1 events and regulation of cell proliferation. *Science* 246:603–608.

Cyclins and Cyclin-dependent Kinases

Draetta, G., Luca, F., Westendorf, J., Brizuela, L., Ruderman, J., and Beach, D. 1989. cdc2 protein kinase is complexed with both cyclin A and B: evidence for proteolytic inactivation of MPF. *Cell* 56:829–838.

Dulik, V., Lees, E., and Reed, S.I. 1992. Association of human cyclin E with a periodic G_1-S phase protein kinase. *Science* 257:1958–1961.

Gautier, J., Minshull, J., Lohka, M., Glotzer, M., Hunt, T., and Maller, J.L. 1990. Cyclin is a component of maturation-promoting factor from *Xenopus*. *Cell* 33:389–396.

Glotzer, M., Murray, A.W., and Kirschner, M.W. 1991. Cyclin is degraded by the ubiquitin pathway. *Nature* 349:132–137.

Kirschner, M. 1992. The cell cycle then and now. *Trends Biochem. Sci.* 17:281–285.

Koff, A., Giordano, A., Desai, D., Yamashita, K., Harper, J.W., Elledge, S., Nishimoto, T., Morgan, D.O., Franza, B.R., and Roberts, J.M. 1992. Formation and activation of a cyclin E-cdk2 complex during the G_1 phase of the human cell cycle. *Science* 257:1689–1694.

Labbé, J.C., Capony, P.-P., Caput, D., Cavadore, J.C., Derancourt, J., Kaghad, M., Lelias, J.-M., Picard, A., and Dorée, M. 1989. MPF from starfish oocytes at first meiotic metaphase is a heterodimer containing one molecule of cdc2 and one molecule of cyclin B. *EMBO J.* 8:3053–3058.

Lewin, B. 1990. Driving the cell cycle: M phase kinase, its partners, and substrates. *Cell* 61:743–752.

Lohka, M.J., Hayes, M.K., and Maller, J.L. 1988. Purification of maturation-promoting factor, an intracellular regulator of early mitotic events. *Proc. Natl. Acad. Sci. USA* 85:3009–3013.

Masui, Y., and Markert, C.L. 1971. Cytoplasmic control of nuclear behavior during meiotic maturation of frog oocytes. *J. Exp. Zool.* 177:129–146.

Matsushime, H., Ewen, M.E., Strom, D.K., Kato, J.-Y., Hanks, S.K., Roussel, M.F., and Sherr, C.J. 1992. Identification and properties of an atypical catalytic subunit ($p34^{PSK-J3}$/cdk4) for mammalian D type G_1 cyclins. *Cell* 71:323–334.

Meyerson, M., and Harlow, E. 1994. Identification of G_1 kinase activity for ckd6, a novel cyclin D partner. *Mol. Cell. Biol.* 14:2077–2086.

Murray, A.W. 1992. Creative blocks: cell-cycle checkpoints and feedback controls. *Nature* 359:599–604.

Nurse, P. 1990. Universal control mechanism regulating onset of M phase. *Nature* 344:503–508.

Ohtsubo, M., and Roberts, J.M. 1993. Cyclin-dependent regulation of G_1 in mammalian fibroblasts. *Science* 259:1908–1912.

Pines, J. 1993. Cyclins and cyclin-dependent kinases: take your partners. *Trends Biochem. Sci.* 18:195–197.

Smith, L.D., and Ecker, R.E. 1971. The interaction of steroids with *Rana pipiens* oocytes in the induction of maturation. *Dev. Biol.* 25:232–247.

Van der Heuvel, S., and Harlow, E. 1993. Distinct roles for cyclin-dependent kinases in cell cycle control. *Science* 262:2050–2054.

Wittenberg, C., Sugimoto, K., and Reed, S.I. 1990. G_1-specific cyclins of *S. cerevisiae*: cell cycle periodicity, regulation by mating pheromone, and association with the p34^{cdc28} protein kinase. *Cell* 62:225–237.

Cyclin D: A Proto-oncogene Inducible by Growth Factors

Baldin, V., Lukas, J., Marcote, M.J., Pagano, M., and Draetta, G. 1993. Cyclin D1 is a protein required for cell cycle progression in G_1. *Genes Dev.* 7:812–821.

Bodrug, S.E., Warner, B.J., Bath, M.L., Lindeman, G.J., Harris, A.W., and Adams, J.M. 1994. Cyclin D1 transgene impedes lymphocyte maturation and collaborates in lymphomagenesis with the *myc* gene. *EMBO J.* 13:2124–2130.

Hinds, P.W., Dowdy, S.F., Eaton, E.N., Arnold, A., and Weinberg, R.A. 1994. Function of a human cyclin gene as an oncogene. *Proc. Natl. Acad. Sci. USA* 91:709–713.

Hunter, T., and Pines, J. 1991. Cyclins and cancer. *Cell* 66:1071–1074.

Koff, A., Cross, F., Fisher, A., Schumacher, J., Leguellec, K., Philippe, M., and Roberts, J.M. 1991. Human cyclin E, a new cyclin that interacts with two members of the *CDC2* gene family. *Cell* 66:1217–1228.

Lew, D.J., Dulic, V., and Reed, S.I. 1991. Isolation of three novel human cyclins by rescue of G_1 cyclin (Cln) function in yeast. *Cell* 66:1197–1206.

Matsushime, H., Roussel, M.F., Ashmun, R.A., and Sherr, C.J. 1991. Colony-stimulating factor 1 regulates novel cyclins during the G_1 phase of the cell cycle. *Cell* 65:701–713.

Motokura, T., Bloom, T., Kim, H.G., Juppner, H., Ruderman, J.V., Kronenberg, H.M., and Arnold, A. 1991. A novel cyclin encoded by a *bcl*1-linked candidate oncogene. *Nature* 350:512–515.

Quelle, D.E., Ashmun, R.A., Shurtleff, S.A., Kato, J., Bar-Sagi, D., Roussel, M.F., and Sherr, C.J. 1993. Overexpression of mouse D-type cyclins accelerates G_1 phase in rodent fibroblasts. *Genes Dev.* 7:1559–1571.

Sherr, C.J. 1993. Mammalian G_1 cyclins. *Cell* 73:1059–1065.

Xiong, Y., Connolly, T., Futcher, B., and Beach, D. 1991. Human D-type cyclin. *Cell* 65:691–699.

Cell Cycle Regulation of Rb

Buchkovich, K., Duffy, L.A., and Harlow, E. 1989. The retinoblastoma protein is phosphorylated during specific phases of the cell cycle. *Cell* 58:1097–1105.

Cao, L., Faha, B., Dembski, M., Tsai, L.-H., Harlow, E., and Dyson, N. 1992. Independent binding of the retinoblastoma protein and p107 to the transcription factor E2F. *Nature* 355:176–179.

Chen, P.-L., Scully, P., Shew, J.-Y., Wang, J.Y.J., and Lee, W.-H. 1989. Phosphorylation of the retinoblastoma gene product is modulated during the cell cycle and cellular differentiation. *Cell* 58:1193–1198.

DeCaprio, J.A., Ludlow, J.W., Lynch, D., Furukawa, Y., Griffin, J., Piwnica-Worms, H., Huang, C.-M., and Livingston, D.M. 1989. The product of the retinoblastoma susceptibility gene has properties of a cell cycle regulatory element. *Cell* 58:1085–1095.

Devoto, S.H., Mudryj, M., Pines, J., Hunter, T., and Nevins, J.R. 1992. A cyclin A-protein kinase complex possesses sequence-specific DNA binding activity: $p33^{cdk2}$ is a component of the E2F-cyclin A complex. *Cell* 68:167–176.

Dowdy, S.F., Hinds, P.W., Louie, K., Reed, S.I., Arnold, A., and Weinberg, R.A. 1993. Physical interaction of the retinoblastoma protein with human D cyclins. *Cell* 73:499–511.

Ewen, M.E., Sluss, H.K., Sherr, C.J., Matsushime, H., Kato, J., and Livingston, D.M. 1993. Functional interactions of the retinoblastoma protein with mammalian D-type cyclins. *Cell* 73:487–497.

Ewen, M.E., Sluss, H.K., Whitehouse, L.L., and Livingston, D.M. 1993. TGFß inhibition of Cdk4 synthesis is linked to cell cycle arrest. *Cell* 74:1009–1020.

Geng, Y., and Weinberg, R.A. 1993. Transforming growth factor ß effects on expression of G_1 cyclins and cyclin-dependent kinases. *Proc. Natl. Acad. Sci. USA* 90:10315–10319.

Goodrich, D.W., Ping Wang, N., Qian, Y.-W., Lee, E.Y.-H.P., and Lee, W.-H. 1991. The retinoblastoma gene product regulates progression through the G_1 phase of the cell cycle. *Cell* 67:293–302.

Hinds, P.W., Mittnacht, S., Dulic, V., Arnold, A., Reed, S.I., and Weinberg, R.A. 1992. Regulation of retinoblastoma protein functions by ectopic expression of human cyclins. *Cell* 70:993–1006.

Kato, J., Matsushime, H., Hiebert, S.W., Ewen, M.E., and Sherr, C.J. 1993. Direct binding of cyclin D to the retinoblastoma gene product (pRb) and pRb phosphorylation by the cyclin D-dependent kinase CDK4. *Genes Dev.* 7:331–342.

Koff, A., Ohtsuki, M., Polyak, K., Roberts, J.M., and Massagué, J. 1993. Negative regulation of G_1 in mammalian cells: inhibition of cyclin E-dependent kinase by TGF-ß. *Science* 260:536–539.

Laiho, M., DeCaprio, J., Ludlow, J., Livingston, D., and Massagué, J. 1990. Growth inhibition by TGF-ß linked to suppression of retinoblastoma protein phosphorylation. *Cell* 62:175–185.

Lees, J.A., Buchkovich, K.J., Marshak, D.R., Anderson, C.W., and Harlow, E. 1991. The retinoblastoma protein is phosphorylated on multiple sites by human cdc2. *EMBO J.* 10:4279–4290.

Lin, B.T.-Y., Gruenwald, S., Moria, A.O., Lee, W.-H., and Wang, J.Y.J. 1991. Retinoblastoma cancer suppressor gene product is a substrate of the cell cycle regulator cdc2 kinase. *EMBO J.* 10:857–864.

Ludlow, J.W., DeCaprio, J.A., Huang, C.-M., Lee, W.-H., Paucha, E., and Livingston, D.M. 1989. SV40 large T antigen binds preferentially to an underphosphorylated member of the retinoblastoma susceptibility gene product family. *Cell* 56:57–65.

Polyak, K., Kato, J.Y., Solomon, M.J., Sherr, C.J., Massagué, J., Roberts, J.M., and

Koff, A. 1994. p27[Kip1], a cyclin-Cdk inhibitor, links transforming growth factor-ß and contact inhibition to cell cycle arrest. *Genes Dev.* 8:9–22.

Shirodkar, S., Ewen, M., DeCaprio, J.A., Morgan, J., Livingston, D.M., and Chittenden, T. 1992. The transcription factor E2F interacts with the retinoblastoma product and a p107-cyclin A complex in a cell cycle-regulated manner. *Cell* 68:157–166.

p53 and MTS1: Tumor Suppressors that Regulate Cdks

El-Deiry, W., Tokino, T., Velculescu, V.E., Levy, D.B., Parsons, R., Trent, J.M., Lin, D., Mercer, W.E., Kinzler, K.W., and Vogelstein, B. 1993. *WAF1*, a potential mediator of p53 tumor suppression. *Cell* 75:817–825.

Gu, Y., Turck, C.W., and Morgan, D.O. 1993. Inhibition of CDK2 activity *in vivo* by an associated 20K regulatory subunit. *Nature* 366:707–710.

Harper, J.W., Adami, G.R., Wei, N., Keyomarsi, K., and Elledge, S.J. 1993. The p21 Cdk-interacting protein Cip1 is a potent inhibitor of G_1 cyclin-dependent kinases. *Cell* 75:805–816.

Hunter, T. 1993. Braking the cell cycle. *Cell* 75:839–841.

Kamb, A., Gruis, N.A., Weaver-Feldhaus, J., Liu, Q., Harshman, K., Tavtigian, S.V., Stockert, E., Day, R.S., III, Johnson, B.E., and Skolnick, M.H. 1994. A cell cycle regulator potentially involved in genesis of many tumor types. *Science* 264:436–440.

Nobori, T., Miura, K., Wu, D.J., Lois, A., Takabayashi, K., and Carson, D.A. 1994. Deletions of the cyclin-dependent kinase-4 inhibitor gene in multiple human cancers. *Nature* 368:753–756.

Serrano, M., Hannon, G.J., and Beach, D. 1993. A new regulatory motif in cell-cycle control causing specific inhibition of cyclin D/CDK4. *Nature* 366:704–707.

Xiong, Y., Hannon, G.J., Zhang, H., Casso, D., Kobayashi, R., and Beach, D. 1993. p21 is a universal inhibitor of cyclin kinases. *Nature* 366:701–704.

Role of p53 in Cell Cycle Arrest and Apoptosis Induced by DNA Damage

Clarke, A.R., Purdie, C.A., Harrison, D.J., Morris, R.G., Bird, C.C., Hooper, M.L., and Wyllie, A.H. 1993. Thymocyte apoptosis induced by p53-dependent and independent pathways. *Nature* 362:849–852.

Dulic, V., Kaufmann, W.K., Wilson, S.J., Tlsty, T.D., Lees, E., Harper, J.W., Elledge, S.J., and Reed, S.I. 1994. p53-dependent inhibition of cyclin-dependent kinase activities in human fibroblasts during radiation-induced G_1 arrest. *Cell* 76:1013–1023.

Kastan, M.B., Zhan, Q., El-Deiry, W.S., Carrier, F., Jacks, T., Walsh, W.V., Plunkett, B.S., Vogelstein, B., and Fornace, A.J., Jr. 1992. A mammalian cell cycle checkpoint pathway utilizing p53 and GADD45 is defective in ataxia-telangiectasia. *Cell* 71:587–597.

Kuerbitz, S.J., Plunkett, B.S., Walsh, W.V., and Kastan, M.B. 1992. Wild-type p53 is a cell cycle checkpoint determinant following irradiation. *Proc. Natl. Acad. Sci. USA* 89:7491–7495.

Lee, J.M., and Bernstein, A. 1993. *p53* mutations increase resistance to ionizing radiation. *Proc. Natl. Acad. Sci. USA* 90:5742–5746.

Lowe, S.W., Ruley, H.E., Jacks, T., and Housman, D.E. 1993. p53-dependent apoptosis modulates the cytotoxicity of anticancer agents. *Cell* 74:957–967.

Lowe, S.W., Schmitt, E.M., Smith, S.W., Osborne, B.A., and Jacks, T. 1993. p53 is required for radiation-induced apoptosis in mouse thymocytes. *Nature* 362:847–849.

Lu, X., and Lane, D.P. 1993. Differential induction of transcriptionally active p53 following UV or ionizing radiation: defects in chromosome instability syndromes? *Cell* 75:765–778.

❖ CHAPTER 19

Development, Differentiation, and Programmed Cell Death

Previous chapters have dealt with the functions of proto-oncogenes and tumor suppressor genes in mitogenic signal transduction and cell cycle regulation. In addition to such roles in normal cell proliferation, one can envision that proto-oncogenes might also function in the transmission of signals that induce other cellular responses, such as differentiation. Indeed, the roles of the *erb*A and *PML/RARα* oncogenes in differentiation of hematopoietic cells were discussed in chapter 16. Moreover, because carefully regulated cell proliferation and differentiation are both critical to normal development, it is a reasonable expectation that some proto-oncogenes function in embryogenesis.

Consistent with these hypotheses, a number of proto-oncogenes have been found to regulate multiple aspects of development, including the meiotic maturation of oocytes and the proliferation and differentiation of a variety of embryonic cell types. In addition, at least one proto-oncogene functions to regulate programmed cell death, which plays a critical role both in embryonic development and in maintenance of adult tissues. Studies of proto-oncogenes and tumor suppressor genes have thus begun to extend beyond analysis of normal and neoplastic cell proliferation to yield new insights into the molecular mechanisms that regulate the fundamental biological processes of development and differentiation.

Proto-oncogenes in Meiosis and Early Embryo Development

The earliest stages of embryonic development (Fig. 19.1) are controlled by maternal RNAs that are synthesized and stored in oocytes that are arrested late in meiotic prophase. Oocytes can remain arrested at this stage of development for long time periods (for example, 40–50 years in humans) until the resumption of meiosis is triggered by hormonal stimulation. The oocyte then proceeds through the first meiotic division, after which meiosis in most species is again arrested until fertilization. The fertilized egg then completes meiosis and begins a series of cell divisions giving rise to the mass of undifferentiated cells that constitute the morula. The first cell determination event in development of mammalian embryos occurs at the transition of the morula to the blastocyst (about the sixteen-cell stage in the mouse). The blastocyst now contains two distinct cell types, the inner cell mass and the trophoblast, which give rise to the embryo and the extra-embryonic membranes, respectively.

In lower organisms, new transcription is not required for these early stages of embryogenesis, which are supported entirely by maternal RNAs. For example, *Xenopus* embryos develop normally in the absence of RNA synthesis to the midblastula stage, at which time a new program of transcription from the embryonic genome is initiated. In contrast, the extent of development supported by maternal RNAs in mammals is limited to meiosis, fertilization, and the first embryonic cell divisions. For example, new transcription is required for development of mouse embryos beyond the two-cell stage and for development of human embryos beyond four cells.

Accumulation of transcripts of a number of proto-oncogenes in oocytes of several species has been described, suggesting the possibility that these proto-oncogenes function as maternal messages in early embryonic development. These genes include c-*mos*, which is specifically expressed in oocytes of a variety of vertebrates, as well as a number of proto-oncogenes encoding growth factors, receptors, and other proteins (for example, Ras and Raf) that generally take part in intracellular signal transduction. In several cases, a developmental role for these proto-oncogenes is not only inferred from their expression as maternal RNAs in oocytes but is directly supported by functional studies.

The c-*mos* proto-oncogene, which encodes a protein-serine/threonine kinase (chapter 15), appears to function in the earliest stage of development—namely, meiotic maturation of the oocyte (Fig. 19.1). The c-*mos* proto-oncogene is unique in that its major sites of transcription are the male and female germ cells. Oocytes in particular accumulate very high levels of c-*mos* mRNA that is translated during meiosis. The biological role of Mos has been studied by microinjection of both *Xenopus* and mouse oocytes with antisense oligonucleotides, leading to degradation of the accumulated maternal c-*mos* mRNA. In both species, the loss of c-*mos* RNA prevents normal meiotic maturation of the

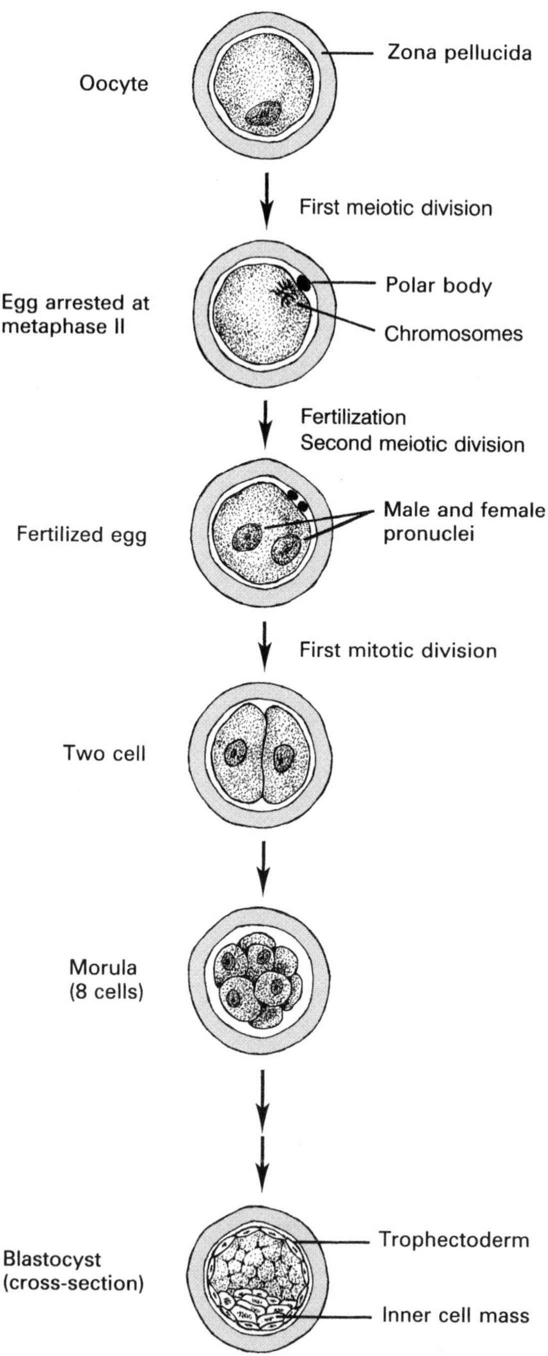

FIGURE 19.1
Oocyte maturation, fertilization, and early stages of embryonic development. Development of a mouse embryo to the blastocyst stage is depicted.

oocytes, and further studies of this system have revealed that Mos plays a unique role in regulating the meiotic cell cycle.

As reviewed in the preceding chapter, studies of amphibian oocytes led to the initial identification of maturation promoting factor (MPF), which is now known to be composed of Cdc2 and cyclin B. Activation of MPF induces chromosome condensation, nuclear membrane breakdown, and the cytoplasmic reorganization associated with entry into the M phase of either mitosis or meiosis. In mitosis, the proteolytic destruction of cyclin B then inactivates MPF, signaling the metaphase to anaphase transition, chromosome decondensation, reformation of a nucleus, and cytokinesis. In contrast, oocyte meiosis includes two successive M phases without reformation of a nucleus, and vertebrate eggs then remain arrested at metaphase II for as long as several days prior to fertilization, suggesting that distinct mechanisms regulate MPF during the meiotic cell cycle (Fig. 19.2). In particular, MPF activity increases upon the resumption

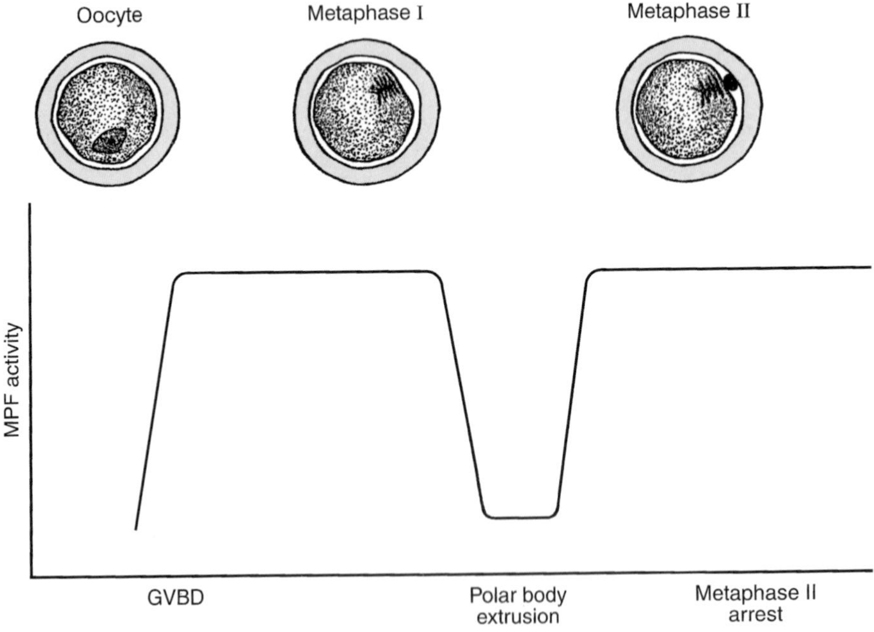

Figure 19.2
Activity of MPF during meiosis. MPF activity, which is low in oocytes, increases upon the resumption of meiosis to induce germinal vesicle (nuclear) breakdown (GVBD). MPF activity remains high until the metaphase/anaphase transition of meiosis I, when it declines to low levels. Following completion of meiosis I and polar body extrusion, MPF activity again increases and remains high during metaphase II arrest.

of meiosis and remains high until the metaphase/anaphase transition of meiosis I, when it declines to low levels. Following completion of meiosis I and extrusion of the first polar body, MPF activity again increases and remains high while the egg is arrested at metaphase II. This maintenance of MPF activity requires an additional factor first detected in amphibian eggs (cytostatic factor, or CSF), which appears to stabilize MPF during metaphase arrest. Fertilization then signals the degradation of cyclin B and exit from metaphase II.

Injection of antisense oligonucleotides indicates that translation of c-*mos* mRNA is required for the initial resumption of *Xenopus* oocyte meiosis. This is not the case for mouse oocytes, which do not require Mos for meiosis I. In both species, however, Mos is required for meiosis II, apparently acting to stabilize cyclin B and allow the increase in MPF activity required for the meiosis I to meiosis II transition. Furthermore, Mos has been identified as a component of cytostatic factor in *Xenopus,* indicating that it continues to prevent cyclin B degradation and maintain MPF activity during metaphase II arrest. However, Mos is not sufficient for maintenance of metaphase II, which also requires the Cdk2 protein kinase.

The biochemical mechanism by which Mos acts to maintain metaphase arrest is not yet fully understood. It is noteworthy, however, that Mos phosphorylates and activates the MEK protein kinase, leading to MAP kinase activation (see Fig. 17.8). This activation of the MAP kinase pathway is likely to account for the oncogenic potential of Mos and may also mediate the action of Mos during meiosis. Consistent with this possibility, MAP kinase is specifically activated in meiosis but not in mitosis. Moreover, a direct role for MAP kinase in meiosis has been indicated by experiments showing that constitutively activated MAP kinase is itself sufficient to induce metaphase arrest. Taken together, these findings indicate that activation of MAP kinase may account for the action of Mos in maintaining metaphase II arrest.

A variety of other proto-oncogenes also are expressed as maternal messages in both mammalian and amphibian eggs, as well as being transcribed after activation of the embryonic genome. In contrast to c-*mos,* these genes include many of the proto-oncogenes commonly expressed in somatic cells. For example, the proto-oncogenes expressed as maternal mRNAs in mouse oocytes include genes encoding growth factors (PDGF and TGF-α), the c-*kit* and c-*abl* protein-tyrosine kinases, members of the *ras* family, c-*raf,* c-*fos,* and c-*myc.* Several growth factors and receptor tyrosine kinases, as well as the *ras* and *raf* genes, are again expressed as transcripts of the embryonic genome (which is activated at the two-cell stage in the mouse), suggesting the possibility that autocrine growth factor signaling may play a role in early embryo development. For example, insulin-like growth factor II (IGF-II) and its receptor are both expressed in early mouse embryos, forming an autocrine loop that appears to contribute to embryonic cell proliferation. In addition, interference with the Ras/Raf signaling pathway by a dominant inhibitory Ras mutant blocks development of mouse embryos at the two-cell stage, suggesting that autocrine growth factor signaling may play a role in the transition between maternal and embryonic programs of gene expression.

In amphibian eggs, maternal messages function significantly beyond meiosis and early cleavages and appear to control determination of cell lineage at the blastula stage—a process in which proto-oncogenes also take part. In particular, the formation of mesoderm in *Xenopus* embryos (a classic example of embryonic induction) has been found to require stimulation of the FGF receptor and activation of the Ras/Raf signaling pathway. In contrast to mammalian eggs, amphibian eggs are polarized, consisting of a dark pigmented animal half and a light pigmented vegetal half (Fig. 19.3). Cells from the animal pole develop into ectoderm, which gives rise to tissues such as skin and the nervous system. Cells from the vegetal pole develop into the second germ layer, endoderm, which gives rise to internal organs such as the liver, stomach, and intestine. Formation of the third germ layer, mesoderm (which gives rise to connective tissues, the hematopoietic system, and muscle), is not intrinsically programmed but takes place through an inductive interaction between animal and vegetal pole cells. That is, animal pole cells will alter their program of differentiation and form mesoderm rather than ectoderm in response to an inductive signal from the vegetal pole cells. It therefore appears that the vegetal pole cells produce a mesoderm-inducing factor to which the animal pole cells respond. The differentiation of animal pole cells to mesoderm can be induced experimentally by synergistic action of the mammalian growth factors FGF and TGF-ß, suggesting the possibility that these growth factors might function as natural inducers of mesoderm differentiation. This hypothesis is supported by the identification of *Xenopus* maternal RNAs encoding basic FGF and members of the TGF-ß family, as well as FGF re-

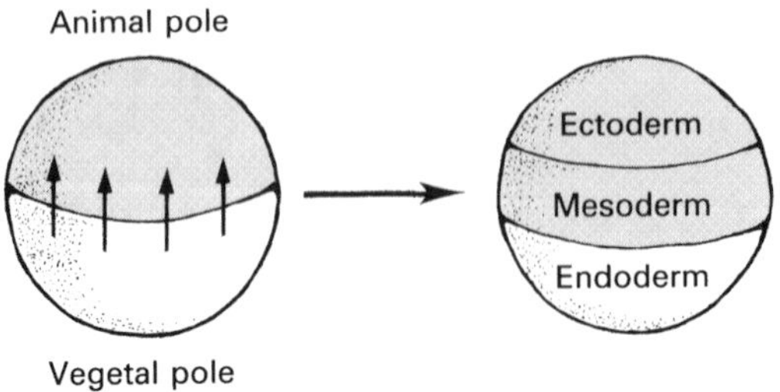

FIGURE 19.3
Mesoderm induction in *Xenopus*. Cells from the animal pole develop into ectoderm, whereas cells from the vegetal pole develop into endoderm. In addition, animal pole cells are induced to develop into mesoderm in response to a factor produced by vegetal pole cells. (Modified from J.M.W. Slack, *From egg to embryo: determinative events in early development*, Cambridge University Press, 1983.)

ceptor. Moreover, dominant negative mutants of the FGF receptor, Ras, and Raf have all been shown to block mesoderm formation, providing direct evidence that growth factor stimulation of the Ras/Raf signaling pathway plays a critical role in this process of embryonic induction.

Role of the Ras/Raf Pathway in Development of *Drosophila* and *C. Elegans*

The extensive genetic analysis that is possible in *Drosophila* and *C. elegans* has provided critical insights into the normal developmental functions of several proto-oncogenes. These include *ras* and *raf*, demonstrating that the Ras/Raf signal transduction pathway transmits intracellular signals leading not only to cell proliferation, but also to differentiation of a variety of cell types (Fig. 19.4).

In *Drosophila,* development of the terminal anterior and posterior regions of the embryo is governed by several genes expressed as maternal messages. One of these genes, *torso,* encodes a receptor tyrosine kinase that is similar to the mammalian PDGF receptor. Another member of this group of genes, *l*(1)*polehole,* is the *Drosophila* homolog of *raf,* which has been demonstrated ge-

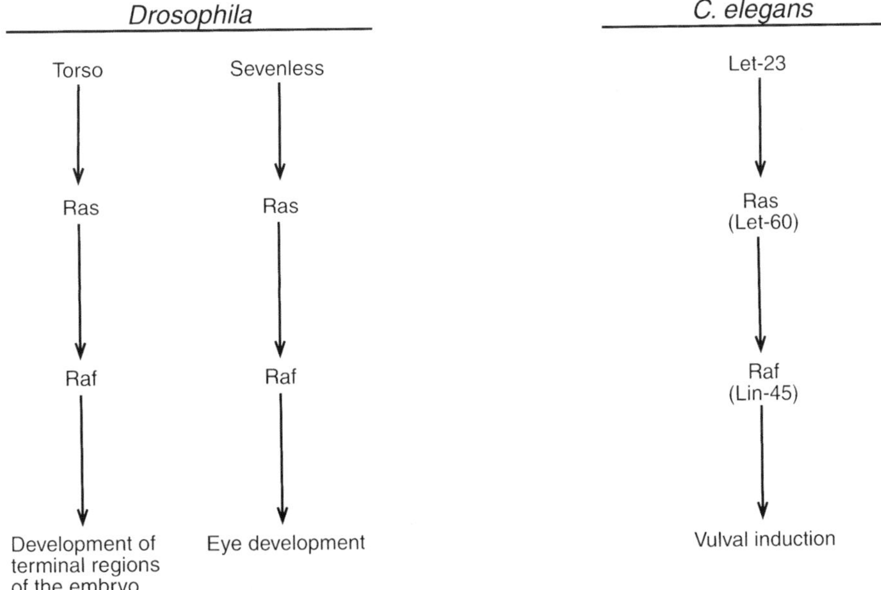

FIGURE 19.4
Roles of the Ras/Raf pathway in *Drosophila* and *C. elegans.* Activation of the Ras/Raf signaling pathway downstream of receptor tyrosine kinases (Torso, Sevenless, and Let-23) leads to formation of the terminal regions of the *Drosophila* embryo, *Drosophila* eye development, and vulval induction in *C. elegans.*

netically to function downstream of *torso*. Further experiments have shown that Ras and the Sos guanine nucleotide exchange factor also are required for Torso signaling and function upstream of Raf. It thus appears that determination of cell fate at the termini of *Drosophila* embryos requires localized stimulation of the Torso receptor tyrosine kinase, leading to activation of the Ras/Raf pathway and transcriptional activation of genes that are specifically expressed in terminal regions of the embryo—a signaling pathway that is entirely parallel to mitogenic signal transduction in mammalian cells.

The Ras/Raf pathway has a similiar role in *Drosophila* eye development. The compound eye of *Drosophila* is composed of about 800 simple eyes, each of which contains eight photoreceptor neurons (called R1–R8). These cells differentiate in a fixed sequence, starting with R8 and ending with R7. Differentiation of the R7 cell is induced by interaction with the neighboring R8 cell, and several genes required for R7 differentiation have been identified. One of these, *sevenless*, encodes a receptor tyrosine kinase related to the vertebrate *ros* proto-oncogene. Another, called *boss*, is specifically expressed on the surface of the R8 cell and encodes the ligand for Sevenless. Thus, a cell–cell interaction with the R8 cell stimulates the Sevenless receptor and leads to differentiation of R7. Further genetic analysis has again established that signaling from Sevenless proceeds through the Ras/Raf pathway, in addition to demonstrating that MAP kinase also is required for R7 development. In this case, signaling through the Ras/Raf pathway clearly leads to cell differentiation in the absence of proliferation, because the R7 cell is already postmitotic.

Vulval induction in *C. elegans* is a third well-characterized developmental pathway in which Ras and Raf play a role. In this system, the developmental fates of vulval precursor cells are determined by an inductive signal from a gonadal anchor cell. The genes required for vulval induction include *let*-23, which encodes a receptor tyrosine kinase related to the EGF receptor. The ligand for the Let-23 receptor, which is the inductive signal for vulval differentiation, is an EGF-like molecule encoded by *lin*-3. Downstream of Let-23, vulval differentiation requires the *let*-60 and *lin*-45 genes, which encode Ras and Raf proteins, respectively. Stimulation of the Ras/Raf pathway by tyrosine kinase receptors is thus a common mechanism of signaling cell differentiation in both *C. elegans* and *Drosophila*.

Proto-oncogenes in Neuronal Differentiation

Although the simpler eucaryotes provide clear advantages for genetic studies of development and differentiation, signaling from tyrosine kinase receptors through Ras and Raf has also been implicated in differentiation of some mammalian cell types. Neuronal differentiation is a particularly well characterized example, studies of which have been facilitated by the use of PC12 pheochromocytoma cells as an *in vitro* model system.

Development of the nervous system is regulated by neurotrophic factors, which signal differentiation and survival of neuronal cell types. The prototype neurotrophin is nerve growth factor (NGF), which is related to three other factors: brain-derived neurotrophic factor (BDNF), neurotrophin-3 (NT-3), and neurotrophin-4 (NT-4). The receptors for these factors are protein-tyrosine kinases encoded by the *trk* family of proto-oncogenes. In particular, Trk is the receptor for NGF, TrkB is the receptor for BDNF and NT-4, and TrkC is the receptor for NT-3. Thus, a family of protein-tyrosine kinases initially identified as oncogene products that induced transformation of fibroblasts normally function to regulate neuronal differentiation.

The PC12 cell line has been widely used as a cell culture model to study the induction of neuronal differentiation by NGF (Fig. 19.5). Upon NGF treatment,

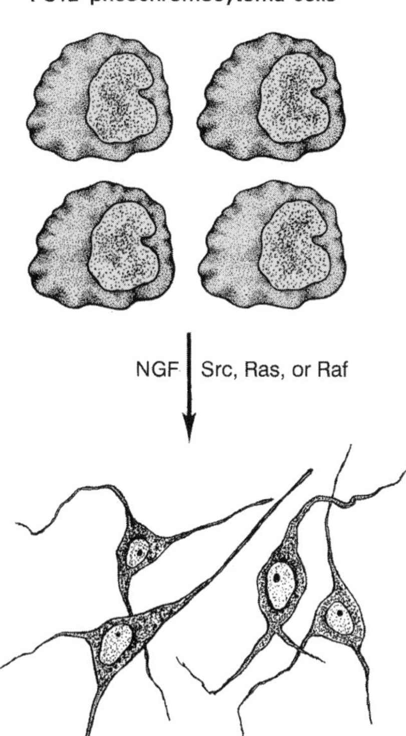

FIGURE 19.5
Differentiation of PC12 pheochromocytoma cells. PC12 cells can be induced to differentiate to postmitotic neuronlike cells by treatment with nerve growth factor (NGF). Similar neuronal differentiation of these cells is induced by the *src, ras,* and *raf* oncogenes.

PC12 cells differentiate to form cells resembling sympathetic neurons, as characterized by expression of a number of neuron-specific genes, extension of neurites, and cessation of cell proliferation. Thus, in this cell culture model, stimulation of a receptor tyrosine kinase induces postmitotic cell differentiation instead of cell division. Furthermore, introduction of activated *src, ras,* or *raf* oncogenes into PC12 cells similarly induces terminal neuronal differentiation. Conversely, interference with Src or Ras function by the use of antibodies or dominant negative mutants has been found to block differentiation induced by NGF, as well as blocking activation of Raf and MAP kinase. Thus, activation of the same Ras/Raf/MAP kinase pathway that signals proliferation of most cells (for example, fibroblasts) instead causes PC12 cells to cease proliferation and differentiate.

It is further noteworthy that NGF induces transcription of many of the same immediate-early genes in PC12 cells (for example, c-*fos,* c-*jun,* and c-*myc*) as are induced by mitogenic growth factors in fibroblasts. Understanding the molecular mechanisms by which the Ras/Raf signaling pathway leads to differentiation of one cell type but proliferation of another remains an open question to be addressed by further research.

Developmental Roles of Other Proto-oncogenes

Whereas the Ras/Raf pathway appears to play a general role in signal transduction processes leading either to cell growth or to differentiation, a number of other proto-oncogenes (like *trk*) encode proteins that are required for specific developmental processes. The roles of these genes have generally been elucidated by studies of mutant organisms lacking gene function. In some cases, proto-oncogenes have been identified as known genetic loci associated with specific developmental defects in flies, mice, or humans. In other cases, the potential developmental roles of proto-oncogenes have been experimentally addressed by using homologous recombination in embryonal stem cells to introduce specific mutations into the mouse germ line (Fig. 19.6).

A notable example of the role of extracellular growth factors in development is provided by the *wnt* genes. The prototype member of this gene family was identified as an oncogene activated by retroviral integration in mouse mammary carcinomas and initially called *int*-1. Its normal role became apparent through studies of *Drosophila* mutants, which established that the *Drosophila* homolog of *int*-1 was the segment polarity gene *wingless.* Studies of the defective development of flies bearing *wingless* mutations indicate that *wingless* encodes a secreted protein that induces pattern formation within groups of cells that form segments of the developing *Drosophila* embryo. These findings were consistent with the action of the Int-1 protein as an extracellular factor and directly established its role in development. Based on its identification as the

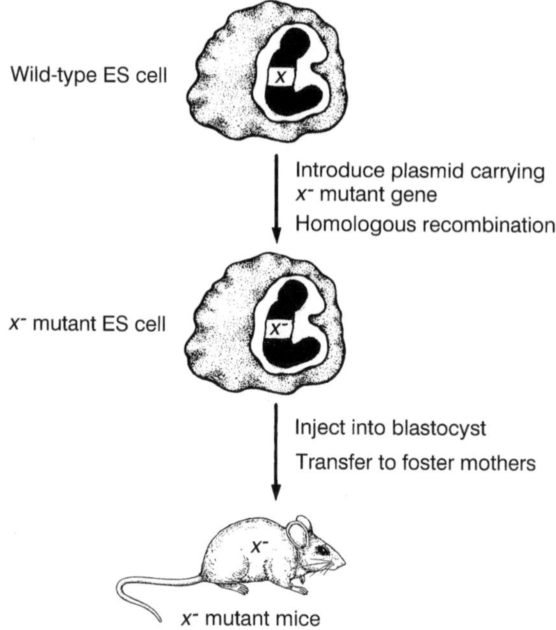

FIGURE 19.6
Homologous recombination in embryonal stem cells. Embryonal stem cells (ES cells), which are derived from mouse blastocysts, can be used to introduce mutations into the mouse germ line. Mutations of any cloned gene (for example, the *x* gene) can be introduced into ES cells by transfection with a plasmid containing a mutated allele, followed by selection of transformed ES cells in which the resident wild-type gene (*x*) has been replaced with the mutant (x^-) by homologous recombination. These cells can then be injected into a blastocyst and will participate in normal development of the embryo. In some cases, the ES cells carrying the x^- allele become incorporated into the germ line, allowing x^- mutant mice to be obtained.

mammalian homolog of *wingless*, the *int*-1 oncogene was subsequently renamed *wnt*-1 (an amalgam between *wingless* and *int*).

Given the role of *wingless* in *Drosophila,* it was noteworthy that *wnt*-1 was specifically expressed in the developing nervous system of mouse embryos. The potential role of *wnt*-1 in mammalian development was therefore investigated by using homologous recombination to disrupt mouse *wnt*-1 genes. These experiments revealed that mouse embryos homozygous for mutated *wnt*-1 alleles failed to develop a midbrain and cerebellum, although the rest of the embryo was normal. Moreover, further studies showed that a previously characterized mouse mutation (*swaying*) affecting development of the cerebellum corresponded to a mutation in *wnt*-1. Thus, *wnt*-1 is required for the development of specific regions of the central nervous system in mice. Other members of the *wnt* family (which includes at least ten distinct genes) also are expressed at specific embryonic sites, including other regions of the nervous system, and are likely to play similar roles in development and differentiation.

Specific developmental roles for two receptor tyrosine kinases, Kit and Ret, have been illuminated from the identification of mutations affecting these genes in both mice and humans. Namely, the *kit* proto-oncogene represents the *W* locus, which was first identified as a mouse coat color mutant more than 50 years ago. In addition to lacking hair pigmentation, mice with mutations at the *W* locus are sterile and anemic. These phenotypes result from defective proliferation or migration of the early embryonic stem cells that give rise to melanocytes, germ cells, and hematopoietic cells, respectively. The *W* locus thus appears to exert a pleiotropic effect on the development of three distinct embryonic stem cell populations. The identification of the *kit* proto-oncogene with the *W* locus therefore implies that this single receptor tyrosine kinase is critical for the development of each of these stem cell populations in the mouse embryo. The ligand for the Kit receptor is encoded at another mouse locus, *steel* (*Sl*), mutations of which result in the same phenotypes as mutations at the W locus. It is also noteworthy that mutations of *kit* are responsible for the inherited human piebald trait, which is similar to *W* locus mutations in mice.

Mutations of the *ret* proto-oncogene in humans are responsible for Hirschprung's disease, which is a relatively common genetic disorder with an incidence of 1 in 5,000 births. Hirschprung's disease is due to defective development of the enteric nervous system, resulting in intestinal obstruction as a consequence of the absence of autonomic ganglion cells in the hindgut. The gene responsible for Hirschprung's disease was found to map to the same chromosomal region as *ret,* leading two groups of researchers to search for and identify mutations of *ret* in Hirschprung's disease patients. Thus, the Ret receptor tyrosine kinase appears to play a normal role in development of the enteric nervous system in human beings. Mice with mutant *ret* genes also have been constructed by homologous recombination and found to display similar defects in enteric nervous system development, as well as being defective in development of the kidney.

Homologous recombination in mice has also been used to investigate the functions of nonreceptor tyrosine kinases, including Abl and members of the Src family. Perhaps surprisingly, defects in the genes encoding these proteins do not result in embryonic lethality, suggesting that other kinases with redundant functions can at least partially compensate for their loss. For example, mice lacking functional *src* genes survive to birth, although they then die as a result of defective osteoclast function. Presumably, the loss of Src activity during embryonic development of these mutant mice is compensated by other members of the Src family, which are redundantly expressed in most cell types.

However, some interesting tissue-specific functions of nonreceptor tyrosine kinases, like the activity of Src in osteoclasts, have been revealed by experiments of this type. For example, studies of mutant mice have implicated *lck* in thymocyte development, *fyn* in T-cell signaling as well as spatial learning and long-term potentiation in the hippocampus, and *abl* in development of the immune system.

An interesting example of a specific developmental role for a proto-oncogene transcription factor is provided by the *Drosophila* gene *dorsal,* which is a homolog of *rel.* The product of the *dorsal* gene, which is expressed as a maternal message, acts to determine cell fate along the dorsoventral axis of the developing embryo. As discussed in chapter 16, the activity of the NF-κB family of transcription factors (including Rel and Dorsal) is regulated by their translocation from the cytoplasm to the nucleus. In the case of Dorsal, the protein is uniformly distributed in eggs and early embryos. However, a gradient of nuclear localization is established during the embryonic cleavage divisions, such that high levels of Dorsal are present in nuclei at the ventral side of the embryo, whereas the protein remains cytoplasmic at the dorsal side. This nuclear concentration gradient of Dorsal results in differential expression of embryonic target genes, thereby determining the fate of cells along the dorsoventral axis of the embryo.

The functions of transcription factors encoded by several proto-oncogenes and tumor suppressor genes also have been investigated by homologous recombination experiments in mice. Perhaps surprisingly, disruption of these genes does not completely block embryonic development. For example, mice lacking c-*fos* genes are capable of developing to adults, although they display defects in development of bone and the hematopoietic system. Thus, c-Fos is not required for the normal development of most cell types, perhaps because its loss can be compensated by the activity of other members of the AP-1 transcription factor family, such as Fos-B, Fra-1, and Fra-2. Inactivation of c-*jun* has a more severe effect, resulting in the death of embryos at mid- to late-gestation. However, analysis of these embryos indicates that lack of c-Jun specifically interferes with liver development, while other cell types develop normally. It thus appears that other family members (for example, Jun-B and Jun-D) are able to compensate for loss of c-Jun in most cell types but not in fetal liver cells. Similar experiments have shown that N-*myc* is required for the development of several organs in mouse embryos (including the nervous system, heart, gut, and lung) and that c-*myb* is required for development of the hematopoietic system.

Mice lacking functional *p53* genes develop normally, although they have a high incidence of tumors, consistent with the function of p53 as a checkpoint regulator of cell cycle progression. In contrast, the *Rb* tumor suppressor gene is required for normal development, with *Rb*-deficient embryos dying in mid-to late-gestation with defects in the hematopoietic and nervous systems. Thus, similar to several proto-oncogenes, *Rb* is required for the development of specific cell types, although it appears to be dispensable for early stages of embryogenesis.

The *bcl*-2 Oncogene and Programmed Cell Death

Development is regulated not only by cell proliferation and differentiation, but also by cell death. The cell death that occurs during normal development (programmed cell death) proceeds by apoptosis, which was discussed in the preceding chapter as an active pathway of cell death triggered by DNA damage. Programmed cell death is a fundamental process during development, serving to eliminate unwanted cells from a variety of different tissues. Dramatic examples include the elimination of tissue between the digits during formation of fingers and toes, as well as the elimination of larval tissues during metamorphosis of insects and amphibians. Programmed cell death also plays an important role in development of the nervous system by eliminating those neurons that fail to make the proper connections. In the immune system, programmed cell death is responsible for the elimination of self-reactive or nonfunctional lymphocytes. In addition, apoptosis plays an important role in the maintenance of many adult tissues, where cell proliferation and cell death are balanced to maintain a constant cell population.

Studies of the *bcl*-2 oncogene, initially detected as the site of a chromosomal translocation in follicular B-cell lymphomas, have provided critical insights into the relation between programmed cell death and cancer. Overexpression of Bcl-2 in hematopoietic cells was found not to stimulate cell proliferation, but to block apoptosis of cells that were deprived of growth factors and would normally undergo programmed cell death. Similarly, transgenic mice expressing Bcl-2 from an immunoglobulin promoter were characterized by extended survival of B lymphocytes. Further studies have shown that Bcl-2 blocks apoptosis induced by a variety of stimuli in multiple different cell types, suggesting that Bcl-2 is a negative regulator of a general apoptosis pathway. It thus appears that *bcl*-2 functions as an oncogene because it prevents apoptosis and extends cell survival rather than because it directly affects cell proliferation. Interestingly, one of the signals that can induce apoptosis of some cells is continued expression of the c-*myc* oncogene under conditions of growth factor deprivation. Bcl-2 blocks such Myc-induced apoptosis, which may account for the cooperative effects of *bcl*-2 and c-*myc* in tumor development (see chapter 11).

Bcl-2 is an integral membrane protein of 25 kd, localized to mitochondria, endoplasmic reticula, and nuclear membranes. Although its mechanism of action remains to be established, Bcl-2 appears to function in an antioxidant pathway to prevent cellular damage resulting from the generation of reactive oxygen species. Other genes related to *bcl*-2 also regulate apoptosis (Fig. 19.7). One such gene, *bax*, encodes a protein of 21 kd that forms heterodimers with

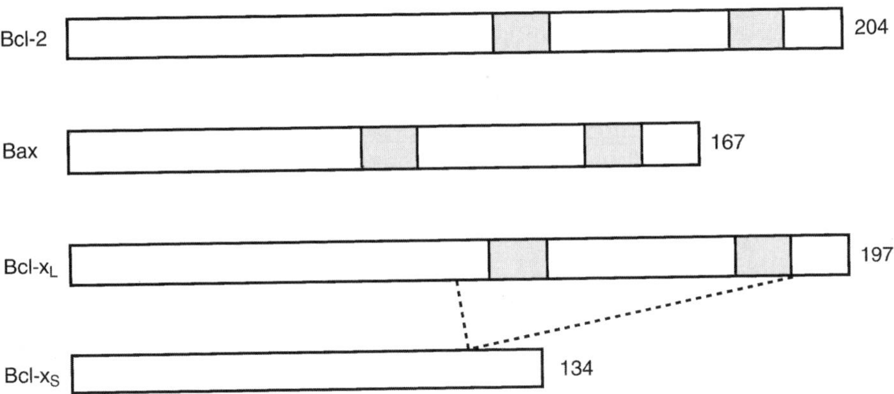

FIGURE 19.7
The *bcl*-2 gene family. Protein sizes are indicated as numbers of amino acids, and two regions of conserved sequence similarity are shown as filled boxes. The region deleted from Bcl-x$_S$ by alternative splicing is indicated by dashed lines.

Bcl-2 and counteracts the effect of Bcl-2 in suppressing apoptosis. Another member of the *bcl*-2 family, *bcl*-x, can encode two functionally distinct proteins as a result of alternative splicing. One of these proteins, Bcl-x$_L$, acts similarly to Bcl-2 to suppress apoptosis. However, the alternatively spliced protein, Bcl-x$_S$, acts instead to facilitate apoptosis by interfering with the action of Bcl-2. Alternative splicing of *bcl*-x may thus serve as an important determinant of apoptosis in certain tissues, such as the nervous system and thymus.

As with several other proto-oncogenes discussed earlier, mice in which *bcl*-2 genes have been inactivated by homologous recombination complete normal embryonic development. Therefore, *bcl*-2 is not required to prevent programmed cell death during embryogenesis, perhaps owing to redundant function of other members of the *bcl*-2 family (for example, *bcl*-x). After birth, however, mice deficient in *bcl*-2 undergo massive apoptosis of the lymphoid system, resulting in loss of both B and T lymphocytes. Thus, *bcl*-2 is necessary to maintain a stable immune system. In addition, *bcl*-2 deficient mice exhibit polycystic kidney disease and abnormalities of hair pigmentation. Both of these pathologies are related to oxidative damage, consistent with the proposed function of Bcl-2 in an antioxidant pathway.

In contrast with the normal embryonic development of *bcl*-2 deficient mice, it is of interest that a homolog of *bcl*-2 is required to regulate programmed cell death during the development of *C. elegans*. During normal nematode development, programmed cell death eliminates 131 somatic cells of a total of 1,090. One of the genes that controls this process is *ced*-9, which is required to prevent apoptosis of those cells that normally survive. Conversely, overexpression of *ced*-9 prevents apoptosis of those cells that normally die. Thus, *ced*-9 acts like *bcl*-2 to inhibit apoptosis. Consistent with this functional similarity, molecular cloning and sequencing have revealed that *ced*-9 is a homolog of *bcl*-2. Moreover, human *bcl*-2 can function in *C. elegans* to rescue *ced*-9 mutants, indicating that the pathway of apoptosis regulated by *bcl*-2 has been conserved from nematodes to human beings.

Summary

The examples discussed in this chapter illustrate the roles of proto-oncogenes in development and differentiation in a variety of organisms. The genes taking part in these processes include representatives of each of the major groups of proto-oncogenes that participate in mitogenic signal transduction, as well as members of the *bcl*-2 gene family, which function to control programmed cell death. It is thus clear that many proto-oncogenes regulate differentiation, rather than proliferation, of appropriate cell types. The identification of proto-oncogenes, initially based on the ability of oncogenes to induce cell transformation, has thus begun to provide a molecular framework applicable to the analysis of regulatory events governing not only cell proliferation, but also the fundamental processes of normal development.

References

Proto-oncogenes in Meiosis and Early Embryo Development

Amaya, E., Musci, T.J., and Kirschner, M.W. 1991. Expression of a dominant negative mutant of the FGF receptor disrupts mesoderm formation in *Xenopus* embryos. *Cell* 66:257–270.

Colledge, W.H., Carlton, M.B L., Udy, G.B , and Evans, M.J. 1994. Disruption of c-*mos* causes parthenogenetic development of unfertilized mouse eggs. *Nature* 370:65–68.

Furuno, N., Nishizawa, M., Okazaki, K., Tanaka, H., Iwashita, J., Nakajo, N., Ogawa, Y., and Sagata, N. 1994. Suppression of DNA replication via Mos function during meiotic divisions in *Xenopus* oocytes. *EMBO J.* 13:2399–2410.

Gabrielli, B.G., Roy, L.M., and Maller, J.L. 1993. Requirement for Cdk2 in cytostatic factor-mediated metaphase II arrest. *Science* 259:1766–1769.

Goldman, D.S., Kiessling, A.A., Millette, C.F., and Cooper, G.M. 1987. Expression of c-*mos* RNA in germ cells of male and female mice. *Proc. Natl. Acad. Sci. USA* 84:4509–4513.

Haccard, O., Sarcevic, B., Lewellyn, A., Hartley, R., Roy, L., Izumi, T., Erikson, E., and Maller, J.L. 1993. Induction of metaphase arrest in cleaving *Xenopus* embryos by MAP kinase. *Science* 262:1262–1265.

Hashimoto, N., Watanabe, N., Furuta, Y., Tamemoto, H., Sagata, N., Yokoyama, M., Okazaki, K., Nagayoshi, M., Takeda, N., Ikawa, Y., and Aizawa, S. 1994. Parthenogenetic activation of oocytes in c-*mos*-deficient mice. *Nature* 370:68–71.

Iwaoki, Y., Matsuda, H., Mutter, G.L., Watrin, F., and Wolgemuth, D.J. 1993. Differential expression of the proto-oncogenes c-*abl* and c-*mos* in developing mouse germ cells. *Exp. Cell Res.* 206:212–219.

Kanki, J.P., and Donoghue, D.J. 1991. Progression from meiosis I to meiosis II in *Xenopus* oocytes requires *de novo* translation of the *mos* protooncogene. *Proc. Natl. Acad. Sci. USA* 88:5794–5798.

Kimelman, D., and Kirschner, M. 1987. Synergistic induction of mesoderm by FGF and TGF-ß and the identification of an mRNA coding for FGF in the early *Xenopus* embryo. *Cell* 51:869–877.

MacNicol, A.M., Muslin, A.J., and Williams, L.T. 1993. Raf-1 kinase is essential for early *Xenopus* development and mediates the induction of mesoderm by FGF. *Cell* 73:571–583.

Manova, K., Nocka, K., Besmer, P., and Bachvarova, R.F. 1990. Gonadal expression of c-*kit* encoded at the *W* locus of the mouse. *Development* 110:1057–1069.

Mutter, G.L., and Wolgemuth, D.J. 1987. Distinct developmental patterns of c-*mos* proto-oncogene expression in female and male mouse germ cells. *Proc. Natl. Acad. Sci. USA* 84:5301–5305.

Nebreda, A.R., and Hunt, T. 1993. The c-*mos* proto-oncogene protein kinase turns on and maintains the activity of MAP kinase, but not MPF, in cell-free extracts of *Xenopus* oocytes and eggs. *EMBO J.* 12:1979–1986.

O'Keefe, S.J., Wolfes, H., Kiessling, A.A., and Cooper, G.M. 1989. Microinjection of antisense c-*mos* oligonucleotides prevents meiosis II in the maturing mouse egg. *Proc. Natl. Acad. Sci. USA* 86:7038–7042.

O'Keefe, S.J., Kiessling, A.A., and Cooper, G.M. 1991. The c-*mos* gene product is required for cyclin B accumulation during meiosis of mouse eggs. *Proc. Natl. Acad. Sci. USA* 88:7869–7872.

Orr-Urtreger, A., Avivi, A., Zimmer, Y., Givol, D., Yarden, Y., and Lonai, P. 1990. Developmental expression of c-*kit*, a proto-oncogene encoded by the *W* locus. *Development* 109:911–923.

Pal, S.K., Crowell, R., Kiessling, A.A., and Cooper, G.M. 1993. Expression of proto-oncogenes in mouse eggs and preimplantation embryos. *Mol. Reprod. Dev.* 35:8–15.

Paria, B.C., Das, S.K., Andrews, G.K., and Dey, S.K. 1993. Expression of the epidermal growth factor receptor gene is regulated in mouse blastocysts during delayed implantation. *Proc. Natl. Acad. Sci. USA* 90:55–59.

Posada, J., Yew, N., Ahn, N.G., Vande Woude, G.F., and Cooper, J.A. 1993. Mos stimulates MAP kinase in *Xenopus* oocytes and activates a MAP kinase kinase *in vitro*. *Mol. Cell. Biol.* 13:2546–2553.

Rappolee, D.A., Brenner, C.A., Schultz, R., Mark, D., and Werb, Z. 1988. Developmental expression of PDGF, TGF-α, and TGF-ß genes in preimplantation mouse embryos. *Science* 241:708–712.

Rappolee, D.A., Sturm, K.S., Behrendtsen, O., Schultz, G.A., Pedersen, R.A., and Werb, Z. 1992. Insulin-like growth factor II acts through an endogenous growth pathway regulated by imprinting in early mouse embryos. *Genes Dev.* 6:939–952.

Rosa, F., Roberts, A.B., Danielpour, D., Dart, L.L., Sporn, M.B., and Dawid, I.B. 1988. Mesoderm induction in amphibians: the role of TGF-ß2-like factors. *Science* 239:783–785.

Sagata, N., Oskarsson, M., Copeland, T., Brumbaugh, J., and Vande Woude, G.F. 1988. Function of c-*mos* proto-oncogene product in meiotic maturation in *Xenopus* oocytes. *Nature* 335:519–525.

Sagata, N., Watanabe, N., Vande Woude, G.F., and Ikawa, Y. 1989. The c-*mos* proto-oncogene product is a cytostatic factor responsible for meiotic arrest in vertebrate eggs. *Nature* 342:512–518.

Slack, J.M.W., Darlington, B.G., Heath, J.K., and Godsave, S.F. 1987. Mesoderm induction in early *Xenopus* embryos by heparin-binding growth factors. *Nature* 326:197–200.

Whitman, M., and Melton, D.A. 1992. Involvement of $p21^{ras}$ in *Xenopus* mesoderm induction. *Nature* 357:252–254.

Yamauchi, N., Kiessling, A.A., and Cooper, G.M. 1994. The Ras/Raf signaling pathway is required for progression of mouse embryos through the two-cell stage. *Mol. Cell. Biol.* 14:6655–6662.

Role of the Ras/Raf Pathway in Development of *Drosophila* and *C. elegans*

Ambrosio, L., Mahowald, A.P., and Perrimon, N. 1989. Requirement of the *Drosophila raf* homologue for *torso* function. *Nature* 342:288–291.

Aroian, R.V., Koga, M., Mendel, J.E., Ohshima, Y., and Sternberg, P.W. 1990. The *let-23* gene necessary for *Caenorhabditis elegans* vulval induction encodes a tyrosine kinase of the EGF receptor subfamily. *Nature* 348:693–698.

Beitel, G.J., Clark, S.G., and Horvitz, H.R. 1990. *Caenorhabditis elegans ras* gene *let-60* acts as a switch in the pathway of vulval induction. *Nature* 348:503–509.

Biggs, W.H., III, Zavitz, K.H., Dickson, B., van der Straten, A., Brunner, D., Hafen, E., and Zipursky, S.L. 1994. The *Drosophila rolled* locus encodes a MAP kinase required in the sevenless signal transduction pathway. *EMBO J.* 13:1628–1635.

Dickson, B., Sprenger, F., Morrison, D., and Hafen, E. 1992. Raf functions downstream of Ras1 in the Sevenless signal transduction pathway. *Nature* 360:600–603.

Hafen, E., Basler, K., Edstroem, J.-E., and Rubin, G.M. 1987. *Sevenless,* a cell-specific homeotic gene of *Drosophila,* encodes a putative transmembrane receptor with a tyrosine kinase domain. *Science* 236:55–63.

Han, M., Golden, A., Han, Y., and Sternberg, P.W. 1993. *C. elegans lin-45 raf* gene participates in *let-60 ras*-stimulated vulval differentiation. *Nature* 363:133–140.

Han, M., and Sternberg, P.W. 1990. *let-60,* a gene that specifies cell fates during *C. elegans* vulval induction, encodes a *ras* protein. *Cell* 63:921–931.

Kramer, H., Cagan, R.L., and Zipursky, S.L. 1991. Interaction of *bride of sevenless* membrane-bound ligand and the *sevenless* tyrosine-kinase receptor. *Nature* 352:207–212.

Lu, X., Chou, T.-B., Williams, N.G., Roberts, T.M., and Perrimon, N. 1993. Control of cell fate determination by $p21^{ras}$/Ras1, an essential component of *torso* signaling in *Drosophila. Genes Dev.* 7:621–632.

Simon, M.A., Bowtell, D.D.L., Dodson, G.S., Laverty, T.R., and Rubin, G.M. 1991. Ras1 and a putative guanine nucleotide exchange factor perform crucial steps in signaling by the Sevenless protein tyrosine kinase. *Cell* 67:701–715.

Sprenger, F., Stevens, L.M., and Nusslein-Volhard, C. 1989. The *Drosophila* gene *torso* encodes a putative receptor tyrosine kinase. *Nature* 338:478–483.

Proto-oncogenes in Neuronal Differentiation

Alema, S., Casalbore, P., Agostini, E., and Tato, F. 1985. Differentiation of PC12 phaeochromocytoma cells induced by v-*src* oncogene. *Nature* 316:557–559.

Bar-Sagi, D., and Feramisco, J.R. 1985. Microinjection of the *ras* oncogene protein into PC12 cells induces morphological differentiation. *Cell* 42:841–848.

Greenberg, M.E., Greene, L.A., and Ziff, E.B. 1985. Nerve growth factor and epidermal growth factor induce rapid transient changes in proto-oncogene transcription in PC12 cells. *J. Biol. Chem.* 260:14101–14110.

Hagag, N., Halegoua, S., and Viola, M. 1986. Inhibition of growth factor-induced differentiation of PC12 cells by microinjection of antibody to *ras* p21. *Nature* 319:680–682.

Kaplan, D.R., Hempstead, B.L., Martin-Zanca, D., Chao, M.V., and Parada, L.F. 1991. The *trk* proto-oncogene product: a signal transducing receptor for nerve growth factor. *Science* 252:554–558.

Klein, R., Jing, S., Nanduri, V., O'Rourke, E., and Barbacid, M. 1991. The *trk* proto-oncogene encodes a receptor for nerve growth factor. *Cell* 65:189–197.

Kremer, N.E., D'Arcangelo, G., Thomas, S.M., DeMarco, M., Brugge, J.S., and Halegoua, S. 1991. Signal transduction by nerve growth factor and fibroblast growth factor in PC12 cells requires a sequence of *src* and *ras* actions. *J. Cell Biol.* 115:809–819.

Kruijer, W., Schubert, D., and Verma, I.M. 1985. Induction of the proto-oncogene *fos* by nerve growth factor. *Proc. Natl. Acad. Sci. USA* 82:7330–7334.

Noda, M., Ko, M., Ogura, A., Liu, D., Amano, T., Takano, T., and Ikawa, Y. 1985. Sarcoma viruses carrying *ras* oncogenes induce differentiation-associated properties in a neuronal cell line. *Nature* 318:73–75.

Raffioni, S., Bradshaw, R.A., and Buxser, S.E. 1993. The receptors for nerve growth factor and other neurotrophins. *Ann. Rev. Biochem.* 62:823–850.

Szeberenyi, J., Cai, H., and Cooper, G.M. 1990. Effect of a dominant inhibitory Ha-*ras* mutation on neuronal differentiation of PC12 cells. *Mol. Cell. Biol.* 10:5324–5332.

Thomas, S.M., DeMarco, M., D'Arcangelo, G., Halegoua, S., and Brugge, J.S. 1992. Ras is essential for nerve growth factor- and phorbol ester-induced tyrosine phosphorylation of MAP kinases. *Cell* 68:1031–1040.

Troppmair, J., Bruder, J.T., App, H., Cai, H., Liptak, L., Szeberenyi, J., Cooper, G.M., and Rapp, U.R. 1992. Ras controls coupling of growth factor receptors and protein kinase C in the membrane to Raf-1 and B-Raf protein serine kinases in the cytosol. *Oncogene* 7:1867–1873.

Wood, K.W., Sarnecki, C., Roberts, T.M., and Blenis, J. 1992. Ras mediates nerve growth factor receptor modulation of three signal-transducing protein kinases: MAP kinase, Raf-1, and RSK. *Cell* 68:1041–1050.

Developmental Roles of Other Proto-oncogenes

Appleby, M.W., Gross, J.A., Cooke, M.P., Levin, S.D., Qian, X., and Perlmutter, R.M. 1992. Defective T cell receptor signaling in mice lacking the thymic isoform of $p59^{fyn}$. *Cell* 70:751–763.

Cabrera, C.V., Alonso, M.C., Johnston, P., Phillips, R.G., and Lawrence, P.A. 1987. Phenocopies induced with antisense RNA identify the *wingless* gene. *Cell* 50:659–663.

Chabot, B., Stephenson, D.A., Chapman, V.M., Besmer, P., and Bernstein, A. 1988. The proto-oncogene c-*kit* encoding a transmembrane tyrosine kinase receptor maps to the mouse *W* locus. *Nature* 335:88–89.

Charron, J., Malynn, B.A., Fisher, P., Stewart, V., Jeannotte, L., Goff, S.P., Robertson, E.J., and Alt, F.W. 1992. Embryonic lethality in mice homozygous for a targeted disruption of the N-*myc* gene. *Genes Dev.* 6:2248–2257.

Clarke, A.R., Maandag, E.R., van Roon, M., van der Lugt, N.M.T., van der Valk, M., Hooper, M.L., Berns, A., and te Riele, H. 1992. Requirement for a functional *Rb-1* gene in murine development. *Nature* 359:328–330.

Donehower, L.A., Harvey, M., Slagle, B.L., McArthur, M.J., Montgomery, C.A., Jr., Butel, J.S., and Bradley, A. 1992. Mice deficient for p53 are developmentally normal but susceptible to spontaneous tumours. *Nature* 356:215–221.

Edery, P., Lyonnet, S., Mulligan, L.M., Pelet, A., Dow, E., Abel, L., Holder, S., Nihoul-Fekete, C., Ponder, B.A.J., and Munnich, A. 1994. Mutations of the *RET* proto-oncogene in Hirschprung's disease. *Nature* 367:378–380.

Fleischman, R.A., Saltman, D.L., Stastny, V., and Zneimer, S. 1991. Deletion of the c-*kit* protooncogene in the human developmental defect piebald trait. *Proc. Natl. Acad. Sci. USA* 88:10885–10889.

Geissler, E.N., Ryan, M.A., and Housman, D.E. 1988. The dominant-white spotting (*W*) locus of the mouse encodes the c-*kit* proto-oncogene. *Cell* 55:185–192.

Giebel, L.B., and Spritz, R.A. 1991. Mutation of the *KIT* (mast/stem cell growth factor receptor) protooncogene in human piebaldism. *Proc. Natl. Acad. Sci. USA* 88:8696–8699.

Grant, S.G.N., O'Dell, T.J., Karl, K.A., Stein, P.L., Soriano, P., and Kandel, E.R. 1992. Imparied long-term potentiation, spatial learning, and hippocampal development in *fyn* mutant mice. *Science* 258:1903–1910.

Holberg, F., Aguzzi, A., Howells, N., and Wagner, E.F. 1993. c- Jun is essential for normal mouse development and hepatogenesis. *Nature* 365:179–181.

Jacks, T., Fazeli, A., Schmitt, E.M., Bronson, R.T., Goodell, M.A., and Weinberg, R.A. 1992. Effects of an *Rb* mutation in the mouse. *Nature* 359:295–300.

Johnson, R.S., Spiegelman, B.M., and Papaioannou, V. 1992. Pleiotropic effects of a null mutation in the c-*fos* proto-oncogene. *Cell* 71:577–586.

Lee, E.Y.-H.P., Chang, C.-Y., Hu, N., Wang, Y.-C.J., Lai, C.-C., Herrup, K., Lee, W.-H., and Bradley, A. 1992. Mice deficient for Rb are nonviable and show defects in neurogenesis and haematopoiesis. *Nature* 359:288–294.

McMahon, A.P., and Bradley, A. 1990. The *Wnt*-1 (*int*-1) proto-oncogene is required for development of a large region of the mouse brain. *Cell* 62:1073–1085.

Molina, T.J., Kishihara, K., Siderovski, D.P., van Ewijk, W., Narendran, A., Timms, E., Wakeham, A., Paige, C.J., Hartmann, K.U., Veillette, A., Davidson, D., and Mak, T.W. 1992. Profound block in thymocyte development in mice lacking $p56^{lck}$. *Nature* 357:161–164.

Mucenski, M.L., McLain, K., Kier, A.B., Swerdlow, S.H., Schreiner, C.M., Miller, T.A., Pietryga, D.W., Scott, W.J., Jr., and Potter, S.S. 1991. A functional c-*myb* gene is required for normal murine fetal hepatic hematopoiesis. *Cell* 65:677–689.

Nusse, R., and Varmus, H.E. 1992. *Wnt* genes. *Cell* 69:1073–1087.

Rijsewijk, F., Schuermann, M., Wagenaar, E., Parren, P., Weigel, D., and Nusse, R. 1987. The *Drosophila* homolog of the mouse mammary oncogene *int*-1 is identical to the segment polarity gene *wingless*. *Cell* 50:649–657.

Romeo, G., Ronchetto, P., Luo, Y., Barone, V., Seri, M., Ceccherini, I., Pasini, B., Bocciardi, R., Lerone, M., Kaarlainen, H., and Martucciello, G. 1994. Point mutations affecting the tyrosine kinase domain of the *RET* proto-oncogene in Hirschprung's disease. *Nature* 367:377–378.

Roth, S., Stein, D., Nusslein-Volhard, C. 1989. A gradient of nuclear localization of the *dorsal* protein determines dorsoventral pattern in the *Drosophila* embryo. *Cell* 59:1189–1202.

Rushlow, C.A., Han, K., Manley, J.L., and Levine, M. 1989. The graded distribution of the *dorsal* morphogen is initiated by selective nuclear transport in *Drosophila*. *Cell* 59:1165–1177.

Schuchardt, A., D'Agati, V., Larsson-Blomberg, L., Costantini, F., and Pachnis, V. 1994. Defects in the kidney and enteric nervous system of mice lacking the tyrosine kinase receptor Ret. *Nature* 367:380–383.

Schwartzenberg, P.L., Stall, A.M., Hardin, J.D., Bowdish, K.S., Humaran, T., Boast, S., Harbison, M.L., Robertson, E.J., and Goff, S.P. 1991. Mice homozygous for the abl^{m1} mutation show poor viability and depletion of selected B and T cell populations. *Cell* 65:1165–1175.

Shackleford, G.M., and Varmus, H.E. 1987. Expression of the proto-oncogene *int*-1 is restricted to postmeiotic male germ cells and the neural tube of mid-gestational embryos. *Cell* 50:89–95.

Soriano, P., Montgomery, C., Geske, R., and Bradley, A. 1991. Targeted disruption of the c-*src* proto-oncogene leads to osteopetrosis in mice. *Cell* 64:693–702.

Stanton, B.R., Perkins, A.S., Tessarollo, L., Sassoon, D.A., and Parada, L.F. 1992. Loss of N-*myc* function results in embryonic lethality and failure of the epithelial component of the embryo to develop. *Genes Dev.* 6:2235–2247.

Stein, P.L., Lee, H.-M., Rich, S., and Soriano, P. 1992. $pp59^{fyn}$ mutant mice display differential signaling in thymocytes and peripheral T cells. *Cell* 70:741–750.

Steward, R. 1987. *Dorsal*, an embryonic polarity gene in *Drosophila*, is homologous to the vertebrate proto-oncogene, c-*rel*. *Science* 238:692–694.

Steward, R. 1989. Relocalization of the *dorsal* protein from the cytoplasm to the nucleus correlates with its function. *Cell* 59:1179–1188.

Thomas, K.R., and Capecchi, M.R. 1990. Targeted disruption of the murine *int*-1 proto-oncogene resulting in severe abnormalities in midbrain and cerebellar development. *Nature* 346:847–850.

Thomas, K.R., Musci, T.S., Neumann, P.E., and Capecchi, M.R. 1991. *Swaying* is a mutant allele of the proto-oncogene *Wnt*-1. *Cell* 67:969–976.

Tybulewicz, V.L.J., Crawford, C.E., Jackson, P.K., Bronson, R.T., and Mulligan, R.C. 1991. Neonatal lethality and lymphopenia in mice with a homozygous disruption of the c-*abl* proto-oncogene. *Cell* 65:1153–1163.

Wang, Z.-Q., Ovitt, C., Grigoriadis, A.E., Mohle-Steinlein, U., Ruther, U., and Wagner, E.F. 1992. Bone and haematopoietic defects in mice lacking c-*fos*. *Nature* 360:741–745.

Wilkinson, D.G., Bailes, J.A., and McMahon, A.P. 1987. Expression of the proto-oncogene *int*-1 is restricted to specific neural cells in the developing mouse embryo. *Cell* 50:79–88.

Witte, O.N. 1990. Steel locus defines new multipotent growth factor. *Cell* 63:5–6.

The *bcl*-2 Oncogene and Programmed Cell Death

Bissonnette, R.P., Echeverri, F., Mahboubi, A., and Green, D.R. 1992. Apoptotic cell death induced by c-*myc* is inhibited by *bcl*-2. *Nature* 359:552–554.

Boise, L.H., Gonzales-Garcia, M., Postema, C.E., Ding, L., Lindsten, T., Turka, L.A., Mao, X., Nuñez, G., and Thompson, C.B. 1993. *bcl*-x, a *bcl*-2 related gene that functions as a dominant regulator of apoptotic cell death. *Cell* 74:597–608.

Evan, G.I., Wyllie, A.H., Gilbert, C.S., Littlewood, T.D., Land, H., Brooks, M., Waters, C.M., Penn, L.Z., and Hancock, D.C. 1992. Induction of apoptosis in fibroblasts by c-myc protein. *Cell* 69:119–128.

Fanidi, A., Harrington, E.A., and Evan, G.I. 1992. Cooperative interaction between c-*myc* and *bcl*-2 proto-oncogenes. *Nature* 359:554–556.

Garcia, I., Martinou, I., Tsujimoto, Y., and Martinou, J.-C. 1992. Prevention of programmed cell death of sympathetic neurons by the *bcl*-2 proto-oncogene. *Science* 258:302–304.

Hengartner, M.O., and Horvitz, H.R. 1994. *C. elegans* cell survival gene *ced*-9 encodes a functional homolog of the mammalian proto-oncogene *bcl*-2. *Cell* 76:665–676.

Hockenberry, D., Nuñez, G., Milliman, C., Schreiber, R.D., and Korsmeyer, S.J. 1990. Bcl-2 is an inner mitochondrial membrane protein that blocks programmed cell death. *Nature* 348:334–336.

Hockenberry, D.M., Oltvai, Z.N., Yin, X.-M., Milliman, C.L., and Korsmeyer, S.J. 1993. Bcl-2 functions in an antioxidant pathway to prevent apoptosis. *Cell* 75:241–251.

Kane, D.J., Sarafian, T.A., Anton, R., Hahn, H., Gralla, E.B., Valentine, J.S., Ord, T., and Bredesen, D.E. 1993. Bcl-2 inhibition of neural death: decreased generation of reactive oxygen species. *Science* 262:1274–1277.

McDonnell, T.J., Deane, N., Platt, F.M., Nuñez, G., Jaeger, U., McKearn, J.P., and Korsmeyer, S.J. 1989. *bcl*-2-immunoglobulin transgenic mice demonstrate extended B cell survival and follicular lymphoproliferation. *Cell* 57:79–88.

Nakayama, K., Nakayama, K., Negishi, I., Kuida, K., Shinkai, Y., Louie, M.C., Fields, L.E., Lucas, P.J., Stewart, V., Alt, F.W., and Loh, D.Y. 1993. Disappearance of the lymphoid system in *bcl*-2 homozygous mutant chimeric mice. *Science* 261:1584–1588.

Oltvai, Z.N., Milliman, C.L., and Korsmeyer, S.J. 1993. Bcl-2 heterodimerizes *in vivo* with a conserved homolog, Bax, that accelerates programmed cell death. *Cell* 74:609–619.

Sentman, A., Harris, A.W., and Cory, S. 1991. *bcl*-2 transgene inhibits T cell death and perturbs thymic self-censorship. *Cell* 67:889–899.

Sentman, C.L., Shutter, J.R., Hockenberry, D., Kanagawa, O., and Korsmeyer, S.J. 1991. *bcl*-2 inhibits multiple forms of apoptosis but not negative selection in thymocytes. *Cell* 67:879–888.

Vaux, D.L., Cory, S., and Adams, J.M. 1988. *bcl*-2 gene promotes haemopoietic cell survival and co-operates with c-*myc* to immortalize pre-B cells. *Nature* 335:440–442.

Vaux, D.L., Weissman, I.L., and Kim, S.K. 1992. Prevention of programmed cell death in *Caenorhabditis elegans* by human *bcl*-2. *Science* 258:1955–1957.

Veis, D.J., Sorenson, C.M., Shutter, J.R., and Korsmeyer, S.J. 1993. Bcl-2-deficient mice demonstrate fulminant lymphoid apoptosis, polycystic kidneys, and hypopigmented hair. *Cell* 75:229–240.

❖ Part V

Clinical Applications

❖ CHAPTER 20

New Prospects for Cancer Prevention and Treatment

The start of our current War on Cancer was declared when the late president Richard M. Nixon signed the National Cancer Act in 1971. As reviewed in the preceding pages, remarkable progress has since been made in understanding cancer from the viewpoint of basic science. Studies emanating from the molecular characterization of tumor viruses led to the identification of cellular oncogenes and tumor suppressor genes, and to an understanding of their roles in human cancer development. Analysis of the functions of these genes has further served to illuminate the molecular mechanisms that control the proliferation, differentiation, and survival of both normal and neoplastic cells. These advances are clearly major scientific accomplishments, which have affected virtually all areas of cell and molecular biology.

So far, however, these advances in understanding cancer have been of little direct benefit to the cancer patient. For many cancers, the success of treatment is primarily determined by early detection, allowing the curative treatment of localized disease by surgery or radiotherapy. Unfortunately, the majority of cancers have already metastasized by the time of diagnosis and cannot be cured by surgery or radiotherapy alone. Major progress has unquestionably been made in cancer chemotherapy, and effective drug combinations are now capable of curing most patients suffering from acute lymphocytic leukemia, Hodgkin's disease, non-Hodgkin's lymphoma, and testicular carcinoma. It is particularly important to note that these cancers occur most frequently among children and young adults, so the development of curative treatments for these diseases is a major advance. Nonetheless, the common adult cancers (such as lung, breast, colon, and prostate carcinomas) remain refractory to available chemotherapeutic drugs. Thus, although substantial progress has clearly been made, the overall cancer death rate has not diminished. Cancer remains a dread disease that eventually claims the lives of one of every four Americans.

This concluding chapter therefore considers the possible contributions that research on oncogenes and tumor suppressor genes may make to the practical issues of cancer prevention and treatment. Can our current insights into the molecular basis of cancer enable us to develop new approaches that will be effective in dealing with the human cancer problem?

Identification of Individuals with Inherited Susceptibilities to Cancer

The most effective way to deal with cancer would, of course, be to prevent development of the disease. A second-best, but still effective, alternative would be to reliably detect early premalignant stages of tumor development that can be easily treated. The major application of molecular biology to prevention and early detection may lie in the identification of individuals with inherited susceptibilities to cancer development. In particular, inherited mutations in tumor suppressor genes (*Rb, p53, APC,* and others), oncogenes (for example, *ret*), or DNA repair genes (for example, *MSH*2) can be detected by genetic analysis. In addition to indicating the use of appropriate measures to prevent cancer development in affected individuals, the ability to reliably detect genetic defects by prenatal diagnosis may be of great importance for those planning a family. In this respect, it is noteworthy that genetic defects can now be detected in early embryos derived from *in vitro* fertilization procedures while they are still in culture, prior to establishment of pregnancy.

Significant progress has been made in the prevention and early detection of some cancers; so the identification of individuals carrying mutations in genes resulting in increased cancer susceptibility can, in many cases, be expected to provide information that will be beneficial to the patient. In colon cancer, for example, polyps can be detected by colonoscopy and removed before their progression to malignancy. Patients with familial adenomatous polyposis (resulting from inherited mutations of *APC*) typically develop hundreds or thousands of polyps within the first two decades of life; so colons of these patients are usually removed before cancers develop. However, patients with the more common hereditary nonpolyposis colon cancer (resulting from mutations in DNA repair genes) develop a smaller number of polyps later in life and might significantly benefit from routine surveillance (for example, colonoscopy) and preventive measures, potentially including pharmacologic treatments as well as polypectomy. As another example, molecular cloning of the *BRCA*-1 tumor suppressor gene is expected to allow identification of some women at high risk for development of breast cancer, which may lead to benefits both from increased surveillance to detect the earliest possible stages of disease and from potential preventive measures, perhaps including hormone therapy to interfere with estrogen-stimulated cell proliferation.

It is important to note that the direct inheritance of cancers resulting from mutations in currently identified genes is relatively rare, corresponding to a small proportion of cancer cases. The most common inherited cancer susceptibility is hereditary nonpolyposis colon cancer, which may affect as many as one in two hundred people and accounts for approximately 15% of total colon cancer incidence. This corresponds to between 20,000 and 25,000 cases each year in the United States, of a total of approximately 1 million cancer cases diagnosed annually. Mutations of the *BRCA*-1 gene are likewise thought to represent a comparatively common inherited predisposition to cancer, being responsible for approximately 5% of all breast cancers (7,000–8,000 of the ~150,000 breast cancer cases diagnosed annually in the United States). Thus, although clearly affecting a significant number of people, inherited mutations in currently identified cancer susceptibility genes are still responsible for a relatively small fraction of total cancer cases. It is possible, however, that additional genes result in increased inherited susceptibilities to a larger number of common adult cancers. For example, it has been estimated that genes conferring increased susceptibility to breast, colon, and lung cancers may be inherited by 10 to 20% of the population, contributing to a substantial fraction of these malignancies. These genes have yet to be isolated and characterized at the molecular level, but this is an important undertaking with clear practical implications. The reliable identification of susceptible individuals, followed by appropriate preventive and early detection measures, might eventually make a major impact on cancer mortality.

Role of Oncogenes in Cancer Diagnosis

Information about the particular oncogenes and tumor suppressor genes responsible for a tumor clearly has the potential of being useful to the physician treating cancer. Indeed, several diagnostic applications of analysis of oncogenes are already being put into clinical practice. In some cases, abnormalities to oncogenes have provided useful assays for monitoring the course of disease during treatment. The translocation of *abl* in chronic myelogenous leukemia is a good example—PCR analysis of *bcr/abl* fusion sequences provides a sensitive means of detecting leukemic cells and is therefore useful for monitoring the response of patients to therapy. Likewise, molecular detection of the *PML/RAR*α oncogene is useful for both diagnosis and monitoring of acute promyelocytic leukemia.

In other cases, abnormalities to specific oncogenes may provide information pertinent to choosing between different therapeutic options. For example, amplification of N-*myc* in neuroblastomas, and *erb*B-2 in breast and ovarian carcinomas, is predictive of rapid disease progression. Therefore, it might be appropriate to treat patients with such amplified oncogenes more aggressively. As another example, discussed in chapter 18, p53 is required for apoptosis following DNA damage induced by irradiation or cancer chemotherapeutic drugs. Consequently, analysis of mutations in *p53* may help to predict the response of tumors to many of the drugs commonly used in chemotherapy. In general, increasingly specific diagnostic information is expected to allow more informed choices between available treatment options, and it is likely that characterization of tumors at the molecular level will play an increasingly important role in cancer diagnosis and treatment.

Development of Drugs Targeted against Oncogenes

The most critical question, however, is whether the discovery of oncogenes and tumor suppressor genes will allow the development of new drugs that act selectively against cancer cells. Most of the drugs currently used in cancer treatment act either by damaging DNA or by inhibiting DNA replication. Consequently, these drugs are toxic not only to cancer cells, but also to normal cells, particularly those normal cells that are undergoing rapid cell division (for example, hematopoietic cells, epithelial cells of the gastrointestinal tract, and hair follicle cells). The action of anticancer drugs against these normal cell populations accounts for most of the toxicity associated with these drugs, and it limits their effective use in cancer treatment. Do oncogenes provide unique targets, against which drugs with increased specificity for cancer cells could be designed?

Unfortunately, from the standpoint of cancer treatment, oncogenes are not unique to tumor cells. Because proto-oncogenes play important roles in normal cells, general inhibitors of oncogene expression or function are likely to act against normal cells as well as tumor cells. The exploitation of oncogenes as targets for anticancer drugs is therefore not a straightforward proposition, but there are reasons to hope that it will not ultimately be an impossible one.

One example of a therapeutic regimen targeted against a specific oncogene already exists—namely, the treatment of acute promyelocytic leukemia by retinoic acid. These leukemic cells, which are characterized by a translocation of the retinoic acid receptor to form the *PML/RARα* oncogene, undergo terminal differentiation in response to retinoic acid, presumably as a result of retinoic acid binding to the RAR portion of the PML-RAR fusion protein (see chapter 16). As a result, retinoic acid induces clinical remission in the majority (~90%) of acute promyelocytic leukemia patients, although these patients eventually relapse and are then resistant to further retinoic acid treatment. The therapeutic activity of retinoic acid against acute promyelocytic leukemia was observed prior to identification of the *PML/RARα* oncogene; so this association between oncogene and treatment was discovered by chance rather than by design. Nonetheless, retinoic acid therapy of acute promyelocytic leukemia is the first example of a cancer treatment specifically targeted against an oncogene protein.

The development of drugs targeted against Ras proteins provides an example of potential therapies directed toward interference with an oncogene protein that is not only involved in a variety of tumors, but also plays a central role in signal transduction pathways in normal cells. The *ras* oncogenes are activated by point mutations in ~15% of cancers, so selective interference with Ras oncogene proteins would be expected to make a significant contribution to cancer chemotherapy. However, Ras proto-oncogene proteins are expressed and appear to function in a wide variety of normal cell types; so general interference with Ras function would also be expected to have toxic side effects. Because the Ras oncogene proteins differ structurally from their normal homologs, it is a theoretical possibility to design drugs that would selectively inhibit proteins encoded by the mutated *ras* oncogenes. Unfortunately, the differences between oncogenic and normal Ras proteins are subtle, resulting from only single amino acid substitutions; so the design of selective inhibitors is an unquestionably difficult undertaking.

On the other hand, recent progress has been made in the development of inhibitors of Ras processing, which appear to display surprising specificity for interference with the growth of *ras*-transformed cells. As discussed in chapter 14, posttranslational modification of Ras proteins by the addition of a fifteen-carbon isoprenyl (farnesyl) group is required for their membrane association and biological activity. Farnesylation is not unique to Ras, and several other cellular proteins (including the nuclear lamins) also are farnesylated. However, because farnesylation is not a common protein modification, several laboratories have investigated the possibility of developing inhibitors of farnesyl-protein transferase as potential Ras-targeted anticancer agents. Several effective inhibitors of this enzyme have now been identified and found to interfere with transformation by *ras* oncogenes. Most interestingly, these inhibitors display considerable selectivity. For example, inhibition of farnesyl-protein transferase inhibits anchorage-independent growth of cells transformed by *ras,* but not of cells transformed by *raf* or *mos.* Moreover, such inhibitors have been reported to reverse the transformed phenotype of *ras*-transformed cells without interfering with normal cell growth. It thus appears that cells transformed by *ras* oncogenes may be more sensitive to interference with Ras processing than their normal counterparts. Although the molecular basis for this selectivity remains to be understood, it clearly offers promise for this approach to treatment of tumors in which *ras* oncogenes are activated.

A wide variety of additional oncogene targets for drug development are possible. For example, antibodies or synthetic peptides might be used to interfere with the binding of growth factors to their receptors—a strategy potentially applicable to tumors in which autocrine growth factor production or amplification of growth factor receptors plays a role (for example, the amplification of *erb*B-2 in breast cancers). In addition to directly affecting the growth of cancer cells, drugs targeted against appropriate growth factors and receptors might also interfere with tumor growth by inhibiting angiogenesis. The design of drugs targeted against oncogenes encoding intracellular signaling molecules (for example, Raf) or transcription factors (for example, Myc) are additional possibilities. Further, the possibility exists of developing agents that would inhibit progression through the cell cycle, potentially compensating for the loss of tumor suppressor genes that function in cell cycle regulation (for example, *p53, Rb,* and *MTS*1).

The roles of p53 and Bcl-2 in apoptosis highlight the importance of cell death, not only in the development of cancer, but also in the response of cancer cells to chemotherapeutic agents that damage DNA. Thus, the possibility of identifying agents that induce apoptosis is another potential avenue for drug development. An interesting example is already available in hormonal therapy of prostate cancer, which includes several possible means of blocking androgen stimulation of the tumor cells. Such androgen ablation induces apoptosis of androgen-dependent prostate cancer cells and substantially impedes tumor progression in most patients. Unfortunately, this hormonal therapy is not curative, and the initial response is generally followed by relapse due to the growth of androgen-independent cells. Hormone therapy of prostate cancer, like the treatment of acute promyelocytic leukemia with retinoic acid, was not specifically designed to induce apoptosis. However, several other possibilities can be envisioned for developing drugs that might selectively induce apoptosis of tumor cells—for example, by inhibiting Bcl-2 or by compensating for the loss of p53 in p53-deficient tumors that fail to undergo apoptosis in response to treatment with DNA-damaging agents.

Gene therapy by the use of antisense oligonucleotides is another possible approach to oncogene-directed cancer treatment, particularly in cases where novel targets are provided by oncogene rearrangements. For example, the transcripts of oncogenes expressed as fusion proteins (such as *bcr/abl* or *PML/RAR*α) are characterized by novel sequences at the junctions between the fusion partners, which could provide tumor-specific targets for antisense oligonucleotides. Indeed, selective inhibition of the proliferation of chronic myelogenous leukemia cells by antisense oligonucleotides directed against the *bcr/abl* oncogene has been reported both *in vitro* and in a mouse model, suggesting the potential feasibility of this approach. Still another possibility is that the novel peptide sequences at the junctions of fusion proteins such as Bcr/Abl may provide unique targets for immunotherapy directed specifically against oncogene proteins.

Summary

Although the identification of oncogenes and tumor suppressor genes has greatly enhanced our understanding of the molecular basis of human cancer, these discoveries have not yet been translated into significant clinical benefits for the cancer patient. However, important applications of advances in molecular biology are now evident in several areas of clinical practice. For example, the identification of individuals with inherited mutations resulting in a high cancer risk may allow preventive measures to be instituted before cancer has a chance to develop, as well as opening a variety of options for family planning. The identification of mutations affecting oncogenes and tumor suppressor genes is also playing an increasing role in cancer diagnosis and monitoring. Perhaps of greatest importance, oncogenes and tumor suppressor genes may provide novel targets for the development of anticancer drugs. The treatment of acute promyelocytic leukemia with retinoic acid and the development of inhibitors of Ras farnesylation provide important examples of drugs targeted against specific oncogenes. Although the impact of this approach to drug discovery remains to be seen, there clearly exists a reasonable basis for the optimistic viewpoint that progress in molecular and cell biology will lead to corresponding advances in cancer treatment. Accomplishing this transition from basic science to clinical practice is a major challenge for the future of cancer research.

References

General References

Cline, M.J. 1994. The molecular basis of leukemia. *N. Engl. J. Med.* 330:328–336.

Grignani, F., Fagioli, M., Alcalay, M., Longo, L., Pandolfi, P.P., Donti, E., Biondi, A., Lo Coco, F., Grignani, F., and Pelicci, P.G. 1994. Acute promyelocytic leukemia: from genetics to treatment. *Blood* 83:10–25.

Karp, J.E., and Broder, S. 1994. New directions in molecular medicine. *Cancer Res.* 54:653–665.

Identification of Individuals with Inherited Susceptibilities to Cancer

Cannon-Albright, L.A., Skolnick, M.H., Bishop, D.T., Lee, R.G., and Burt, R.W. 1988. Common inheritance of susceptibility to colonic adenomatous polyps and associated colorectal cancers. *N. Engl. J. Med.* 319:533–537.

Handyside, A.H., Lesko, J.G., Tarin, J.J., Winston, R.M.L., and Hughes, M.R. 1992. Birth of a normal girl after *in vitro* fertilization and preimplantation diagnostic testing for cystic fibrosis. *N. Engl. J. Med.* 327:905–909.

Henderson, B.E., Ross, P.K., and Pike, M.C. 1993. Hormonal chemoprevention of cancer in women. *Science* 259:633–638.

King, M.-C. 1992. Breast cancer genes: how many, where and who are they? *Nature Genet.* 2:89–90.

Powell, S.M., Petersen, G.M., Krush, A.J., Booker, S., Jen, J., Giardiello, F.M., Hamilton, S.R., Vogelstein, B., and Kinzler, K.W. 1993. Molecular diagnosis of familial adenomatous polyposis. *N. Engl. J. Med.* 329:1982–1987.

Sellers, T.A., Bailey-Wilson, J.E., Elston, R.C., Wilson, A.F., Elston, G.Z., Ooi, W.L., and Rothschild, H. 1990. Evidence for Mendelian inheritance in the pathogenesis of lung cancer. *J. Natl. Cancer Inst.* 82:1272–1279.

Skolnick, M.H., Cannon-Albright, L.A., Goldgar, D.E., Ward, J.H., Marshall, C.J., Schumann, G.B., Hogle, H., McWhorter, W.P., Wright, E.C., Tran, T.D., Bishop, D.T., Kushner, J.P., and Eyre, H.J. 1990. Inheritance of proliferative breast disease in breast cancer. *Science* 250:1715–1720.

Winawer, S.J., Zauber, A.G., Ho, M.N., O'Brien, M.J., Gottlieb, L.S., Sternberg, S.S., Waye, J.D., Schapiro, M., Bond, J.H., Panish, J.F., Ackroyd, F., Shike, M., Kurtz, R.C., Hornsby-Lewis, L., Gerdes, H., Stewart, E.T., and the National Polyp Study Workshop. 1993. Prevention of colon cancer by colonoscopic polypectomy. *N. Engl. J. Med.* 329:1977–1981.

Role of Oncogenes in Cancer Diagnosis

Biondi, A., Rambaldi, A., Alcalay, M., Pandolfi, P.P., Lo Coco, F., Diverio, D., Rossi, U., Mencarelli, A., Longo, L., Zangrilli, D., Masera, G., Barbui, T., Mandelli, F., Grignani, F., and Pelicci, P.G. 1991. *RAR-*α rearrangements as a genetic marker for diagnosis and monitoring in acute promyelocytic leukemia. *Blood* 77:1418–1422.

Delage, R., Soiffer, R.J., Dear, K., and Ritz, J. 1991. Clinical significance of *bcr-abl* gene rearrangement detected by polymerase chain reaction after allogeneic bone marrow transplantation in chronic myelogenous leukemia. *Blood* 78:2759–2767.

Look, A.T., Hayes, F.A., Shuster, J.J., Douglass, E.C., Castleberry, R.P., Bowman, L.C., Smith, E.I., and Brodeur, G.M. 1991. Clinical relevance of tumor cell ploidy and N-*myc* gene amplification in childhood neuroblastoma: a pediatric oncology group study. *J. Clin. Oncol.* 9:581–591.

Miller, W.H., Kakizuka, A., Frankel, S.R., Warrell, R.P., DeBlasio, A., Levine, K., Evans, R.M., and Dmitrovsky, E. 1992. Reverse transcriptase polymerase chain reaction for the rearranged retinoic acid receptor α clarifies diagnosis and detects minimal residual disease in acute promyelocytic leukemia. *Proc. Natl. Acad. Sci. USA* 89:2694–2698.

Sawyers, C.L., Timson, L., Kawasaki, E.S., Clark, S.S., Witte, O.N., and Champlin, R. 1990. Molecular relapse in chronic myelogenous leukemia patients after bone marrow transplantation detected by polymerase chain reaction. *Proc. Natl. Acad. Sci. USA* 87:563–567.

Development of Drugs Targeted against Oncogenes

Berges, R.R., Furuya, Y., Remington, L., English, H.F., Jacks, T., and Isaacs, J.T. 1993. Cell proliferation, DNA repair, and p53 function are not required for programmed death of prostatic glandular cells induced by androgen ablation. *Proc. Natl. Acad. Sci. USA* 90:8910–8914.

Castaigne, S., Chomienne, C., Daniel, M.T., Ballerini, P., Berger, R., Fenaux, P., Castaigne, S., and Degos, L. 1990. All-*trans* retinoic acid as a differentiation therapy for acute promyelocytic leukemia. I. Clinical results. *Blood* 76:1704–1709.

Chen, W., Peace, D.J., Rovira, D.K., You, S.G., and Cheever, M.A. 1992. T-cell immunity to the joining region of p210 BCR-ABL protein. *Proc. Natl. Acad. Sci. USA* 89:1468–1472.

Gibbs, J.B., Oliff, A., and Kohl, N.E. 1994. Farnesyltransferase inhibitors: Ras research yields a potential cancer therapeutic. *Cell* 77:175–178.

Huang, M., Ye, Y., Chen, S., Chai, J., Lu, J.-X., Zhoa, L., Gu, L., and Wang, Z. 1988. Use of all *trans*-retinoic acid in the treatment of acute promyelocytic leukemia. *Blood* 72:567–572.

James, G.L., Goldstein, J.L., Brown, M.S., Rawson, T.E., Somers, T.C., McDowell, R.S., Crowley, C.W., Lucas, B.K., Levinson, A.D., and Marsters, J.C., Jr. 1993. Benzodiazepine peptidomimetics: potent inhibitors of Ras farnesylation in animal cells. *Science* 260:1937–1942.

Kim, K.J., Li, B., Winer, J., Armanini, M., Gillett, N., Phillips, H.S., and Ferrara, N. 1993. Inhibition of vascular endothelial growth factor-induced angiogenesis suppresses tumour growth *in vivo*. *Nature* 362:841–844.

Kohl, N.E., Mosser, S.D., deSolms, S.J., Giuliani, E.A., Pompliano, D.L., Graham, S.L., Smith, R.L., Scolnick, E.M., Oliff, A., and Gibbs, J. 1993. Selective inhibition of *ras*-dependent transformation by a farnesyltransferase inhibitor. *Science* 260:1934–1937.

Lippman, M.E. 1993. The development of biological therapies for breast cancer. *Science* 259:631–632.

Millauer, B., Shawver, L.K., Plate, K.H., Risau, W., and Ullrich, A. 1994. Glioblastoma growth inhibited *in vivo* by a dominant-negative Flk-1 mutant. *Nature* 367:576–579.

Skorski, T., Nieborowska-Skorska, M., Nicolaides, N.L., Szczylik, C., Iversen, P., Iozzo, R.V., Zon, G., and Calabretta, B. 1994. Suppression of Philadelphia[1] leukemia cell growth in mice by *BCR-ABL* antisense oligodeoxynucleotide. *Proc. Natl. Acad. Sci. USA* 91:4504–4508.

Szczylik, C., Skorski, T., Nicolaides, N.C., Manzella, L., Malaguarnera, L., Venturelli, D., Gewirtz, A.M., and Calabretta, B. 1991. Selective inhibition of leukemia cell proliferation by *BCR-ABL* antisense oligodeoxynucleotides. *Science* 253:562–565.

❖ Glossary

abl The oncogene of Abelson murine leukemia virus that is activated by the Philadelphia translocation in human chronic myelogenous and acute lymphocytic leukemias and encodes a nonreceptor protein-tyrosine kinase.

acutely transforming virus A retrovirus that rapidly induces tumors in infected animals and efficiently transforms cells in culture. The genomes of acutely transforming viruses contain one or more oncogenes.

adenoviruses A family of DNA tumor viruses with genomes of ~35 kb.

adenylate cyclase An enzyme that catalyzes the synthesis of cAMP from ATP.

akt A murine retrovirus oncogene that encodes a protein-serine/threonine kinase.

alk An oncogene activated by chromosome translocation in lymphomas that encodes a receptor tyrosine kinase.

allele One of two or more alternative forms of a gene.

***Alu* sequences** A family of short interspersed repetitive sequences present approximately once every 5 kb in human genomic DNA.

***aml*1** An oncogene activated by chromosome translocation in acute myeloid leukemias that encodes a transcription factor.

anchorage independence The ability of transformed fibroblasts and epithelial cells in culture to proliferate without attachment to a surface.

aneuploidy An abnormal number of chromosomes.

AP-1 A transcription factor composed of the *jun* and *fos* gene products.

APC A tumor suppressor gene responsible for familial adenomatous polyposis and frequently inactivated in noninherited colorectal carcinomas.

apoptosis An active process of cell death characterized by cell shrinkage, chromosome condensation, and DNA fragmentation.

arg An *abl*-related gene that encodes a nonreceptor protein-tyrosine kinase.

autocrine growth stimulation Production of a growth factor by a responsive cell, resulting in continual stimulation of cell proliferation.

autophosphorylation Phosphorylation of a protein by its own kinase activity.

autoregulation Regulation of the expression of a gene by its own product.

axl A human oncogene detected by gene transfer assays that encodes a receptor tyrosine kinase.

B lymphocyte An antibody-producing lymphocyte.
Bax A Bcl-2 related protein that counteracts the effect of Bcl-2 in suppressing apoptosis.
***bcl*-2** An oncogene activated by chromosome translocation in B-cell follicular lymphomas that encodes a protein that suppresses apoptosis.
***bcl*-3** An oncogene activated by chromosome translocation in chronic B-cell leukemias that encodes a protein related to I-κB.
***bcl*-6** An oncogene activated by chromosome translocation in diffuse B-cell lymphomas that encodes a transcription factor.
***bcl*-x** A gene related to *bcl*-2 that encodes two proteins, derived by alternative splicing, that suppress or promote apoptosis, respectively.
bcr Breakpoint cluster region, a gene on human chromosome 22 that is fused to *abl* as a result of the Philadelphia translocation.
benign tumor A tumor that remains confined to its normal location and neither invades surrounding normal tissue nor spreads to other organ sites.
Blk A member of the Src family of nonreceptor tyrosine kinases.
***bmi*-1** An oncogene activated by retroviral insertion in mouse leukemias that encodes a putative transcription factor.
BRCA1 A tumor suppressor gene responsible for inherited breast cancers.
Burkitt's lymphoma A human B-cell lymphoma associated with Epstein-Barr virus infection and characterized by translocation of c-*myc* to an immunoglobulin gene locus.

carcinogen Any cancer-inducing agent.
carcinoma A neoplasm arising from epithelial cells of either endodermal or ectodermal origin.
carcinoma *in situ* A small localized epithelial tumor that has not invaded surrounding normal tissue.
catalytic domain Region of an enzyme responsible for its catalytic activity.
cbl The oncogene of Cas NS-1 murine retrovirus.
Cdc2 A protein-serine/threonine kinase that is a component of maturation promoting factor and controls cell entry into M phase of both mitosis and meiosis.
Cdc25 A Ras guanine nucleotide exchange factor.
Cdk Cyclin-dependent kinase; a family of protein-serine/threonine kinases that regulate cell cycle progression.
cDNA A DNA molecule synthesized by copying RNA with reverse transcriptase.
cDNA clone A molecular clone of a cDNA.
***ced*-9** A *C. elegans* gene related to *bcl*-2 that suppresses programmed cell death.

chromosome banding pattern Characteristic pattern of bands in stained metaphase chromosomes.

chromosome translocation Exchange of segments between nonhomologous chromosomes.

***cis*-acting regulatory sequence** A DNA sequence that regulates expression of an adjacent gene.

clonality Origin of a group of cells or a tumor from a single progenitor cell.

coding sequence A nucleotide sequence that specifies the amino acid sequence of a protein.

colony-stimulating factor-1 (CSF-1) A hematopoietic growth factor that stimulates macrophage proliferation and can be activated as an oncogene by insertional mutagenesis.

colony-stimulating factor-1 receptor The protein-tyrosine kinase receptor for CSF-1.

complementation Ability of two viruses (or other organisms) bearing mutations in two different genes to compensate for each other's defects.

c-*onc* General term for a cellular oncogene or proto-oncogene.

contact inhibition The inhibition of movement of normal fibroblasts that results from cell contact.

cot A human oncogene activated in gene transfer experiments that encodes a protein-serine/threonine kinase.

crk The oncogene of avian sarcoma virus CT10 that encodes an adapter protein containing SH2 and SH3 domains.

Csk A nonreceptor tyrosine kinase that phosphorylates negative regulatory sites of Src subfamily kinases.

cyclic AMP (cAMP) Adenosine monophosphate in which the phosphate group is covalently bound to both the 3' and 5' carbon atoms, forming a cyclic structure; an important second messenger in the response of cells to a variety of hormones.

cyclic AMP-dependent protein kinase A protein kinase that is activated by cAMP.

cyclins A family of proteins that regulate the activity of Cdks and control progression through the cell cycle.

cytoskeleton Structural framework of the cytoplasm.

cytosol Solution phase of the cytoplasm.

cytostatic factor (CSF) Protein(s) that maintain metaphase II arrest in meiosis of vertebrate oocytes.

dbl A human oncogene activated by DNA rearrangements in gene transfer that encodes a guanine nucleotide exchange factor for a Rho family protein.

DCC A tumor suppressor gene frequently deleted in colorectal carcinomas.

dek/can An oncogene activated by chromosome translocation in acute myeloid leukemia.

delayed response gene A gene whose transcription is stimulated by growth factors subsequent to the initial induction of immediate-early genes.

deletion Loss of genetic material.

density-dependent inhibition The characteristic cessation of normal cell growth at a finite cell density in culture.

diacylglycerol A second messenger, generated by hydrolysis of phosphatidylinositol 4,5-bisphosphate or phosphatidylcholine, that activates protein kinase C.

DNA amplification Increased number of DNA copies in a cell resulting from repeated replication.

DNA binding domain The region of a DNA binding protein that interacts directly with DNA.

dorsal The *Drosophila* homolog of the *rel* proto-oncogene that is required for development of embryonic dorsal-ventral polarity.

double minute A small chromosome-like structure containing amplified DNA.

dysplasia An early preneoplastic stage in the development of carcinomas characterized by loss of cellular regularity in an epithelial tissue.

E1A* and *E1B Oncogenes of adenoviruses.

E2A/pbx1 An oncogene activated by chromosome translocation in acute pre-B-cell leukemia that encodes a chimeric transcription factor.

E2F A family of transcription factors that activate genes taking part in cell proliferation and are regulated by formation of complexes with Rb.

E5*, *E6*, and *E7 Oncogenes of papillomaviruses.

early gene A gene expressed early after infection of a cell by a virus.

***ect*-2** An oncogene encoding a protein with a putative guanine nucleotide exchange factor domain.

ectoderm Germ layer giving rise to tissues that include the skin and nervous system.

Elk-1 A member of the Ets family of transcription factors that is activated by MAP kinase and stimulates c-*fos* transcription.

embryonal stem cell An embryonic cell derived from the blastocyst inner cell mass that can be grown in culture and reintroduced into mouse embryos.

endoderm Germ layer giving rise to internal organs, including the liver, stomach, and intestine.

enhancer A *cis*-acting regulatory sequence that increases transcription of an adjacent gene independent of position or orientation.

env Retroviral gene encoding the envelope glycoproteins.

epidermal growth factor (EGF) A mitogenic polypeptide of 53 amino acids.

epidermal growth factor receptor The protein-tyrosine kinase receptor for EGF that is also the product of the *erb*B proto-oncogene.

epithelial cells Cells that form continuous sheets covering the surface of the body and lining the internal organs.

Epstein-Barr virus A herpesvirus that is the etiologic agent of Burkitt's lymphoma.

***erb*A** An oncogene of avian erythroblastosis virus that encodes a transcriptional regulatory protein derived from the thyroid hormone receptor.

***erb*B** An oncogene of avian erythroblastosis virus that encodes a protein-tyrosine kinase derived from the epidermal growth factor receptor.

***erb*B-2** An oncogene activated in rat neuroblastomas (originally called *neu*) and frequently amplified in human breast and ovarian carcinomas that encodes a receptor protein-tyrosine kinase related to *erb*B.

erg An oncogene activated by chromosome translocation in myeloid leukemias that encodes a transcription factor belonging to the Ets family.

Erk-1 and Erk-2 Extracellular signal regulated protein kinases; *see* MAP kinase.

erythropoietin A growth factor that stimulates proliferation and maturation of erythroid cells.

erythropoietin receptor Cytokine receptor for erythropoietin that can act as an oncogene in mouse erythroleukemias.

established cell line *See* immortal cell line.

ets An oncogene of avian erythroblastosis virus E26 that encodes a transcription factor.

***evi*-1** An oncogene activated by retroviral insertional mutagenesis in mouse myeloid leukemias that encodes a transcription factor.

***ews/fli*1** An oncogene activated by chromosome translocation in Ewing's sarcomas that encodes a chimeric transcription factor.

exon A region of a gene that is retained in processed mRNA.

extracellular matrix An array of secreted products in which cells are embedded.

familial adenomatous polyposis A hereditary form of colon cancer resulting from inherited mutations of the *APC* tumor suppressor gene.

fes The oncogene of Gardner-Arnstein feline sarcoma virus (feline homolog of *fps*) that encodes a nonreceptor protein-tyrosine kinase.

***fgf*-5** A human oncogene activated in gene transfer assays that encodes a member of the fibroblast growth factor family.

fgr The oncogene of Gardner-Rasheed feline sarcoma virus that encodes a member of the *src* subfamily of nonreceptor protein-tyrosine kinases.

fibroblast A connective tissue cell commonly grown in culture.

fibroblast growth factor (FGF) A family of polypeptide growth factors, including acidic and basic FGF, keratinocyte growth factor, and the products of the *hst*, *int*-2, and *fgf*-5 oncogenes.

fibroblast growth factor receptor Protein-tyrosine kinase receptor for FGF.

fibronectin A glycoprotein component of the extracellular matrix.

fli-1 An oncogene activated by retroviral insertion in mouse erythroleukemias and by chromosome translocation in Ewing's sarcomas that encodes a transcription factor belonging to the Ets family.

fms The oncogene of McDonough feline sarcoma virus that encodes a protein-tyrosine kinase derived from the CSF-1 receptor.

focus A group of transformed cells that can be distinguished from normal cells by altered morphology and growth.

fos The oncogene of FBJ murine osteogenic sarcoma virus that encodes a component of the AP-1 transcription factor.

*fos-***B** A member of the *fos* gene family.

fps The oncogene of Fujinami avian sarcoma virus (chicken homolog of *fes*) that encodes a nonreceptor protein-tyrosine kinase.

Fra-1 and Fra-2 Fos-related antigens-1 and -2; members of the Fos family.

fusion protein A protein that consists of coding sequences originally derived from two different genes.

Fyn A member of the Src subfamily of nonreceptor protein-tyrosine kinases.

$\mathbf{G_0}$ A quiescent state in which cells are not proliferating.

$\mathbf{G_1}$ The stage of the cell cycle between the end of mitosis and the beginning of DNA synthesis.

$\mathbf{G_2}$ The stage of the cell cycle between the end of DNA synthesis and the beginning of mitosis.

G proteins A family of guanine nucleotide binding proteins that function to couple cell surface receptors to target enzymes.

gag Retroviral gene encoding the major virion structural proteins.

gene family A set of genes that encode related proteins.

gene linkage Localization of two genes on the same chromosome.

gene transfer Transfer of genetic information, usually in the form of purified DNA, from one cell to another; also called *transfection*.

genomic clone A molecular clone of genomic DNA.

genomic library A collection of genomic clones extensive enough to include most or all of the DNA sequences in the genome of an organism.

$\mathbf{G_i}$ A G protein that inhibits adenylate cyclase.

gli An oncogene amplified in glioblastomas that encodes a putative transcription factor.

$\mathbf{G_q}$ A G protein that stimulates phospholipase C.

granulocyte-macrophage colony-stimulating factor (GM-CSF) A hematopoietic growth factor that stimulates proliferation of granulocyte and macrophage precursors and can be activated as an oncogene.

Grb2 An adapter molecule containing SH2 and SH3 domains that takes part in Ras activation.

growth factor A protein that stimulates cell proliferation.
G_s A G protein that stimulates adenylate cyclase.
GTPase An enzyme that catalyzes the hydrolysis of GTP to GDP and inorganic phosphate.
GTPase activating protein (GAP) A protein that stimulates the GTPase activity of Ras or related proteins.
GTPase superfamily Large family of proteins, including the G proteins and Ras, whose activity is regulated by GTP binding and hydrolysis.
guanine nucleotide exchange factor (GEF) A protein that promotes the exchange of GDP bound to Ras or a related protein for GTP.

Hck A member of the Src subfamily of nonreceptor protein-tyrosine kinases.
helix-loop-helix (HLH) A dimerization motif, found in a number of transcription factors, in which two amphipathic α helices are separated by a loop.
hematopoietic cells Cells giving rise to mature blood cells.
hepatitis B viruses A family of small DNA viruses (genomes of ~3 kb) that can cause liver carcinomas.
hereditary nonpolyposis colorectal cancer (HNPCC) A relatively common form of hereditary colorectal cancer resulting from inherited defects in DNA repair.
herpesviruses A family of large DNA viruses (genomes of 100–200 kb), some of which induce tumors.
heterozygous Containing two different alleles of a gene.
homogeneous staining region A chromosome region containing amplified DNA that stains homogeneously rather than yielding a normal banding pattern.
homozygous Containing two identical alleles of a gene.
***hox*2.4** A homeobox gene activated as an oncogene by retroviral insertion in mouse leukemias.
***hox*11** A homeobox gene activated as an oncogene by chromosome translocation in T-cell leukemias.
hrx An oncogene activated by chromosome translocations in acute leukemias that encodes a transcription factor.
hst A human oncogene activated by DNA rearrangements in gene transfer experiments that encodes a member of the fibroblast growth factor family.

I-κB An inhibitory subunit of NF-κB transcription factors.
immediate-early gene A gene whose transcription is rapidly and transiently induced by growth factors without a requirement for protein synthesis.
immortal cell line A cell line with an unlimited lifespan in culture; also called *established cell line.*

inositol triphosphate (IP_3) A second messenger, formed by hydrolysis of phosphatidylinositol 4,5-bisphosphate, that stimulates release of calcium from intracellular stores.

insertional mutagenesis Genetic alteration resulting from integration of proviral DNA.

insulin receptor Protein-tyrosine kinase receptor for insulin.

***int*-1** *See wnt*-1.

***int*-2** An oncogene activated by retroviral insertional mutagenesis in mouse mammary carcinomas that encodes a member of the fibroblast growth factor family.

integral membrane protein A protein that is inserted into the lipid bilayer of a membrane.

interleukin-2 *See* T-cell growth factor.

interleukin-3 *See* multipotential colony-stimulating factor.

intracisternal A particle A type of defective retrovirus-like particle.

intron Region of a gene that is removed from the primary transcript by splicing and therefore is not present in processed mRNA.

Jak A subfamily of nonreceptor tyrosine kinases associated with cytokine receptors.

JNK A protein-serine/threonine kinase, related to MAP kinase, that phosphorylates and activates Jun.

jun The oncogene of avian sarcoma virus 17 that encodes a component of transcription factor AP-1.

***jun*-B and *jun*-D** Members of the *jun* gene family.

karyotype The chromosome complement of a cell.

kb Kilobase, one thousand nucleotides.

kd Kilodalton, one thousand daltons.

kinase An enzyme that transfers phosphate groups from one molecule, usually from ATP, to another.

kinase domain The catalytic domain of a kinase.

kit The oncogene of Hardy-Zuckerman-4 feline sarcoma virus that encodes a receptor protein-tyrosine kinase derived from the mouse *W* locus.

K-*sam* An oncogene amplified in stomach cancers that encodes a member of the FGF receptor family.

late gene A gene expressed late in virus infection.

latent period Time between exposure to a carcinogen and development of a tumor.

Lck A member of the Src subfamily of nonreceptor protein-tyrosine kinases.

leucine zipper A dimerization domain found in many transcriptional regulatory proteins, including members of the Fos and Jun families, that consists of four or five leucine residues, each separated by six other amino acids.

leukemia Cancer arising from the precursors of circulating blood cells.

leukosis Cancer of white blood cells.

Li-Fraumeni cancer family syndrome A rare inherited cancer susceptibility, leading to the development of multiple kinds of tumors, that results from mutations of the *p53* tumor suppressor gene.

ligand binding domain Region of a protein that binds a ligand—for example, the portion of a growth factor receptor that binds the growth factor.

***LMP*-1** An oncogene of Epstein-Barr virus.

L-*myc* A member of the *myc* gene family that is amplified in human lung carcinomas.

long terminal repeat (LTR) Sequences of several hundred nucleotides present at both termini of retroviral DNAs that contain the viral transcriptional promoter and enhancer sequences.

***lyl*-1** An oncogene activated by chromosome translocations in acute T-cell leukemias that encodes a transcription factor.

lymphocyte A white blood cell that functions in immune responses.

lymphoma A cancer of lymphoid cells.

Lyn A member of the Src subfamily of nonreceptor protein-tyrosine kinases.

***lyt*-10** An oncogene activated by chromosome translocations in B-cell lymphomas that is related to NF-κB.

M phase The stage of the cell cycle during which mitosis takes place.

Mad A protein that forms dimers with Max.

maf An avian sarcoma virus oncogene that encodes a transcription factor.

malignant tumor A cancer that invades adjacent tissues and metastasizes to other organ sites.

MAP kinase Mitogen-activated protein kinase, a protein-serine/threonine kinase that is activated by threonine and tyrosine phosphorylation in mitogen-stimulated cells.

MAP kinase kinase A dual specificity protein kinase that activates MAP kinase by phosphorylation on tyrosine and threonine residues.

mas A human oncogene activated in gene transfer experiments that encodes a receptor coupled to phosphatidylinositol turnover by a G protein.

maternal RNA An RNA that is transcribed in oocytes and translated during oocyte maturation or early development.

maturation promoting factor (MPF) A complex of Cdc2 and cyclin B that promotes entry into the M phase of either mitosis or meiosis.

Max A protein that forms dimers with Myc.

***mdm*-2** An oncogene amplified in sarcomas that encodes an inhibitor of p53.

meiotic maturation The first meiotic division of an oocyte.

MEK *See* MAP kinase kinase.

MEKK MEK kinase; a protein serine/threonine kinase that phosphorylates and activates MEK.

mesoderm Germ layer giving rise to connective tissues, the hematopoietic system, and muscle.

met A human oncogene activated in a chemically transformed osteosarcoma cell line that encodes a protein-tyrosine kinase receptor for hepatocyte growth factor.

metaphase The stage of mitosis or meiosis during which the condensed chromosomes are aligned in the center of the spindle.

metastasis Spread of cancer cells through the blood or lymphatic system to other organ sites.

mil *See raf.*

mitogen An agent that induces cell proliferation.

***MLH*1** A gene encoding an enzyme functioning in mismatch repair of DNA that is mutated in some cases of hereditary nonpolyposis colorectal cancer.

molecular clone A specific DNA fragment inserted into a replicating DNA molecule—for example, a bacterial plasmid or virus—so that the inserted DNA can be propagated as a recombinant molecule.

mos The oncogene of Moloney murine sarcoma virus that encodes a protein-serine/threonine kinase.

mpl A murine leukemia virus oncogene that encodes a cytokine receptor for thrombopoietin.

mRNA An RNA molecule that serves as a template for protein synthesis.

***MSH*2** A gene encoding an enzyme functioning in mismatch repair of DNA that is mutated in most cases of hereditary nonpolyposis colorectal cancer.

***MTS*1** A tumor suppressor gene inactivated in melanomas and a variety of other tumors that encodes an inhibitor of cyclin-dependent kinases.

multipotential colony-stimulating factor A growth factor that stimulates proliferation of stem cells giving rise to multiple hematopoietic cell lineages and can be activated as an oncogene by insertional mutagenesis; also called *interleukin-3.*

myb The oncogene of avian myeloblastosis virus that is activated by insertional mutagenesis in murine and avian leukemias and lymphomas and encodes a transcription factor.

myc The oncogene of avian myelocytomatosis virus-29—activated by insertional mutagenesis in avian, murine, and feline lymphomas, by chromosome translocation in mouse plasmacytomas and human Burkitt's lymphomas, and by amplification in a variety of human neoplasms—that encodes a transcription factor.

Nck An adapter molecule containing SH2 and SH3 domains that can act as an oncogene.

neoplasm An abnormal growth of cells.

nerve growth factor (NGF) A polypeptide that induces sympathetic neuron differentiation.

neu See *erb*B-2.

neurofibromin The product of the *NF1* tumor suppressor gene.

NF1 A tumor suppressor gene inactivated in neurofibromatosis type 1 that encodes a Ras GTPase activating protein.

NF2 A tumor suppressor gene inactivated in neurofibromatosis type 2.

NF-κB A transcription factor whose activity is controlled by translocation from the cytoplasm to the nucleus.

NIH 3T3 cells A nontransformed mouse cell line frequently used as recipient cells in gene transfer assays.

N-*myc* A member of the *myc* gene family that is amplified in human neuroblastomas and lung carcinomas.

nonreceptor tyrosine kinase An intracellular protein-tyrosine kinase.

nov An oncogene activated by retroviral insertion in avian nephroblastoma.

nucleic acid hybridization Hydrogen bond association between two strands of nucleic acid (either DNA or RNA) through complementary base pairing.

nude mouse An immunodeficient mouse, frequently used for tumorigenicity assays, in which T lymphocytes do not develop.

oncogene A gene capable of inducing one or more aspects of the neoplastic phenotype.

ovc A human oncogene activated in gene transfer experiments.

p16 An inhibitor of cyclin-dependent kinases encoded by the *MTS*1 tumor suppressor gene.

p21 An inhibitor of cyclin-dependent kinases whose transcription is induced by p53.

p27 An inhibitor of cyclin-dependent kinases induced by TGFß.

p53 A tumor suppressor gene that is frequently inactivated in a variety of human neoplasms and encodes a transcription factor.

p107 A protein related to Rb.

papillomaviruses A family of small DNA tumor viruses with genomes of ~8 kb.

permissive cell A cell in which a particular type of virus can replicate.

permissive temperature Temperature at which the product of a temperature-sensitive gene can function normally.

phage library A library of molecular clones prepared in a bacteriophage vector.

Philadelphia chromosome An abnormal human chromosome 22, identified in chronic myelogenous leukemias, that results from a reciprocal translocation between chromosomes 9 and 22 in which the *abl* proto-oncogene is activated.

phorbol esters A class of tumor promoters that activate protein kinase C.

phosphatidylinositol bisphosphate (PIP_2) Membrane phospholipid that is hydrolyzed by phospholipase C to yield the second messengers diacylglycerol and inositol triphosphate.

phosphatidylinositol 3-kinase (PI 3-kinase) An enzyme that phosphorylates phosphatidylinositides at the 3 position of the inositol ring.

phospholipase C An enzyme that hydrolyzes phosphatidylinositol 4,5-bisphosphate to yield the second messengers diacylglycerol and inositol triphosphate or that hydrolyzes phosphatidylcholine to yield diacylglycerol and phosphocholine.

***pim*-1** An oncogene activated by insertional mutagenesis in murine T-cell lymphomas that encodes a protein-serine/threonine kinase.

platelet-derived growth factor (PDGF) A polypeptide mitogen for fibroblasts that is also the product of the *sis* proto-oncogene.

platelet-derived growth factor receptor Protein-tyrosine kinase receptor for PDGF.

PML-RARα An oncogene activated by chromosome translocation in acute promyelocytic leukemias that encodes a recombinant fusion protein consisting of the retinoic acid receptor and a putative transcription factor.

point mutation Alteration of a single nucleotide pair.

pol Retroviral gene encoding reverse transcriptase.

polyomavirus A small DNA tumor virus with a genome of ~5 kb.

posttranslational processing Conversion of a polypeptide into the mature form of the protein by modifications that include glycosylation, phosphorylation, and cleavage.

poxvirus A family of large DNA viruses (genomes of ~200 kb), some of which induce benign tumors.

***PRAD*-1** An oncogene activated by chromosome translocations in B-cell leukemias and parathyroid adenomas and by amplification in breast and squamous cell carcinomas that encodes cyclin D1.

preneoplastic cell A cell that displays increased proliferative capacity and is capable of progressing to the full neoplastic phenotype as a result of further alterations.

primary cell culture The initial culture of cells from an *in vivo* source, such as an embryo, a tissue, or a tumor.

primary transcript An RNA molecule that has not yet been modified by splicing or polyadenylation.

probe A radiolabeled nucleic acid used to detect complementary molecules by nucleic acid hybridization.

programmed cell death An active process of cell death that occurs as a normal part of development. *See also* apoptosis.

promoter A *cis*-acting regulatory sequence at which RNA polymerase binds to initiate transcription.

promoter insertion A form of insertional mutagenesis in which a retroviral LTR acts as a promoter to drive transcription of an adjacent cellular gene.
protein kinase An enzyme that phosphorylates a protein.
protein kinase C A family of protein-serine/threonine kinases that function in intracellular signal transduction.
protein-serine/threonine kinase A protein kinase that phosphorylates serine or threonine residues on its substrate proteins.
protein-tyrosine kinase A protein kinase that phosphorylates tyrosine residues on its substrate proteins.
proto-oncogene A normal cell gene that can be converted into an oncogene.
provirus The integrated DNA form of a retrovirus genome.

qin An avian sarcoma virus oncogene that encodes a transcription factor.
quiescent cell A cell that is not proliferating.

Rab, Rac, Ral, Rap, and Rho Ras-related low molecular weight GTP- binding proteins.
raf An oncogene of 3611 murine sarcoma and MH2 avian carcinoma viruses (also called *mil*) that is frequently activated in gene transfer experiments and encodes a protein-serine/threonine kinase. A-*raf* and B-*raf* are additional members of the gene family.
***RAR*α** Gene encoding the retinoic acid receptor α, which is activated as an oncogene in acute promyelocytic leukemia.
ras Oncogenes of Harvey (*ras*H) and Kirsten (*ras*K) rat sarcoma viruses. The *ras*H and *ras*K oncogenes, as well as a third member of the *ras* gene family (*ras*N), are frequently activated in human neoplasms and encode guanine nucleotide binding proteins.
Rb A tumor suppressor gene identified by genetic analysis of retinoblastoma, and also frequently inactivated in sarcomas and lung carcinomas, that encodes a transcriptional regulatory protein.
receptor protein-tyrosine kinases Membrane-spanning protein-tyrosine kinases that are receptors for extracellular ligands.
recombinant fusion protein *See* fusion protein.
regulatory domain Region of an enzyme which regulates its catalytic activity.
regulatory sequence A DNA sequence that controls expression of an adjacent gene.
rel The oncogene of avian reticuloendotheliosis virus that is homologous to the *Drosophila* gene *dorsal* and encodes a transcription factor related to NF-κB.
repetitive sequence A DNA sequence that is present multiple times in a genome.
restriction fragment-length polymorphism (RFLP) A genetic marker based on inherited differences in restriction endonuclease cleavage sites surrounding a cloned DNA segment.

ret A human oncogene activated in gene transfer experiments and in thyroid carcinomas that encodes a receptor protein-tyrosine kinase.

retinoblastoma An eye tumor of children that is frequently inherited.

retrovirus An RNA virus whose genome is reverse transcribed into DNA in infected cells.

reverse transcriptase An enzyme that catalyzes the synthesis of DNA from an RNA template.

***rhom*-1 and *rhom*-2** Oncogenes activated by chromosome translocations in acute T-cell leukemias that encode transcription factors.

ros The oncogene of UR2 avian sarcoma virus that is also activated in gene transfer assays and encodes a receptor protein-tyrosine kinase.

RSK A protein-serine/threonine kinase that is phosphorylated and activated by MAP kinase.

S phase The period of the cell cycle during which DNA replication occurs.

sarcoma A neoplasm arising from cells of mesodermal origin.

sea The oncogene of avian erythroblastosis virus S13 that encodes a receptor protein-tyrosine kinase.

second messenger A compound whose metabolism is modified as a result of a ligand-receptor interaction and which then functions as a signal transducer by regulating other intracellular processes.

secretory leader or signal sequence A series of amino acids that direct the entry of a protein into the secretory pathway.

Sem5 An adapter molecule containing SH2 and SH3 domains that is the *C. elegans* homolog of Grb2.

senescence Limited proliferative capacity of normal cells.

serum response element (SRE) Regulatory sequence that mediates induction of *c-fos* and other immediate-early genes by serum and growth factors.

serum response factor (SRF) A transcription factor that binds to the SRE and regulates expression of *c-fos* and other genes.

sevenless A *Drosophila* gene required for development of the R7 photoreceptor cell that encodes a receptor protein-tyrosine kinase.

SH2 domain A protein domain of approximately 100 amino acids that binds phosphotyrosine-containing peptides.

SH3 domain A protein domain of approximately 60 amino acids that binds proline-rich peptides.

Shc A protein containing SH2 domains that takes part in Ras activation and can function as an oncogene.

sis The oncogene of simian sarcoma virus that encodes platelet-derived growth factor.

ski The oncogene of avian SK virus that encodes a putative transcription factor.

***Sl* locus** A mouse locus affecting development of melanocytes, germ cells, and hematopoietic cells that encodes the ligand for Kit.

somatic cell hybridization Fusion of two different types of somatic cells—for example, a tumor cell and a normal cell.

Sos A Ras guanine nucleotide exchange factor.

***spi*-1** An oncogene activated by retroviral insertion in mouse erythroleukemias that encodes a transcription factor belonging to the Ets family.

splicing RNA processing reaction in which introns are removed from a primary transcript.

src The oncogene of Rous sarcoma virus that encodes a nonreceptor protein-tyrosine kinase.

STATs A family of transcription factors that contain SH2 domains and are activated by tyrosine phosphorylation, which promotes their translocation from the cytoplasm to the nucleus.

stem cell A mitotically active cell that divides to form one new stem cell and another cell that differentiates.

SV40 Simian virus 40, a small DNA tumor virus with a genome of ~5 kb.

Syp A protein-tyrosine phosphatase that contains an SH2 domain and associates with receptor tyrosine kinases.

T antigens The early proteins of polyomavirus and SV40 that induce cell transformation.

T lymphocyte A lymphocyte that functions in cell-mediated immune responses.

***tal*-1 and *tal*-2** Oncogenes activated by chromosome translocations in acute T-cell leukemias that encode transcription factors.

***tan*-1** An oncogene activated by chromosome translocation in acute T-cell leukemias.

tax An oncogene of human T-lymphotropic virus.

T-cell growth factor A hematopoietic growth factor that stimulates proliferation of T lymphocytes and can be activated as an oncogene by insertional mutagenesis; also called interleukin-2.

temperature-sensitive A conditional mutation that results in normal function of a gene product at only certain temperatures.

TGF-α A polypeptide growth factor related to epidermal growth factor.

TGF-ß A family of polypeptides that regulates growth and differentiation of a variety of cell types.

torso A *Drosophila* gene required for development of terminal regions of the embryo that encodes a receptor protein-tyrosine kinase.

TPA The phorbol ester tumor promoter 12-*O*-tetradecanoyl-phorbol-13-acetate that activates protein kinase C as an analog of diacylglycerol.

***tpl*-2** An oncogene activated by retroviral insertion in rat lymphomas that encodes a protein-serine/threonine kinase.

***trans*-activating domain** The region of a transcription factor that interacts with an adjacent protein to stimulate transcription.

transcription Synthesis of RNA from a DNA template.
transcription factor A protein that regulates transcription.
transcription unit A gene together with its transcriptional regulatory sequences.
transfection *See* gene transfer.
transformation Conversion of a normal cell in culture into a cell displaying one or more phenotypic characteristics of a tumor cell.
transgenic An organism carrying a germ line insertion of foreign DNA.
translation Protein synthesis directed by mRNA.
transmembrane domain Region of a protein that spans a membrane lipid bilayer.
transmembrane protein An integral membrane protein with regions exposed at both sides of the membrane.
tre A human oncogene activated in gene transfer experiments.
trk A human oncogene detected by gene transfer assays that encodes the protein-tyrosine kinase receptor for nerve growth factor.
TrkB The protein-tyrosine kinase receptor for the neurotrophins BDNF and NT-4.
TrkC The protein-tyrosine kinase receptor for NT-3.
tumor initiation The first step in development of a tumor.
tumor progression The development of increasing malignancy during the pathogenesis of a neoplasm.
tumor promoter A compound that leads to neoplasm development by stimulating the proliferation of cells that have already sustained carcinogen-induced mutations.
tumor suppressor gene A cellular gene whose loss of function leads to tumor development.

vav A human oncogene activated in gene transfer that encodes a putative guanine nucleotide exchange factor.
VHL A tumor suppressor gene inactivated in von Hippel-Lindau family cancer syndrome.
v-*onc* General term for a viral oncogene.

***W* locus** The *kit* proto-oncogene; a mouse locus that affects development of melanocytes, germ cells, and hematopoietic cells.
weakly oncogenic virus A retrovirus that induces neoplasms with long latent periods in animals and does not transform cells in culture. The weakly oncogenic viruses do not contain oncogenes.
wingless The *Drosophila* homolog of *wnt*-1 that functions in embryonic pattern formation.

***wnt*-1** An oncogene activated by retroviral insertion in mouse mammary carcinomas that encodes an extracellular protein and is the mammalian homolog of the *Drosophila* gene *wingless*.
***wnt*-3** An oncogene belonging to the *wnt* family that is activated by retroviral insertion in mouse mammary carcinomas.
WT1 A tumor suppressor gene inactivated in Wilms tumors that encodes a transcriptional repressor.

***X* gene** An oncogene of hepatitis B virus.

yes The oncogene of Y73 avian sarcoma virus that encodes a member of the Src subfamily of nonreceptor protein-tyrosine kinases.
Yrk A member of the Src subfamily of protein tyrosine kinases.

❖ Index